C000220770

THE DICTIONARY OF CONTEMPORARY POLITICS OF SOUTHERN AFRICA

Dictionaries of Contemporary Politics

Dictionary of Contemporary Politics of Southern Africa
Dictionary of Contemporary Politics of Central America
Dictionary of Contemporary Politics of South America

THE DICTIONARY OF CONTEMPORARY POLITICS OF SOUTHERN AFRICA

Gwyneth Williams
Brian Hackland

ROUTLEDGE
London

First published in 1988 by
Routledge
11 New Fetter Lane, London EC4P 4EE

Set in Linotron Times and Helvetica
by Input Typesetting Ltd, London
and printed in Great Britain
by T J Press (Padstow) Ltd
Padstow, Cornwall

© Gwyneth Williams
and Brian Hackland 1988

No part of this book may be reproduced in
any form without permission from the publisher
except for the quotation of brief passages
in criticism

British Library Cataloguing in Publication Data
Williams, Gwyneth
The dictionary of contemporary politics
of Southern Africa—(Dictionaries of
contemporary politics).
1. Southern Africa. Politics
I. Title II. Hackland, Brian III. Series
320.968

ISBN 0–415–00245–1

Contents

List of illustrations

Introduction

The states of southern Africa could hardly be more disparate. Some are committed to Marxist-Leninist policies, others follow the road of 'African socialism', and still others are pursuing capitalist paths. South Africa, in the face of bitter opposition from the other states, has continued with its apartheid practices.

The demands of geography and economics, in contrast, pull the countries of the region closer together. The economic and military might of South Africa allows it to dominate and manipulate its neighbouring states. This has provided the impetus for the formation of SADCC, the Southern African Development Coordination Conference, by countries determined to reduce their dependence on their powerful neighbour. Politically too, most governments in the region have been united by their opposition to South Africa's controversial system of government, but they still cannot break away from the thrall of its economic might. They continue to rely on South African food, manufactured goods and technology and many of their citizens work in South Africa's mines. Such dependence works in both directions. South Africa imports hydro-electric power from Mozambique and Angola, and water from Lesotho; it needs southern African markets for its produce; and it relies on foreign migrant workers to operate its mines. This interdependence is increased by the colonial heritage of road and railway networks which funnel the bulk of trade through South African ports. Such complex inter-relationships make it impossible to understand the problems of any one country without an understanding of the region as a whole.

Despite the worldwide focus of media and academic attention on southern Africa, many people have difficulty disentangling its politics. This dictionary seeks to provide a guide through the confusion, identifying and explaining in a clear and practical way the political figures, organisations, systems and terminology of politics in the region. No book of its length could be exhaustive, and the choice of what should be left out has often been a difficult one. To those who find that it does not give as much detail as they need, we recommend the considerable literature on the region, a selection from which is listed at the end.

To give the book a wide relevance we have included interesting and controversial terms as well as the essential references. Brief biographies of central figures are provided, a short history is given of each country, and key geographic, demographic and economic data are listed. Important non-English terms, acronyms, abbreviations and basic details of party policies, structures and political systems are included.

The dictionary begins with a map of the region, illustrating the primary infrastructural and geographic features linking the countries. The main body of the book then takes the form of a cross-referenced alphabetical listing. It concludes with a short selection of books suggested for those interested in reading further. Entries in the alphabetical section are placed under the most commonly used name or acronym, and cross-referenced from less common versions.

Choice of terminology, in South Africa especially, often depends on political perspective. Apartheid laws may classify people according to one set of categories, while the subjects categorise themselves in completely different ways. To make this clear, and to avoid confusion, we have placed official apartheid terms in inverted commas and have begun them with capital letters. For instance, we have used 'Black', 'White', 'Coloured' and 'Indian' to denote the population groups identified by the South African state and labelled according to the language of apartheid. Thus 'Black', in official parlance, is used to mean the African population, excluding the groups known as 'Coloured', 'White' and 'Indian'; 'White' is used to refer to those of pale skin colour and considered by the state to be of European origin; 'Coloured' is applied to people of 'mixed' descent; and 'Indian' refers to people who are classified as of Asian or Indian descent under South Africa's racial laws. To refer to the black or white population generally we have used lower case without inverted commas. By black we mean those people who identify themselves as part of the black community, and by white we mean those who regard themselves as part of the white community.

The Dictionary of Contemporary Politics of Southern Africa fills a gap in the growing literature on the region, providing an accurate and objective source of reference. It is written in an accessible style. We hope you will find it interesting, useful and, above all, enjoyable.

Gwyneth Williams
Brian Hackland

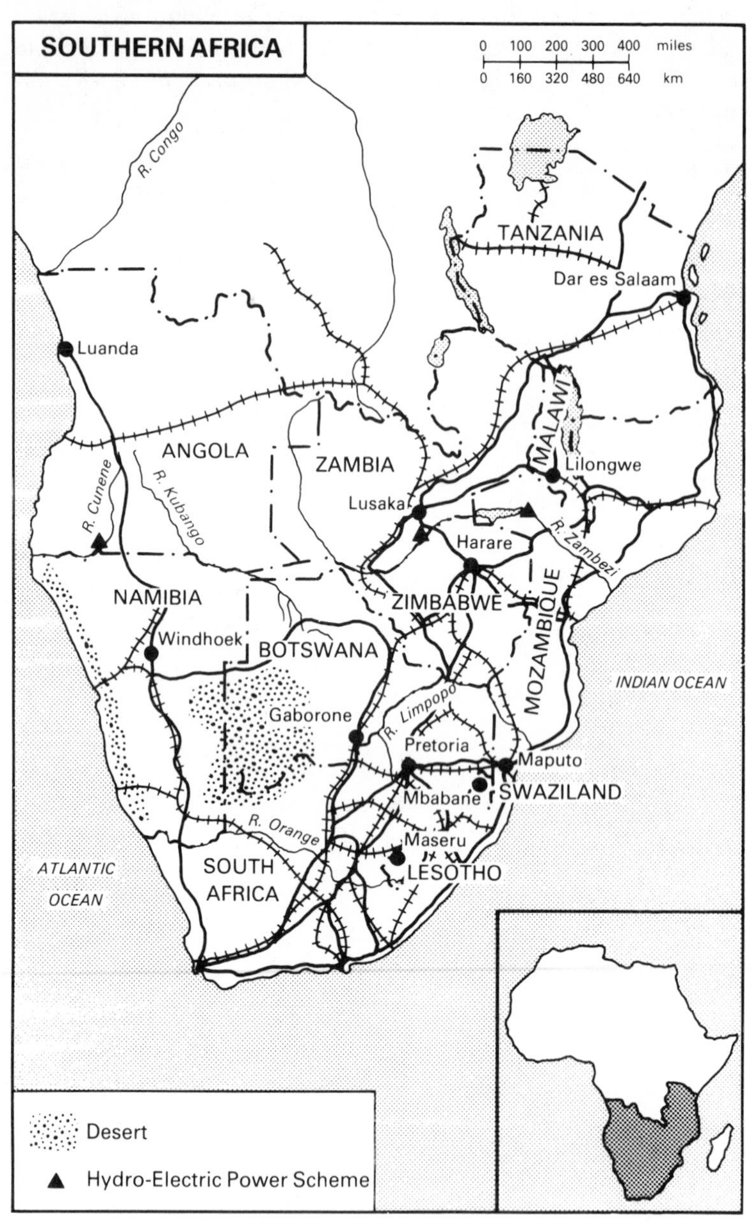
SOUTHERN AFRICA
0 100 200 300 400 miles
0 160 320 480 640 km
R. Congo
TANZANIA
Dar es Salaam
Luanda
ANGOLA
ZAMBIA
MALAWI
Lilongwe
R. Cunene
R. Kubango
Lusaka
Harare
R. Zambezi
NAMIBIA
ZIMBABWE
MOZAMBIQUE
Windhoek
BOTSWANA
INDIAN OCEAN
Gaborone
R. Limpopo
Pretoria
Maputo
Mbabane
SWAZILAND
R. Orange
Maseru
LESOTHO
SOUTH AFRICA
ATLANTIC OCEAN
Desert
Hydro-Electric Power Scheme

AA (*Tanzania*) See **African Association**.

aamati (*Namibia*)
Among black Namibians in the north of the country this name, an Ovambo word meaning 'the boys', is applied to soldiers of PLAN. By contrast, the common word used to refer to members of the South African military and paramilitary forces, and to Namibians serving in them, is *omakakunya*, an Ovambo word meaning 'creatures which gnaw at bones', 'bonepickers', or, more freely translated, 'scavengers'.
See also **PLAN; SWATF**.

Abdullah, Sayyid Jamshid ibn (*Tanzania*)
Sayyid Jamshid ibn Abdullah was the Sultan of Zanzibar. He ruled Zanzibar as a British protectorate from 1890 until he was overthrown in 1964: see **Zanzibar**.

Abdurahman, Abdulla (*South Africa*)
Dr Abdurahman was founder and president of the African People's Organisation, the first 'Coloured' political movement. He argued for black unity in the struggle against white domination. Despite opposition within the APO from younger more radical elements, including his daughter Zainunnissa 'Cissie' Gool (who left to form the Non-European United Front) he remained president of the APO until his death in 1940.
See also **African People's Organisation**.

Acção Nacional Popular (*Mozambique*)
The *Acção Nacional Popular* (ANP) was the official fascist party in Mozambique during the colonial period.

accord between Malawi and Mozambique, 1986 (*Malawi, Mozambique*)
A Joint Security Commission was set up between Malawi and Mozambique after a meeting in September 1986 in Blantyre during which President Machel of Mozambique and President Kaunda of Zambia both warned President Banda that they would consider closing their borders with Malawi unless he acted against the Mozambican rebel MNR movement operating from Malawi. Robert Mugabe, Zimbabwe's president, was also present.

Meetings of the Security Commission culminated, on 18 December

1986 in an accord on mutual co-operation and a protocol covering defence, state security and public order. Just before this Malawi announced that it had accepted proposals from the Mozambican government intended to 'normalise' relations between Malawi and Mozambique. It was reported that the proposals included Malawi's re-routing of its trade away from South Africa's transport system through the Mozambican system, to meet with the aims of SADCC, to which Malawi belongs.
See also **Banda; Machel; MNR; SADCC**.

Action Front for the Retention of Turnhalle Principles (*Namibia*) See **AKTUR**.

Action Own Future (*South Africa*) See ***Aksie Eie Toekoms***.

Action Save White South Africa (*South Africa*) See ***Aksie Red Blank Suid-Afrika***.

Action Save White South West Africa (*Namibia*)
A right wing organisation established to resist 'reform' in Namibia: see **vigilante organisations**.

Active Revolt (*Angola*)
Active Revolt (*Revolta Activa*) was an oppositionist faction of the MPLA which broke away in May 1974, at a time of division and weakness in the movement. Opposed to the leadership of Agostinho Neto, it was a movement dominated by intellectuals of the left. It never attracted a large following, but had among its members leading figures from the early years of the MPLA, including brothers Mario and Joaquim Pinto de Andrade.
See also **MPLA; Neto; Pinto de Andrade, Mario and Joaquim**.

ADMARC (*Malawi*) See **Agricultural Development and Marketing Corporation**.

Administrator General (*Namibia*)
A South African judge, Justice Marthinus Steyn, was appointed by South Africa as the first administrator general in Namibia on 6 July 1977. The post was created to provide an interim system of administration in Namibia until the holding of UN supervised elections prior to independence. It followed agreement between South Africa and the Contact Group of western countries on the principle of free elections for a constituent assembly in Namibia. In fact South Africa has made several attempts since to set up interim or transitional governments in Namibia, in contravention of the agreement, and has used the post as a means of direct rule in Namibia.

Steyn held office until August 1979 when he was replaced by Professor Gerrit Viljoen, formerly head of the *Broederbond* in South Africa. In September 1980 Danie Hough became AG, and was replaced by Willem van Niekerk in February 1983. Van Niekerk was in turn succeeded by Louis Pienaar in May 1985.

See also ***Broederbond*; Contact Group; DTA: MPC: Transitional Government of National Unity; Viljoen.**

AET (*South Africa*) ***Aksie Eie Toekoms.***

African Advisory Council (*Botswana*)
The African Advisory Council was founded in 1920 as the Native Advisory Council to discuss matters seen by the colonial authorities as affecting African interests. The Council provided a platform for people to argue against incorporating Bechuanaland (later Botswana) into the Union of South Africa. A Joint Advisory Council was proposed as a supplement to the African and European Advisory Councils, but all councils ceased to exist in 1961 when the Legislative Council was created.
See also **European Advisory Council.**

African Association (*Tanzania*)
The African Association (AA) was founded in 1929 to oppose the closer union with Kenya and Uganda which was sought by white settlers in Tanganyika (mainland Tanzania). It became known as *Chama cha Umoja wa Watu wa Afrika* – the Association of the Unity of the People of Africa. The organisation was designed to foster unity amongst Africans and its motto was 'Unity is Strength'. Its first president was *Mwalimu* (or teacher) Cecil Matola, a distinguished teacher.

As time went on the association was accused of elitism, but various branches were formed and gradually its character changed and broadened. In 1945 the Tanganyika African Welfare Society, a student group led by Julius Nyerere (later to become president of Tanzania) converted its organisation into a branch of the African Association. In 1948 the association broke its links with Zanzibar after a dispute and renamed itself the Tanganyika African Association, often known as the TAA. Soon the TAA began to decline in popularity and in 1954 Julius Nyerere converted the organisation into TANU, the Tanganyika African National Union, which was to become the country's main nationalist political party campaigning for independence.
See also **Nyerere; TANU.**

Africanisation (*southern Africa*)
Africanisation is the policy followed by most newly independent African countries in which expatriates from the former colonial power are replaced by local people in state and para-statal jobs. In Zambia this policy was sometimes more specifically called 'Zambianisation'. Its aim is to exercise a greater control over the country as well as provide jobs for the population and train local people in skills which have often been denied to them under the colonial regime.
See also **Tanganyika.**

Africanism (*South Africa*)
'Africanism' is a political philosophy developed by a pressure group within the Congress Youth League during the 1940s. One of its initiators was Anton Muziwakhe Lembede, a founder of the Youth League. The

philosophy behind Africanism was an assertive nationalism which emphasised local leadership and the liberation of Africans by Africans. It argued that Africans should assert themselves and their rights and that mass struggles were necessary to overthrow white domination. The Youth League aimed to inspire the ANC with 'the spirit of African nationalism', and it became a dominant influence in the organisation through its Programme of Action which was adopted by the ANC in 1949.

Africanism conflicted with the ANC's developing policy of a multiracial, class-oriented approach to liberation, influenced, though not dominated, by the Communist Party of South Africa. This was consolidated during the 1950s under what the Africanists called the 'leftist' or 'Charterist' (after the ANC's Freedom Charter) leadership of Chief Lutuli. Africanists were opposed to the Congress Alliance, which brought together democrats of all races in a common front during the 1950s. In 1959 the Africanists split from the ANC and formed the Pan-Africanist Congress under the leadership of Robert Sobukwe.

Africanism had its influence on the later development of black consciousness during the 1960s and 70s, but it differed profoundly from the modern movement. Black consciousness, for instance, included 'Coloureds' and 'Indians' in its definition of black, and it also conceded differences within the black community which it often attributed to class differences.

See also **African National Congress of South Africa; Black Consciousness; Congress Youth League; Freedom Charter; Lutuli**.

African Mineworkers' Union (*Zambia*)

The African Mineworkers' Union (AMU) was formed in 1949 and led by the charismatic leader Lawrence Katilungu. It joined with other unions to form the Trades Union Congress in 1950. The AMU renamed itself the Zambia Mineworkers' Union in 1965, and after a merger, the Mineworkers' Union of Zambia in 1967. This is the most powerful trade union in the country, representing the mineworkers of the Copperbelt.

See also **Mineworkers' Union of Zambia**.

African National Congress (*Zambia*)

In 1948 the Federation of African Welfare Societies formed itself into the Northern Rhodesia African Congress. This in turn renamed itself the African National Congress in 1951 under the presidency of Harry Nkumbula. The ANC was to become Zambia's first militant nationalist organisation.

In the period 1952–4 the ANC led the struggle against the imposition of the Central African Federation and organised boycotts of premises which discriminated against Africans. In 1958 the ANC agreed to support a new constitution for Northern Rhodesia and Harry Nkumbula prepared to stand for election. Kenneth Kaunda and other prominent members split from the ANC over this issue and formed another party, the Zambia African National Congress, which was to evolve into Zambia's main nationalist party, UNIP.

See also **Central African Federation; Federation of Welfare Societies; Kaunda; Nkumbula; Northern Rhodesia African Congress**.

African National Congress (*Zimbabwe*)

The African National Congress (ANC) was formed in 1957 from the Congress Youth League which began two years earlier. It should not be confused with the African National Council, also known as the ANC, which was led from 1971 to 1976 by the Zimbabwean politician, Bishop Abel Muzorewa. The President of the African National Congress was Joshua Nkomo and the general secretary, George Nyandoro. The ANC drew its main support from the urban areas and the trade union movement. Its slogan was 'Forward ever, backward never', and it proclaimed a policy of national unity and opposition to tribalism and racialism. It was pledged to encourage trade unions, fight for the repeal of the pass laws and work for the introduction of universal adult suffrage. It also demanded the repeal of the Land Apportionment Act by which land rights for black Zimbabweans were restricted to the overcrowded 'reserves'.

The first ANC in Zimbabwe was formed after the 1914–18 war in response to the growth of the ANC of South Africa, but this was a minority movement and did not survive for long. In February 1959 a State of Emergency was declared by the Prime Minister of what was then Southern Rhodesia (later to become Zimbabwe). Under this the ANC was banned and many of its members detained. In January 1960 the National Democratic Party (NDP) was formed as a caretaker party for the ANC.

See also **African National Congress of South Africa; African National Council; Muzorewa; National Democratic Party; Nkomo; Nyandoro**.

African National Congress of South Africa (*South Africa*)

The ANC or African National Congress leads the liberation struggle in South Africa. It was founded as the South African Native National Congress on 8 January 1912 at a conference in Bloemfontein, and in 1923 changed its name to the African National Congress. The first president of the organisation was John L. Dube, leader of the Natal Native Congress and the General Secretary was Sol T. Plaatje, who had been active in the Cape-based 'Coloured' organisation, the African People's Organisation.

The ANC was a modern nationalist movement. It opposed tribalism. 'We', its founders asserted, 'are one people'. Nevertheless, at the beginning it was conservative. At first the organisation was formed primarily to promote the interests of the growing but small African professional middle class: doctors, teachers, journalists and lawyers. It opposed the colour bar and sought to extend African democratic rights and advance Africans within the society generally. This it aimed to accomplish through peaceful lobbying and protest.

The Congress's first major campaign took place in 1913 with the passing of the Land Act which deprived Africans of land rights outside the 'reserves' set aside for them by the White state. The campaign failed but peaceful protest continued throughout the world war of 1914–18. Then a growing militancy became evident, the result of economic changes after the war and of a more radical Transvaal leadership. The organisation still had only a few thousand subscribing members but it began to support striking workers on the Witwatersrand. Soon Congress was openly backing mineworkers just before the strike of 1920. An attempt by J. T. Gumede, President-General of the ANC, to radicalise the organisation further failed and he lost his office in 1930, largely because of his overt association with the Communist Party. The close association between the Communist Party and the ANC has been a constant source of debate within the nationalist movement.

It was not until the 1940s that the ANC began to broaden its appeal and grow into a mass protest movement. This paralleled growing industrialisation and political consciousness among the black population during the years of the 1940–45 war. A new constitution and political programme was adopted in 1943 including a demand for full political rights. The Congress Youth League was formed in 1943 with a radical Programme of Action demanding strikes, boycotts and civil disobedience. This was adopted by Congress in 1949, and with the passing of the Suppression of Communism Act by the state – an act which ruthlessly suppressed organised opposition – the ANC became more militant in its political struggle against white oppression. In 1950 the Communist Party was banned and dissolved itself, taking the decision to work within the Congress movement.

The new radicalism of the ANC took its first form in the 1952 Defiance Campaign Against Unjust Laws, which was designed to demonstrate to the people the force of mass political action. The campaign increased ANC membership to approximately 100,000. Soon afterwards the Congress Alliance was formed to coordinate policies among the ANC, the South African Indian Congress, the Coloured People's Congress, the Congress of Democrats (a white group) and later, the trade union group, SACTU. In 1953 a decision was taken to convene a national convention, the Congress of the People, and to draw up a Freedom Charter for a future democratic South Africa. The idea came from Professor Z. K. Matthews, an ANC leader. The Congress, which took place in June 1955 in an open space near a 'Coloured' township called Kliptown, near Johannesburg, was broken up by police at the end of its second day. A year later the state charged 156 delegates with treason and a five year treason trial began. The charges were later dropped by the state.

The Freedom Charter produced by the Congress of the People has become the basic document of the ANC and sets out its main aims and philosophy. It reaffirms the multiracial character of South Africa and promises equal status for 'all national groups', as well as various welfare provisions, equal opportunities and the nationalisation of the mines, banks and monopoly industry. It brought to a head a central ideological difference within the ANC, between the Africanists, led by Robert Sobukwe, and the Charterists, led by the ANC's then president Chief Albert Lutuli. The Africanists claimed that the ANC had betrayed the nationalist struggle to win South Africa for black South Africans and were instead pursuing a left-wing, communist-inspired class struggle. In November 1958 they split from the ANC and founded the Pan-Africanist Congress in April 1959.

In April 1960 following the killings at Sharpeville, the ANC was declared an 'unlawful organisation', and forced to continue its activities underground. The relentless state persecution of opposition movements finally brought the ANC to the point where it abandoned peaceful protest and turned to the armed struggle. Its armed wing, *Umkhonto we Sizwe*, was formed in agreement with the South African Communist Party in 1960, with Nelson Mandela as its commander-in-chief.

Umkhonto began with a series of sabotage attacks within South Africa. However, this campaign did not last long, for the state soon exposed the organisation and captured most of the leaders at its secret Rivonia headquarters. The 'Rivonia Trial' resulted in the sentencing to life imprisonment of Nelson Mandela, Walter Sisulu, Govan Mbeki and others. The ANC networks had been destroyed and the focus turned towards regional alliances (as for instance with SWAPO of Namibia and ZAPU of Zimbabwe) and the propaganda battle abroad.

After the Soweto uprising in 1976 and the exodus of thousands of young people from South Africa new opportunities presented themselves to the ANC. The political situation inside the country was changing, becoming more unstable and more repressive at the same time as there was talk of 'reforming' apartheid by the White government. The developing trade unions gave a sharp edge to black workers' demands in South Africa and the ANC was involved directly in this through SACTU and other agencies. Furthermore, the exiled leadership was helped by the emergence of new, sympathetic black governments in the surrounding states: Mozambique, Angola and later, Zimbabwe. The ANC had decided to admit all races to membership after the Morogoro (Tanzania) conference held in 1969.

The 1980s have seen a dramatic change in the image of the ANC, both within South Africa and abroad. Within South Africa it is widely perceived as the legitimate symbol of all the years of protest; it represents a long tradition of resistance to the oppressive White state and it has become a rallying point for young people. Despite the fact that it is still officially banned, unquoted, and has most of its leaders in prison, it has become the focus for the seething discontent in the townships and campaigns like the 'Release Mandela Campaign' have gained widespread support. Even in white liberal circles the organisation has gained credibility and some respectability, to the extent that liberal South African business leaders met and talked to ANC leaders in September 1985 in Zambia. As the international community has imposed sanctions against the apartheid regime, so it has extended recognition to the ANC. Oliver Tambo, the ANC president, has had an official meeting with the British foreign secretary, Sir Geoffrey Howe and the US secretary of state, George Schultz. After years of support from the Soviet Union and

Eastern Europe, the West has at last also been forced to turn to the ANC as the legitimate representative of black South Africa.
See also **'Africanism'; African People's Organisation; Coloured People's Congress; Congress Alliance; Congress of Democrats; Congress Youth League; Defiance Campaign of 1952; Dube; Freedom Charter; Gumede; Land Act of 1913; Mandela, Nelson; Mandela, Winnie; Matthews; Mbeki; Pan-Africanist Congress; Plaatje; 'Rivonia Trial'; SACTU; Sharpeville; Sisulu, Walter; Sobukwe; South African Communist Party; South African Indian Congress; Suppression of Communism Act; Tambo; 'Treason Trial'; *Umkhonto we Sizwe*.**

African National Council (*Zimbabwe*)

The African National Council (also known as the ANC) was formed in 1971 with a specific purpose – to oppose the Anglo-Rhodesian Agreement made in that year between the white Rhodesian leader, Ian Smith, and the British government. This was designed to end the Rhodesian conflict and Zimbabwean opinion was to be tested in the country by Lord Pearce and the Pearce Commission. The new organisation (an amalgamation of ZANU and ZAPU – the main nationalist groups) was launched less than a month before the test of acceptability was due to begin. Its name and initials were designed to reawaken African support for the banned African National Congress. Bishop Abel Muzorewa, the president, announced its aims: 'The ANC', he declared, 'exists only to mobilize opinion against the settlement proposals'.

Despite the Rhodesian state's hostility to the new organisation, the ANC scored a massive and unexpected victory – the country overwhelmingly rejected the 1971 agreement. The ANC continued to exist after this campaign but it split in 1976 when Bishop Muzorewa broke away and formed the UANC.
See also **African National Congress (Zimbabwe); Anglo-Rhodesian Agreement; Muzorewa; Pearce Commission; UANC**.

African National Independence Party (*Zambia*)

The African National Independence Party (ANIP) was a small splinter group which broke away from Kenneth Kaunda's Zambia African National Congress. It was one of the factions which united in 1959 to form Zambia's main political party, UNIP.
See also **UNIP; Zambia African National Congress**.

African People's Organisation (*South Africa*)

The African People's Organisation (APO) was founded in 1902 by Dr Abdurahman. It was the oldest 'Coloured' political organisation and was in favour of economic action against the white state by all black South Africans. Together with the South African Native National Congress the APO opposed the proposed constitution for the Union of South Africa.
See also **Abdurahman; Plaatje; South African Native National Congress; Union of South Africa**.

African Settlement Convention (*Zimbabwe*)

The African Settlement Convention and the Rhodesian Settlement Forum

were black organisations formed to give support to the 1971 Anglo-Rhodesian agreement. They enabled Ian Smith, the leader of the white Rhodesian regime, to argue that there was black support for the settlement proposals. However, the Pearce Commission tested majority opinion in the country and found it overwhelmingly against the agreement.
See also **Anglo-Rhodesian agreement; Pearce Commission**.

African Socialism (*Tanzania, Zambia*)
African Socialism is a concept which was formulated in the 1950s by Africa's nationalist leaders. It has exasperated many political scientists as it is a phrase used to describe widely differing systems of government from strictly socialist to capitalist and mixed economies. Its real significance lies in the idea it represents. It conveys a desire by leaders to retain the good things from the traditional African societies they govern while at the same time moving their countries forward as fast as possible into the modern world. It is also a phrase which is used with effect to mobilise and unite the people.

Whatever its exact meaning, African Socialism has been used throughout the continent to describe something quintessentially African and something which is, at the same time, modern. For instance, it is part of President Nyerere of Tanzania's guiding philosophy of *ujamaa* and it is used by President Kaunda of Zambia in his approach to government which he describes as 'humanism.'
See also **'humanism'; Kaunda; Nyerere; *Ujamaa***.

Afrikaanse Handelsinstituut (*South Africa*)
Founded by the Afrikaner secret society, the *Broederbond*, in 1942 to organise small Afrikaner traders, the *Afrikaanse Handelsinstituut* (Afrikaans Commercial Institute) was set up as part of the drive to mobilise Afrikaner capital. Its aim was to break the virtual monopoly held by whites of English origin. After a long period of internal conflict, large business enterprises became dominant in the organisation during the 1970s and began to advocate reforms in the labour market to suit their interests.

By the end of the 1970s the *Afrikaanse Handelsinstituut*, despite having been instrumental in developing the system of apartheid, was supporting the 'modernisation' of the apartheid system in South Africa, and the biggest corporations became closely identified with P. W. Botha's 'reformism'. This period coincided with the beginning of a decline in the influence of the *Handelsinstituut*. This was brought about by a growing recognition among multinational companies of common interests, whether having originated as English or Afrikaans capital. Large corporations also found they were able to command the attention of government directly. Despite its more liberal image, however, the *Handelsinstituut* remained hostile to the activities of independent non-racial trade unions in South Africa and sought to limit union membership for the black workers employed by its members to the more compliant unions.
See also **ASSOCOM; Chamber of Mines; FCI; 'reforms'; SEIFSA**.

Afrikaanse Studentebond (*South Africa*) See **ASB**.

Afrikaner Broederbond (*South Africa*)
The *Afrikaner Broederbond* (Afrikaner League of Brothers) is a secret, elitist, exclusively male and Protestant Afrikaner society set up in 1918. New members undergo secret rites and swear binding oaths never to divulge information about the organisation to outsiders. It draws its members from the top echelons of the civil and public service, business and the Afrikaans churches. It is highly influential within the ruling National Party and was the driving force in the development of Afrikaner nationalism. The *Broederbond* exercises considerable power in South Africa, and all the country's prime ministers since 1948 have been members. Its present leader is Professor J. P. ('Piet') de Lange, former head of the Rand Afrikaans University. Its youth wing, the *Ruiterwag* (Mounted Guard or Cavalry Guard) had about 4,000 members in 1983, between the ages of 18 and 33.

Following a major split in the National Party in 1982, a rival organisation to the *Broederbond* was formed in 1984. Headed by a former head of the *Broederbond*, the *Afrikaner Volkswag* (Afrikaner People's Watch) opposed the 1983 constitution and the strategy of P. W. Botha. As a result of the split, the *Broederbond* lost about ten per cent of its members. It has also undergone a marked change in its politics, and its leader has described apartheid as 'a simplistic approach based on extremely naive political assumptions'. The policy of separate development, he said at the same time, was 'as dead as a dodo'.
See also ***Afrikaner Volkswag*; constitution of 1983; National Party**.

Afrikaner nationalism (*South Africa*)
The ideological basis of the National Party of South Africa, based on white supremacy and the concept of an Afrikaner 'nation': see ***volk***.

Afrikaner Party (*South Africa*)
A breakaway from the *Herenigde Nasionale Party of Volksparty* in 1941 which later reunited with it and other elements to form the present National Party: see **National Party**.

Afrikaner Resistance Movement (*South Africa*) See **AWB**.

Afrikaner Volkswag (*South Africa*)
The *Afrikaner Volkswag* (AV) or Afrikaner People's Guard was established in 1984 to challenge the *Afrikaner Broederbond*. Led by a former Chairman of the *Afrikaner Broederbond*, Carel Boshoff, its founding conference heard speeches of support from the leaders of the Conservative Party (Andries Treurnicht, the present leader, was himself a past chairperson of the *Broederbond*), the HNP and the AWB. The avowed intent of the organisation was to combat 'liberal influences' which had created a 'cultural crisis' in Afrikanerdom. Although closely associated with the neo-Nazi AWB in the public mind at the time of its formation, the AV later attempted to soften this image.
See also **AWB; Conservative Party; *Herstigte Nasionale Party***.

Afrikaner Weerstandbeweging (*South Africa*) See **AWB**.

Afro-Shirazi Party (*Tanzania*)
In 1957 the Afro-Shirazi Union was formed. This was an alliance between the African Association on Zanzibar and the Shirazi Association on the island of Pemba. Both could trace their origins back to the 1930s but had functioned mainly as social clubs. The new union was modelled on TANU (the Tanganyika African National Union), Tanganyika's campaigning nationalist party, and it maintained close links with TANU's leaders.

In the 1957 elections the Afro-Shirazi Union won five of the six seats in Zanzibar, and transformed itself into a more formal party, the Afro-Shirazi Party. In 1959 the Pemba wing (the members of the old Shirazi Association) broke away and formed the Zanzibar and Pemba People's Party. This set up an alliance with the Afro-Shirazi Party's rival, the Zanzibar National Party, and caused the Afro-Shirazi Party to lose the decisive 1963 pre-independence elections. However, the Zanzibar revolution took place the following year and power passed to the Afro-Shirazi Party. Soon after this the Afro-Shirazi Party and the Umma Party merged. After some internal power struggles Abeid Karume became President of Zanzibar and began negotiating a union with mainland Tanganyika.

The United Republic of Tanzania came into being on 23 April 1964. Although there was cooperation between Tanganyika and Zanzibar the two remained administratively separate. In April 1977 the two parties, TANU and the Afro-Shirazi Party, joined to form a new party, *Chama cha Mapinduzi*, which became Tanzania's only political party.
See also **African Association; Chama cha Mapinduzi; Karume; TANU; Umma Party; Zanzibar; Zanzibar and Pemba People's Party; Zanzibar National Party**.

Afro-Shirazi Union (*Tanzania*)
An alliance between the African Association on Zanzibar and the Shirazi Association on the island of Pemba, and closely linked to TANU: see **Afro-Shirazi Party**.

AG (*Namibia*) See **Administrator General**.

AG8 (*Namibia*)
The South African Administrator General rules by decree. Proclamation AG8 of 1980 made provision for a middle tier of government (between municipal and national levels), the so-called representative authorities. These were set up to provide separate representation for each of the eleven apartheid 'racial' categories in Namibia. They were given responsibility for areas such as health, education and personal income taxation.

The effect of AG8 has been to entrench the power and privileges of the 'White' community in Namibia. Its provisions allow the maintenance of segregated education and give the 'White' 'representative authority' effective control of the bulk of tax income (most of which is spent on services for the privileged 'White' community). Following controversy about the operation of the representative authorities, a commission of

inquiry under Justice Thirion (a South African judge) was set up to investigate allegations of corruption and maladministration in all levels of government in Namibia.
See also **Administrator General**.

a

AG9 (*Namibia*)
Proclamation AG9 of November 1977 (as amended) is the most controversial of decrees passed by South African administrators general in Namibia. It replaced and extended similar legislation. Under its terms any area in Namibia may be declared a security district. Thereafter, it provides for the detention without trial of individuals for periods of 30 days which may be repeated *ad infinitum*. Detainees may be denied access to lawyers, human rights representatives, relatives and even medical personnel, and may be held at 'any suitable place'. AG9 has been used primarily against officials and suspected supporters of SWAPO, and most cases of torture exposed in the courts have been of detainees under AG9. More than half of Namibia has been declared a security district with the result that the proclamation's provisions apply to 80 per cent of the population. In June 1979, AG9 was used to impose a dusk-to-dawn curfew in the Ovambo region, prohibiting night-time travel without official permission.
See also **Administrator General; AG8; *Koevoet***.

Aggett, Neil (*South Africa*)
Dr Neil Aggett was the Transvaal regional secretary of the Food and Canning Workers' Unions. He was arrested in late 1981 and killed in February 1982 in detention. The response to Dr Aggett's death was united and impressive. Over 100,000 workers stopped work for 30 minutes in protest. His funeral was attended by over 15,000 people, the majority of them black, a tribute to the non-racial character of the progressive wing of South Africa's trade unionism, since Neil Aggett was white.
See also **Food and Canning Workers' Union**.

Agricultural Development and Marketing Corporation (*Malawi*)
The Agricultural Development and Marketing Corporation (ADMARC) is a state-owned corporation which controls the buying and selling of agricultural products. It tries to increase agricultural production on state-owned land mainly through keeping prices stable.
See also **Malawi**.

Ahtisaari, Marti (*Namibia*)
UN Commissioner for Namibia from 1977 to 1982, and the UN Secretary General's Special Representative for Namibia, Ahtisaari is a Finnish diplomat who, previous to his period as commissioner, had been Finland's ambassador to Tanzania.
See also **Carlsson; MacBride; Mishra; UN Commissioner for Namibia; UN Council for Namibia**.

Ai Gams Action Committee (*Namibia*)

Committee set up by signatories of Ai Gams declaration to implement its decision to organise nationwide rallies in support of implementation of UN Security Council Resolution 435: see **Ai Gams declaration**.

Ai Gams declaration (*Namibia*)
This was a declaration by 16 organisations at a meeting convened by the Namibia Council of Churches in April 1986 to oppose the MPC 'internal government'. The groups, including SWAPO, demanded the end of South Africa's occupation of Namibia and agreed to 'embark on a campaign of positive action aimed at bringing about the immediate and unconditional implementation of Resolution 435'. The Ai Gams Action Committee was set up to organise rallies throughout Namibia demanding implementation of UN Security Council Resolution 435. Tens of thousands of Namibians subsequently participated in rallies organised by the committee or by SWAPO.
See also **churches; MPC; UN Security Council Resolution 435**.

Aksie Eie Toekoms (*South Africa*)
A small right wing organisation set up in 1981, *Aksie Eie Toekoms* (AET) or Action Own Future, was formed by members of the *Broederbond* to oppose the new direction of the National Party. Some of its members contested the 1981 general election, but it was not formally constituted as a political party until after the election. In 1982 it dissolved itself so that its members could join the newly-formed Conservative Party.
See also ***Broederbond*; Conservative Party; National Party**.

Aksie Red Blank Suid-Afrika (*South Africa*)
Aksie Red Blank Suid-Afrika (Action Save White South Africa) was set up in 1981 to coordinate far-right opposition to the policies of the then prime minister, P. W. Botha. The alliance included the National Conservative Party, AWB, and *Aksie Eie Toekoms*. The *Herstigte Nasionale Party* boycotted the founding conference, despite the attendance of its former leader, Albert Hertzog. The functions of *Aksie Red Blank Suid-Afrika* (ARBSA) have been largely taken over by the Conservative Party, which successfully attracted the support of most of the organisations previously involved in ARBSA.
See also ***Aksie Eie Toekoms*; AWB; Conservative Party; *Herstigte Nasionale Party*; National Conservative Party**.

Aksie Red Blank Suidwes Afrika (*Namibia*)
Action Save White South West Africa, a right wing organisation established to resist 'reform' in Namibia: see **vigilante organisations**.

AKTUR (*Namibia*)
AKTUR, the Action Front for the Retention of Turnhalle Principles, was formed in 1977 as an alliance of the right wing of the SWANP (South West Africa National Party) and a group of ultra-right black politicians. Its formation followed a split in the SWANP. AKTUR campaigned actively against the DTA, winning 6 out of 50 seats in the 'constituent assembly' of 1978 compared with the DTA's 41. With the collapse of

the DTA puppet administration, AKTUR was quietly shelved and the SWANP re-emerged.
See also **DTA; Republican Party; SWANP**.

a

aldeamentos (*Angola*)
The *aldeamentos* were so-called 'strategic villages' introduced by the Portuguese to counter the armed liberation struggle. People in the rural areas were gathered into the *aldeamentos* ostensibly to provide defensible settlements against guerrilla attack. In fact, the primary objective was to deny peasant support to guerrilla fighters and to assist the Portuguese colonial forces in their control of local people. The policy often severely disrupted local social and economic structures, forcing some peasant farmers to walk long distances to their fields.

ALIAZO (*Angola*)
ALIAZO, the *Alliance des Ressortissants de Zombo*, was a Christian-based organisation campaigning peacefully for an autonomous Zombo region in the north of Angola within a wider Angolan federation. It was the precursor to the PDA, and grew out of the ASSOMIZO, the *Association Mutuelle des Ressortissants de Zombo*.
See also **PDA; UPA**.

Alliance des Ressortissants de Zombo (*Angola*) See **ALIAZO**.

Alves, Nito (*Angola*)
Nito Alves rose through the ranks of the MPLA to join the Central Committee in September 1974. A powerful orator, he became the Minister of the Interior in the first post-independence government. Alves used his position to build a personal following in the MPLA, organising opposition to the rest of the leadership on the grounds that the movement was dominated by a *mestiço* coterie, and playing on growing resentments against the food shortages of the early years of independence. Alves was sacked as Interior Minister in 1976 and expelled from the Central Committee of the MPLA in May 1977. On 27 May 1977 he led an abortive coup against the Government.
See also **coup attempt, 1977; _mestiço_; MPLA; Nitistas**.

Alvor Agreement (*Angola*)
Following the 1974 coup in Portugal and the decision to concede independence to its colonies, the Portuguese reached an agreement with the three main nationalist organisations in Angola, the MPLA, FNLA and UNITA, on the basic principles of government in the lead-up to independence. Signed at Alvor, in Portugal, on 15 January 1975, this set a date for independence, provided for a transitional government and established a ceasefire in the liberation war. The Alvor Agreement was suspended by Portugal in August 1975 after the outbreak of civil war in Angola.
See also **FNLA; MFA; MPLA; UNITA**.

AMANGOLA (*Angola*)
Amigos do Manifesto Angolano, or AMANGOLA, was a December 1964 breakaway from Holden Roberto's UPA. A precursor to Jonas Savimbi's

UNITA, AMANGOLA called on Angolans living in exile to return to their country to participate in the armed struggle.
See also **FNLA; Roberto; Savimbi; UNITA; UPA**.

Amigos do Manifesto Angolano (*Angola*) See **AMANGOLA**.

AMU (*Zambia*) See **African Mineworkers' Union**.

ANC (*South Africa*) See **African National Congress of South Africa**.

ANC (*Zambia*) See **African National Congress (Zambia)**.

ANC (*Zimbabwe*) See **African National Congress (Zimbabwe); African National Council**.

Andrade (*Angola*) See **Pinto de Andrade**.

Anglo (*South Africa*)
This is the term commonly used to refer to the Anglo American Corporation of South Africa, the dominant multinational corporation in the country: see **Anglo American Corporation**.

Anglo-American proposals (*Zimbabwe*)
The Anglo-American proposals for a Rhodesian settlement were announced in September 1977. Soon afterwards David Owen, then the British Foreign Secretary, and Andrew Young, then the United States Ambassador to the United Nations, visited Zimbabwe to present the proposals to the illegal Rhodesian regime. They included the following conditions: surrender of power by the illegal regime and a return to legality; an orderly transition to independence in 1978; free elections on the basis of universal suffrage; the British government to establish a transitional regime; a United Nations presence including a peace force during the transitional period; a constitution providing for certain basic freedoms; a development fund to revive the economy. The proposals named Lord Carver as the proposed resident commissioner during the transitional period, and talks took place between Lord Carver and General Prem Chand of India, the special United Nations representative, and Ian Smith, the rebel Rhodesian leader. They ended in disagreement and Ian Smith publicly rejected the talks and the proposals.
See also **Carver; Smith**.

Anglo American Corporation (*South Africa*)
Anglo American totally dominates the private sector of the South African economy. It owns over two thirds of mining capital, and is the central presence in many other fields. In 1981 Anglo American became the largest foreign investor in the USA as it shifted part of its vast capital resource into safer economies to offset its huge stake in what was by then one of the shakiest economies in the world.

Despite being the largest private employer of migrant labour in South Africa, Anglo has a reputation as a leader of liberal opinion. This derives from the decision of its former chair and chief executive, Harry Oppenheimer, to support the liberal wing of the United Party when it broke away in 1959 in protest at the party's conservative position on 'race'. Oppenheimer was, for years thereafter, the sole significant financial supporter of the Progressive Party, and there were repeated exchanges of top personnel between the Anglo American Corporation and the Progressive Party.

More recently, Oppenheimer's successor as chair, Gavin Relly, was among the group of business leaders who visited Lusaka for talks with leaders of the African National Congress (ANC) about the future of South Africa. In a quirky reversal of roles, the ANC participants all wore sober suits while the business representatives dressed down in casual clothes. Perhaps more important symbolically, since the ANC representatives were banned from South Africa, it was the representatives of capital who were forced to go to the ANC.

See also **ANC; migrant labour; Oppenheimer; Progressive Federal Party; Relly; United Party**.

Anglo-Rhodesian agreement (*Zimbabwe*)

The Anglo-Rhodesian agreement came about after the 1969 constitution introduced by the rebel Rhodesian leader, Ian Smith. The constitution repudiated the first and most important of the five principles formulated by the British government in 1964 as the basis of any settlement – that there should be unimpeded progress to majority rule. Instead, the 1969 constitution limited African representation to parity in the foreseeable future. Furthermore, Ian Smith argued that the British principles had become irrelevant because Rhodesia was now in a stronger position than in earlier negotiations.

Discussion with the British government led to visits by Lord Goodman, sent by the British prime minister to Rhodesia to try to reach a settlement. Eventually Lord Goodman devised a plan based not on the principle of one-person, one-vote, but on a vote for the majority of Zimbabweans with certain property and educational qualifications. This put off the idea of black majority rule for long enough to be acceptable to Ian Smith. Lord Goodman returned to Salisbury (now Harare) in September 1971 to negotiate. Ian Smith appeared to be willing to make some amendments to the constitution, including a Declaration of Rights.

On 24 November 1971, the British Foreign Secretary, Sir Alec Douglas-Home, and Ian Smith announced their agreement. It was widely interpreted as a victory for the rebel Rhodesian regime. The proposals allowed for a continuation of white rule until at least the end of the century (one expert estimated this to be until the year 2035). In addition, sanctions were to be lifted and the 1969 constitution, which Ian Smith called the 'world-beater' would remain in force with few amendments. There were no external safeguards to prevent the regime from altering the agreement at a later date. Neither was there provision for repealing the security laws or guarantees that racial discrimination would be abandoned. In exchange Ian Smith gave up the idea of parity, enshrined in the 1969 constitution, and accepted majority rule – but only in the dim and distant future.

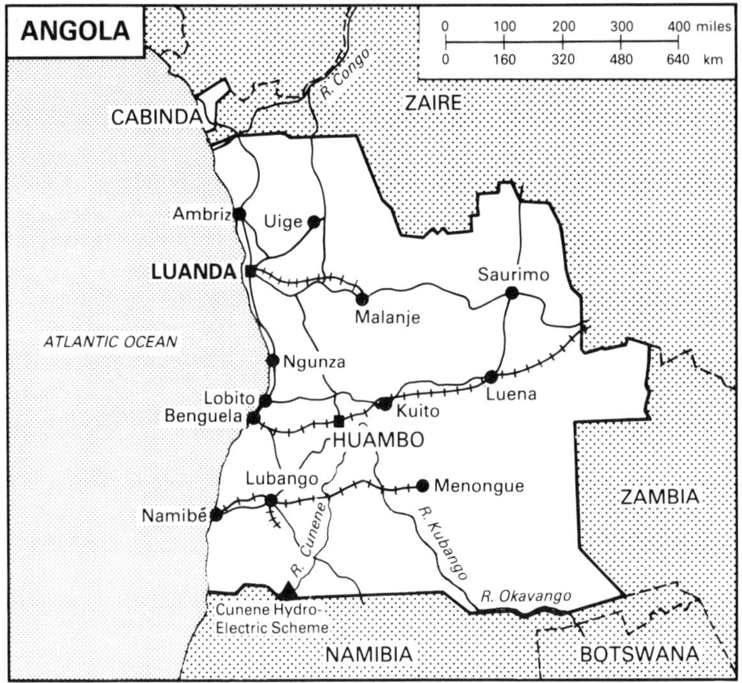

Within hours of the agreement there were reports that the shops in Salisbury had sold out of champagne. However, the celebrations of the rebel Rhodesians were shortlived, for the Pearce Commission later tested majority black opinion and in 1972 the proposals were rejected.
See also **constitution of 1969; Pearce Commission**.

Angola

Official title: People's Republic of Angola
Head of state and government: President José Eduardo dos Santos
Area: 1,246,700 sq. km
Population: 8,573,493 (official estimate, 1985)
Capital: Luanda
Official languages: Portuguese (several African languages also spoken)
GDP per capita: US$456 (1984)
Major exports: petroleum and oil products (67.7 per cent of total value), coffee (16.9 per cent), diamonds (10.7 per cent) (figures for 1979 – oil may now account for as much as 90 per cent)
Currency: Kwanza (Kw) = 100 lwei; Kw30.65 per US$ (July 1987)
Political parties: *Movimento Popular de Libertação de Angola-Partido de Trabalho* (MPLA-PT), the ruling party. Main opposition organisations are the *Frente de Libertação do Enclave*

de Cabinda (FLEC) and the *União Nacional para a Independência Total de Angola* (UNITA).

Angola is probably one of the countries in Africa richest in mineral wealth, the bulk of it unexploited. It is bordered on the west by the Atlantic Ocean, in the north and north-east by Zaire and in the south-east by Zambia. The enclave of Cabinda, an oil-rich part of Angola, is completely surrounded by Zairean territory. Angola's southern border is shared with Namibia, a fact which has allowed repeated invasions of the country by the South African occupation forces in Namibia. While infrastructure is reasonable along the coast, much of the interior of the country has poor or non-existent roads. Its main railway line, the Benguela railway, was an important route for the export of Zairean and Zambian copper, but repeated sabotage by UNITA and possibly South Africa has disrupted this.

UNITA and South African destruction of infrastructure and production equipment, coupled with the loss of most of Angola's trained personnel at independence and the destruction of vehicles and equipment by the fleeing Portuguese settlers, has devastated the economy. For a potentially rich country, most people are desperately poor, relying on subsistence agriculture to stay alive. The one bright spot in the economy is the oil industry, relatively safe in the north of the country. Reduced diamond and coffee production account for main export earnings after oil. Despite being well-watered, disruption of subsistence and commercial agriculture has made Angola a net food importer.

For fifteen years Portugal used force to try to combat the nationalist movements fighting for independence in Angola. After the 1974 coup in Portugal, however, independence became merely a matter of time. Nonetheless, with no clearly dominant liberation movement, the country was forced to go through a civil war to establish who should form the government. The *República Popular de Angola* or People's Republic of Angola was declared by the MPLA on 11 November 1975 after it had successfully repulsed invasions by the FNLA and Zairean forces from the north and the South Africans and UNITA from the south. The RPA's first president was Dr Agostinho Neto, and the first prime minister was Lopo do Nascimento. Separate governments were also declared simultaneously by the FNLA and UNITA with Holden Roberto and Jonas Savimbi, respectively, as their heads. The FNLA and UNITA governments, based in the north and south of the country, subsequently formed an uneasy 'union', the Democratic People's Republic of Angola (DPRA).

While the OAU recognised the RPA in February 1976, and further international recognition followed swiftly, no such support was forthcoming for the DPRA. By the end of March 1976 any hope of recognition had disappeared, with the FNLA forces routed and the South African invasion force, with UNITA in tow, forced out of the country.

Angola is a secular, socialist, one-party state ruled by the MPLA-PT. Popular participation occurs through the institutions of people's power headed by the elected People's Assembly. The supreme body of the

government of the RPA was the Council for Revolution, but was replaced in 1980 by a newly elected People's Assembly.
See also **FLEC; FNLA; MPLA; Neto; UNITA**.

ANIP (*Zambia*) See **African National Independence Party**.

ANP (*Mozambique*)
The *Acção Nacional Popular* (ANP) was the official fascist party in Mozambique during the colonial period.

apartheid (*South Africa*)
'Apartheid' is a word of Afrikaans derivation which means 'apart-ness' or separation. It refers to South Africa's policy of 'racial' separation enshrined in the laws of that country since the ruling National Party came to power in 1948. It is also known by government supporters as 'separate development'. It is resented and opposed by black people in South Africa and has made the country an international pariah.

In 1948 when it first came to power the National Party advocated complete apartheid between white and black people in the social, residential, industrial and political fields. Ten years later in 1958, under Prime Minister H. F. Verwoerd, classical apartheid was given a more definite form and direction. He believed that Africans should be divided into different ethnic communities each of which should be regarded as a 'national unit'. This gave rise to the 'Homelands' policy. 'Coloureds' and 'Indians', he argued, should manage their own affairs as much as possible but be subordinate to the 'White' parliament.

More recently, under the leadership of President P. W. Botha, apartheid has become an unacceptable term. The president has introduced various modifications to the policy which he calls 'reforms', in an attempt to convince the international community that apartheid is dead. Nevertheless, many of the seven so-called 'pillars of apartheid' remain (1987): separate voters' rolls, African 'Homelands', separate education, 'group areas', as well as many separate amenities. Since 1985 the Immorality Act and the Mixed Marriages Act have been abolished and some public amenities have been opened to all races. The much resented 'pass laws' have also been replaced with a new form of identification and South African citizenship has been restored to some people from the 'independent homelands'. There is an attempt underway to devise a method of giving limited political participation to urban 'Blacks' through a system involving local government, and there is talk of replacing the Group Areas Act with something less overtly racial. Nevertheless, apartheid continues to blight the lives of South Africa's people.
See also **Bantu Education; Group Areas; 'Homelands'; Immorality Act; pass laws; 'reforms'; sanctions; urban 'Blacks'**.

APO (*South Africa*) See **African People's Organisation**.

ARBSWA (*Namibia*)
Aksie Red Blank Suidwes Afrika, a right wing organisation established to resist 'reform' in Namibia: see **vigilante organisations**.

Armed Forces Movement (*Angola, Mozambique*) See **MFA**.

Arusha Declaration (*Tanzania*)
'The Arusha Declaration marked a turning point in Tanzanian politics', said President Nyerere in 1968. It defined the country as having a socialist ideology and socialist policy objectives. 'The policy of TANU (Tanzania's sole political party) is to build a socialist state', the declaration begins.

The draft was written by President Nyerere and the ideas and strategy were largely his. The National Executive Committee of TANU amended and approved it and it was published on 5 February 1967. It was officially incorporated into the constitution some years later in the Interim Constitution Amendment Bill. The declaration outlined the underlying politics and then quoted the 'principles of socialism' from the TANU constitution. The first seven concerned individual human rights and the last two defined the role of the state. The state, it said, 'must have effective control over the principal means of production', and then: 'The state should intervene actively in the economic life of the nation'.

The third part of the declaration was an analysis of Tanzania's economic position. Called 'The Policy of Self-Reliance', it concluded that the only basis for development was hard work by the people. The declaration contained five 'leadership conditions', which commit TANU and government leaders to the socialist principles of high standards of personal conduct and service to the people.
See also **Interim Constitution Amendment Bill; Nyerere; TANU; *Ujamaa***.

ASB (*South Africa*)
The *Afrikaanse Studentebond* (ASB) or Afrikaans Student Federation is an exclusivist organisation of Afrikaans students. The student bodies at all Afrikaans universities in South Africa, except the University of Stellenbosch, are members. The original ASB was established in 1916, but was eclipsed by the formation of the National Union of South African Students (NUSAS) in 1924. A separatist Afrikaner student organisation, the *Afrikaans-Nasional Studentebond* (ANS) was formed later, but became tainted with support for National Socialism during the period of the Second World War. In 1948 the ANS was dissolved and the ASB was brought back to life.

Although ostensibly a cultural organisation, the ASB religiously reflected National Party politics until recently, when splits in Afrikaner nationalism produced divisions in the ASB. In recent years, as it has battled to accommodate divisions between supporters of the National Party and the Conservative Party, the ASB has once again emphasised its role as a cultural organisation and attempted to avoid taking an overt party-political stance. At the same time it has been developing a relationship with the conservative Zulu *Inkatha* organisation.
See also **Conservative Party; *Inkatha*; National Party; NUSAS; POLSTU**.

ASP (*Tanzania*) See **Afro-Shirazi Party**.

Assembleia Popular (*Mozambique*)
The People's Assembly, the supreme organ of the state in Mozambique: see **People's Assembly**.

Assembleias do Provo (*Mozambique*)
Local and provincial people's assemblies or councils which provide a democratic input under the Mozambican constitution.
See also **People's Assembly**.

assimilado (*Angola, Mozambique*)
Meaning, literally, someone 'assimilated' into the Portuguese colonial system, this term was applied to people in the colonies who were not of Portuguese origin, but who could speak and write Portuguese and who subscribed to Portuguese cultural norms. Until the early 1960s it had a legal basis, with many privileges confined to *assimilados* and Portuguese settlers. Non-*assimilados*, or *indigenas*, were denied education above a very elementary level, were subject to forced labour requirements, and were forbidden to organise themselves economically, socially, culturally or politically.

The denial of access to education to the vast majority of Portugal's colonial subjects ensured that only a very small proportion of the population ever attained *assimilado* status. In Angola the proportion of the African population which did so was about 0.75 per cent, while in Mozambique it was even lower.
See also ***indigena; mestiço***.

Association Mutuelle des Ressortissants de Zombo
(*Angola*) See **ASSOMIZO**.

Associations (*Malawi*)
Various political organisations, known collectively as Associations, were formed in Nyasaland (later Malawi) during the early part of the century, and acted as pressure groups on the colonial administration. The first was the North Nyasa Native Association formed in 1912. Their memberships tended to be elitist, consisting of teachers, civil servants or leaders of the church. The Associations initially pressed the colonial administration for reform, but in 1944 they formed themselves into the Nyasaland African Congress, the country's first nationalist movement.
See also **Nyasaland African Congress**.

Association of Chambers of Commerce (*South Africa*) See **ASSOCOM**.

ASSOCOM (*South Africa*)
The Association of Chambers of Commerce (ASSOCOM) is the national body which brings together local chambers of commerce from throughout the country. Although most Afrikaner enterprises are organised outside ASSOCOM, as are most black enterprises, it remains the most representative body of private companies, reflecting the continued domination of the capitalist sector by the white English-speaking population.

Established in 1892, ASSOCOM moved from a primarily commercial base to a membership which included both commerical and industrial enterprises. Having fought for the interests of indigenous capital for most of the century, by the 1970s it was advocating reform and the liberalisation of apartheid laws. It was initially highly supportive of P. W. Botha's programme, but later became critical of the slow pace of his 'reforms'. See also ***Afrikaanse Handelsinstituut*; Chamber of Mines; FCI; NAFCOC; 'reforms'; SEIFSA**.

ASSOMIZO (*Angola*)

The *Association Mutuelle des Ressortissants de Zombo*, a social organisation in the north of Angola which was transformed into a political organisation, ALIAZO in 1959.
See also **ALIAZO**.

austerity programme (*Zambia*)

Zambia has been forced to introduce a severe austerity programme under the auspices of the International Monetary Fund (IMF) in order to satisfy its external creditors. The economy is almost totally dependent on copper (copper and cobalt account for over 95 per cent of export earnings) and world prices have now dropped to historically low levels. Add to this the fact that most experts predict that Zambia's copper reserves will be exhausted early in the next century, and it is obvious that the economy urgently needs to be diversified. Real income per head in Zambia was halved between 1974 and 1980, and by 1985 was less than one-third of the 1974 level. This compares with the first six years after independence when Zambia was one of the richest and fastest-growing countries in Africa – GDP increased by an estimated 13 per cent per annum from 1964 to 1969.

In February 1986 a rigorous new reform programme was agreed with the IMF. One of the main aspects of the programme is a foreign exchange auctioning system which has led to the devaluation of the *kwacha* by at least 70 per cent. There were also cuts in basic good subsidies, interest rate rises and promised reductions in civil service staffing and perks.

The public reacted strongly to the austerity measures. Earlier, less drastic measures provoked a student riot in 1984 leading to the dismissal of over 2,000 students and the closure of Lusaka University. There were angry demonstrations in Lusaka in October 1985 and in December that year students protested against the reintroduction of boarding fees for pupils. The university was again closed after boycotts in May 1986.

Then in December 1986 the country experienced its worst rioting since independence. Four days of serious food riots led to an official death toll of 15 and an unofficial estimate of 25 deaths with over 150 injuries. Shops and property were burnt and destroyed. The riots spread throughout the Copperbelt affecting all seven towns in the area. The crisis was triggered by the government's decision to end subsidies on basic food commodities. There was an immediate increase of over 100 per cent in the price of the country's main staple, maize meal. A bag of 50kg was selling at some 37 *kwacha* before the announcement and it shot up to 82 *kwacha* overnight. The monthly income of a domestic worker is 70 *kwacha;* inflation is

running officially at 50 per cent but unofficially some estimates put it as high as 600 per cent. Dr Kaunda revoked the increase in maize prices, and in early 1987 he was forced to abandon the programme – and IMF money. The country needed a new strategy, he declared, which 'while ensuring austerity, recognises the need to protect the poor and vulnerable members of our society'.
See also **Copperbelt**.

'Authorised Person' (*Swaziland*)
The 'Authorised Person' has an extremely important role within the traditionalist power structure of Swazi society. It is a heriditary office, the most well-known recent incumbent being Prince Sozisa Dlamini. The 'Authorised Person', together with the Elders of the Nation, oversees Swazi politics and is the spokesperson of the elders. The office is actually superior to that of the monarchy, but rarely becomes directly involved in the political process. However, the period since the death of King Sobhuza in 1982 has proved an exception and shown the extent of the power vested in the office.

In this instance the 'Authorised Person' acted to remove the queen regent, Dzeliwe, from power because she refused in August 1983 to sign a document transferring many of her powers to the office of the 'Authorised Person', at the bidding of the *Liqoqo* or the Supreme Ruling Council. A week later an issue of the *Government Gazette*, signed by the 'Authorised Person' announced Dzeliwe's dismissal. She was forcibly removed from her residence and Queen Regent Ntombi – who signed the controversial document – was installed in her place. The result has been the strengthening of the already powerful office of the 'Authorised Person'.

The recent trend has been to reduce the power of the *Liqoqo* and the *Government Gazette* of 10 October 1986 took away the right of the *Liqoqo* to appoint the 'Authorised Person'. This has revoked the King's Proclamation Amendment Decree of 1982 which gave the *Liqoqo* that power.
See also **Dzeliwe; Elders of the Nation; *Liqoqo*; Sozisa**.

AV (*South Africa*) See ***Afrikaner Volkswag***.

AWB (*South Africa*)
The *Afrikaner Weerstandbeweging* (AWB) or Afrikaner Resistance Movement is an extreme, neo-Nazi organisation of white South Africans opposed to the policies of the National Party under P. W. Botha. Formed in 1979, it sports a three-armed swastika-like emblem, and calls its 'military wing' the *Stormvalke* (Storm Falcons). Its leader, Eugene Terre-Blanche, an ex-policeman, is an accomplished demagogue, surrounding himself with black-uniformed, jackbooted bodyguards during his speeches. The AWB violently disrupted National Party meetings during by-elections in 1985 and 1986, fighting pitched battles before television cameras for control of the venues. Members of its so-called *Blitzkommando* have been at the forefront of such events. In 1981 the AWB formed the *Blanke Volksstaat Party* (White People's State Party) to

pursue its political ends. This was dissolved in 1982 when the AWB supported the formation of the Conservative Party. The Government has, from time to time, threatened to take action against the AWB to control its worst excesses.
See also ***Aksie Red Blank Suid-Afrika***; **Conservative Party; TerreBlanche**.

AZACTU (*South Africa*)
AZACTU, the Azanian Confederation of Trade Unions, joined with CUSA, (the Council of Unions of South Africa) in 1976 to form CUSA-AZACTU (renamed NACTU, the National Confederation of Trade Unions, in 1987). This is a black consciousness grouping of trade unions in contrast to COSATU, the largest union federation, which is non-racial.

AZACTU's roots were in the black consciousness movement and it admitted only black members, claiming a total of 70,000 in 1986. BAMCWU (the Black Allied Mining and Construction Workers' Union) was the only really large union affiliated to AZACTU. AZACTU's General Secretary was Pandelani Nefolovhodwe, now one of the two main leaders of NACTU.
See also **COSATU; CUSA; NACTU**.

Azania (*South Africa*)
Azania was the new name adopted for South Africa by the Pan-Africanist Congress during the 1960s. It was a term used by early cartographers for part of East Africa and derived from an Arabic expression. Its use now indicates a leaning towards black consciousness ideology, for instance in AZAPO (the Azanian People's Organisation), a black consciousness grouping. An exception to this is AZASO which is non-racial.
See also **AZAPO; AZASO; Black Consciousness Movement; PAC**.

Azanian Confederation of Trade Unions (*South Africa*) See **AZACTU**.

Azanian People's Organisation (*South Africa*) See **AZAPO**.

Azanian Students' Movement (*South Africa*) See **AZASM**.

Azanian Students' Organisation (*South Africa*) See **AZASO**.

AZAPO (*South Africa*)
AZAPO, the Azanian Peoples' Organisation, was founded in May 1978 to fill the political vacuum left by the banning of black consciousness organisations in 1977. It is currently (1987) the leading black consciousness group within South Africa. Its motto, 'One people, one Azania', placed it firmly within this tradition. However, AZAPO has sought to take the philosophy further and direct itself towards the black working class. At its second conference at Roodepoort in September 1979 a policy paper stated: 'AZAPO has taken Black Consciousness beyond the phase of Black awareness into class struggle'.

The organisation recognised the importance of trade unions in the liberation struggle and it also accepted that some black people would collaborate with the 'White' state because it would be in their class

interest to do so. Despite this attempt to broaden the movement, AZAPO's influence continues to be strongest among the intelligentsia, particularly in Soweto. The organisation has been heavily suppressed by the state and its leaders constantly banned and arrested. Some of the most prominent are George Wauchope, the general secretary, Nkosi Molala, the current president (1987) and Saths Cooper, a former president.
See also **Black Consciousness Movement; National Forum**.

AZASM (*South Africa*)
AZASM, the Azanian Students' Movement, was inaugurated in Pietersburg in July 1983 at a conference addressed by a former executive member of SASO, the black students' movement from the 1960s. AZASM's president, K. Lengene, said that its establishment was a reaction to the movement of the rival AZASO away from black consciousness to non-racialism. In constrast, AZASM was adopting a firm black consciousness stance. It recognised AZAPO as the authentic liberation movement for black people and pledged support for the formation of black-consciousness-orientated trade unions. In 1984 and 1985 there were clashes between AZASM and AZASO.
See also **AZASO; COSAS; SASO**.

AZASO (*South Africa*)
AZASO, the Azanian Students' Organisation, works closely with COSAS, the Congress of South African Students. It split from the main black consciousness organisation, AZAPO, in 1980 after it opposed the suspension of AZAPO's president, Curtis Nkondo. AZASO differs from AZAPO, according to a former president, Joe Phaahla, in that it adopts a class analysis of society. However, he claimed AZASO was willing to cooperate with AZAPO to get rid of apartheid. In 1984 and 1985 there were clashes between the two organisations. At its 1985 congress AZASO's priorities were stated to be a campaign for democratic student representative councils, an education charter campaign and a commitment to build organisational structures. Billy Ramokgopa, a medical student at the University of Natal, was elected president.
See also **AZASM; COSAS**.

B52s (*Angola*)
Mosquito-borne malaria is a major problem in Angola and has been exacerbated by the disruption caused by years of war. B52s is the humorous and popular name given to the mosquitos which plagued those fighting in the north of the country to resist the invasions by the CIA-backed FNLA and Zairean forces during the civil war.

baasskap (*Namibia*)
This is an Afrikaans term used in Namibia and South Africa to denote white domination or control of the political process. It is a noun meaning, literally, a situation of 'mastery' or control.
See also **apartheid**.

Babu, Abdul Rahman (*Tanzania*)
In 1963 Abdul Rahman Babu formed the Umma Party with a socialist philosophy. He became a minister in the union government in 1964 and proved to be one of President Nyerere's most vocal critics on the left, demanding more aggressive socialist and nationalist policies. In 1972 he was suspected of being involved in the assassination of Zanzibar's president, Abeid Karume, and detained on the mainland. He was released in an amnesty declared by President Nyerere in April 1978.
See also **Karume; Nyerere; Umma Party**.

bairro (*Angola*)
The neighbourhood, or *bairro*, is an important unit of organisation under the Angolan constitution. Through the concept of *poder popular*, committees organised for each *bairro* (neighbourhood committees) take responsibility for organising essential services such as schools, local health facilities, food distribution cooperatives and information and political work.
See also ***Poder Popular***.

Bamangwato (*Botswana*)
The Bamangwato, or Ngwato, group is the largest ethnic group in Botswana, making up about a third of the population and owning a fifth of the land. Sir Seretse Khama was their hereditary chief and his marriage in 1948 to an English woman provoked a political crisis which resulted in his not taking up his traditional position.
See also **Khama**.

Banana, Canaan Sodindo (*Zimbabwe*)

Rev. Canaan Banana became the first President of Zimbabwe in April 1980. He is a Methodist clergyman and was detained and restricted during the nationalist struggle by the white Rhodesian regime. He was a founder member of Bishop Muzorewa's African National Council and became a member of the ruling party, ZANU, in 1976.

See also **African National Council; Muzorewa; ZANU**.

b

Banda, Hastings Kamuzu (*Malawi*)

Dr Hastings Banda is the Life President of Malawi and president of the country's only legal political party, the MCP (Malawi Congress Party). He is a controversial figure in southern Africa because his regime is politically out of tune with the other 'frontline' states of the region. Unlike them, his philosophy is conservative and he maintains friendly relations with white-ruled South Africa. In 1967 diplomatic ties were opened with South Africa and the next year Malawi accepted a South African loan to finance the first stage of the new capital city officially established at Lilongwe. In May 1970 the South African Prime Minister, B. J. Vorster, visited the country and diplomatic contacts were raised to ambassadorial level. Then, to underline his different approach, Dr Banda himself visited South Africa in August 1971.

This cordiality has not always been extended to his other neighbours, Tanzania and Zambia. Opposition Malawian politicians have fled to neighbouring states for refuge and in 1965 Dr Banda accused Tanzania of helping to organise subversive activities against him. In 1968 there was a boundary dispute over Lake Malawi in which Malawi laid claim to the northern half, including the Tanzanian shores. However, during the 1970s the neighbouring states made an effort to improve their relations with Dr Banda. The first overtures were made by Zambia's president, Kenneth Kaunda; in July 1971 he opened a high commission in Malawi. By 1980 the situation had improved considerably and Malawi joined SADCC (the Southern African Development Coordination Conference), an organisation of southern African states committed to developing alternative economic networks not dependent on South Africa. In 1985 full diplomatic relations were established with Zambia and Tanzania, and an agreement was signed with Mozambique providing for defence and security cooperation. Despite this, Malawi is still accused of supporting the Mozambique rebels known as the MNR.

Dr Banda pursued his university education abroad, in the USA and Britain, where he qualified as a medical doctor. He did not return to Malawi until July 1958 at the request of Henry Chipembere of the Nyasaland African Congress. The consensus was that a charismatic leader was needed to head the party. Dr Banda campaigned successfully for independence and formally became prime minister in February 1963. Malawi became a republic in 1966 and Dr Banda was elected president. These were the last presidential elections and in 1971 parliament declared Dr Banda the country's Life President.

The main challenge to Dr Banda's increasingly despotic rule came in

1964 from his own cabinet. Several ministers were dismissed and others resigned after they failed to make the president pursue more radical policies. After the cabinet crisis Dr Banda became more conservative and repressive. He expanded traditional courts (the president regards British justice as too permissive), and he even introduced regulations on tight trousers and short skirts. News and information are strictly controlled. Although President Banda is now one of Africa's oldest rulers – he was born in 1898 – he has controlled power so tightly in Malawi that no clear successor has been allowed to emerge. One name mentioned in this regard, however, is John Tembo, Secretary-General of the MCP and Governor of the Central Bank of Malawi.
See also **Chirwa; Chiume; 'frontline' states; MCP; MNR; Nyasaland African Congress; SADCC; Tembo**.

banning order (*Namibia, South Africa*)
Banning orders have been used by South Africa as a means of silencing and punishing political opponents. They are executive orders which cannot be challenged in court, for which no reasons need be given and which variously restrict their subjects' rights of abode, of association and rights to work. They also make it an offence to quote banned people or to publish material written by them. Banning is a highly flexible form of house arrest which has been used to great effect to suppress opposition to apartheid. Orders are issued under the Internal Security Act. In Namibia banning orders have been used against SWAPO leaders in the main, including vice-president Nathaniel Maxuilili, SWAPO Women's Council leader Gertrude Kandanga, and SWAPO Secretary for Labour Jason Angula.
See also **detention without trial; Internal Security Act**.

bantfanenkosi (*Swaziland*) See ***mtfanenkosi***.

Bantu (*southern Africa*)
Bantu is the name given to a group of languages spoken by almost all the people living in southern Africa, south of a line between Cameroon and central Kenya. Bantu languages spoken south of the Zambezi River are usually called 'southeast African Bantu'. However, the linguistic term is commonly used in southern Africa as an ethnic or racial identification. In particular 'Bantu' was adopted by the white South African state to refer to all Africans and it became a derogatory, racist term for African people generally.
See also **'Black'**.

Bantu Education Act (*South Africa*)
'Bantu education' is one of the most hated aspects of apartheid. It was the immediate cause of the Soweto uprising of 1976 and has created deep resentment in the black population over many years. The Bantu Education Act was introduced in 1953 after a Commission on Native Education chaired by Dr W. W. M. Eiselen had reported in 1951. The commission's report formed the basis for the system of 'Bantu education' introduced in 1954. It acknowledged that Africans it had consulted

showed 'an extreme aversion to any education specifically adapted for the Bantu'. Nevertheless, the commission argued, 'Bantu education' should be separate. 'Educational practice must recognise that it has to deal with a Bantu child, that is a child trained and conditioned in Bantu culture, endowed with a knowledge of a Bantu language, and imbued with values, interests, and behaviour patterns learned at the knee of a Bantu mother. These facts must dictate to a very large extent the content and methods of his early education.' The commission considered that Bantu education should be an integral part of a carefully planned policy of socio-economic development for the 'Bantu', emphasising the value of a school as an institution for the transmission and development of the 'Bantu' cultural heritage.

This philosophy was strongly opposed by the government's critics. One, a leading African educationalist, Dr D. G. S. M'Timkulu, in an article published by the South African Institute of Race Relations in 1959, argued that Africans 'seek for integration into the democratic structure and institutions of the country. To them, one of the most effective ways of achieving this is by education – an education essentially in no way different from, or inferior to, that of the other sections of the community.'

The government spends far less on black education than on white. For instance in 1984–5 per capita expenditure on 'Black' pupils (in 'White' areas and non-independent 'Homelands') was R293.86; on 'Coloured' pupils R708.32; on 'Indian' pupils R1,182.00 and on 'White' pupils R1,926.00. The pupil-teacher ratios are also very different. In 1984 the ratios were: 'White', 18.7 to 1; 'Indian', 22.5 to 1; 'Coloured', 25.4 to 1; 'Black', 41.2 to one. Education is compulsory for 'Whites', 'Indians' and 'Coloureds' up to the age of sixteen. Compulsory education for Africans has been introduced at those schools where the school committees have requested it. The government has said that school attendance should be compulsory for six years.
See also **apartheid; Soweto uprising of 1976**.

bantustans (*South Africa*)
Bantustans or 'Homelands' or 'Black States' are those areas of land set aside under the South African government's policy for different groups of the African population in South Africa. The 'Homelands' policy is a cornerstone of apartheid.
See also **'Black'; 'Homelands'**.

Basotho (*Lesotho*)
The people living in Lesotho who speak the Sesotho language are called Basotho. Some Basotho people live in Botswana and others in South Africa.
See also **Orange Free State**.

Basotho National Party (*Lesotho*)
The Basotho National Party (BNP) was formed by Chief Leabua Jonathan in 1959. It held power through all the early years of independence from 1965 until 1970 when Chief Jonathan declared a state of

emergency and suspended the constitution. Its philosophy was conservative, traditionalist and anti-communist.

The BNP did not do well in the 1960 elections despite the support of the Roman Catholic church which is very powerful in Lesotho. However, by 1965 the party was strongly supported by South Africa and won a majority of seats in the elections. Chief Jonathan was the only political leader allowed to campaign inside South Africa among the thousands of Basotho migrant workers. Although he declared himself opposed to apartheid, Chief Jonathan also argued that he would be able to achieve the best deal with South Africa, while maintaining Lesotho's independence.
See also **Jonathan**.

Basutoland (*Lesotho*)
Basutoland was the name given to Lesotho before its independence on 4 October 1966. It had become a British colony on 12 March 1868. In 1871 the territory was formally annexed to the Cape Colony. In 1883 the British government again took over control of Basutoland and administered it as one of the 'High Commission Territories' (together with Swaziland and Bechuanaland) until independence.
See also **Lesotho**.

Basutoland Congress Party (*Lesotho*)
The Basutoland Congress Party (BCP) was formed in 1952 as the Basutoland African Congress. It was Lesotho's first modern political party inspired by African nationalism and modelled on the ANC of South Africa. Ntsu Mokhehle was its founder and first president. The party spread its message inside the country through its publication, *The Warrior* or *Mohlabani*. In 1960 the renamed Basutoland Congress Party won the 1960 elections with a large majority. However, splits developed within the party and there were some defections. The crucial 1965 pre-independence elections were won by the opposition party of Chief Leabua Jonathan. The BCP vowed to fight back in the 1970 elections, and the results seemed to be to their advantage. But Chief Jonathan suddenly declared the elections null and void and announced a state of emergency, arresting many oppostion leaders including Ntsu Mokhehle.

Since 1970 the BCP has been in disarray. A rump lead by Gerard Ramoreboli sat in Chief Leabua Jonathan's Interim National Assembly, but it was officially expelled by the party leadership. In 1974 violence erupted and Ntsu Mokhehle fled the country. Trials led to jail sentences for 35 BCP supporters convicted of high treason and contravention of internal security laws in connection with the abortive coup in January 1974. Clashes between the BCP's military wing, the Lesotho Liberation Army (widely thought to be backed by South Africa), and the Lesotho government have continued over the years. But the party has remained in exile and its influence has gradually diminished.
See also **Jonathan; Mokhehle; Ramoreboli**.

Basutoland Freedom Party (*Lesotho*)
The Basutoland Freedom Party was formed in 1960 following a split from the Basutoland Congress Party by B. M. Khaketla, the deputy president. The new party attacked the BCP for antagonising the chiefs, largely over religion; B. M. Khaketla was a devout Anglican while the BCP leaders tended not to be religious. The party later merged with the Marema-Tlou Party.
See also **Basutoland Congress Party; Marema-Tlou Party**.

b

Basutoland National Party (*Lesotho*) See **Basotho National Party**.

Batswana (*Botswana*)
The Batswana, or Tswana, are the citizens of Botswana. The local language is Setswana.

BCP (*Lesotho*) See **Basutoland Congress Party**.

BDP (*Botswana*) See **Botswana Democratic Party**.

Bechuanaland (*Botswana*)
The name of Botswana until its independence in 1966: see **Botswana**.

Bechuanaland Democratic Party (*Botswana*)
The Botswana Democratic Party was called the Bechuanaland Democratic Party when it was first formed in 1961 by members of the Legislative Council, including Sir Seretse Khama who was to be its leader and President for many years.
See also **Botswana Democratic Party; Khama**.

Bechuanaland People's Party (*Botswana*)
The Botswana People's Party was called the Bechuanaland People's Party when it was first formed in 1960. It was Botswana's (then Bechuanaland) first really modern political party with a nationalist approach. Its founder was Kgaleman Motsete.
See also **Botswana People's Party; Motsete**.

Bechuanaland Protectorate Federal Party (*Botswana*)
The Bechuanaland Protectorate Federal Party (BPFP) was Botswana's first political party. It was started in 1959 by Leetile Disang Raditladi and campaigned for support in the 1961 elections for a Legislative Council. The party wanted the chiefs to be removed from party politics but was otherwise very conservative and opposed a black majority in the council. By 1962 it had ceased to exist.
See also **Bechuanaland People's Party**.

Bemba (*Zambia*)
Bemba-speakers are the largest group in Zambia and make up some 34 per cent of the population. They have always been strongly represented in Zambia's main political party, UNIP, and within the party they were led by Simon Kapwepwe, Zambia's vice-president.

In 1971 President Kaunda dismissed several Bemba-speaking ministers

from his government and this alienated the group within UNIP. In response Simon Kapwepwe formed the United Progressive Party, drawing its support mainly from Bemba-speakers within the Copperbelt. The Government feared that its popularity would spread and banned the UPP in 1972, declaring Zambia a one-party state.
See also **Copperbelt; Kapwepwe; UNIP; United Progressive Party**.

Benguela railway (*Angola*)
Potentially one of the most important trade routes for southern Africa, and one of the few routes to the sea which does not run through South African territory, the Benguela railway has been a major target for opponents of the MPLA government. During the independence struggle the guerrilla forces avoided attacks on the line, on the whole, as their host countries, Zaire and Zambia, were dependent on it as a route for exports and imports.

After independence, however, UNITA made it a particular target, both as a way of increasing economic pressure on Angola and as a way of increasing the dependence of other southern African countries on UNITA's sponsor, South Africa. A working Benguela railway is essential to the effective operation of the SADCC, Zambian and Zimbabwean independence of South African controlled trade routes.

The railway runs east from the Atlantic port of Benguela to the Zairean border and there connects to the Zairean and then Zambian railway networks. It is the natural export route for copper produced in the Copperbelt and the spasmodic nature of its operation since Angolan independence has been costly for both countries.
See also **MPLA; SADCC; UNITA**.

Bevryder Demokratiese Party (*Namibia*)
This is a coalition of the small, ethnically based *Rehoboth Bevrydingsparty* and the Rehoboth Democratic Party.
See also **DTA; Liberation Front; National Convention of Namibia**.

Biko, Stephen (*South Africa*)
Steve Biko was the most prominent leader of the black consciousness movement in South Africa, and is sometimes known as the 'father of Black Consciousness'. He was the driving force behind the formation of the first black consciousness organisation, SASO (the South African Student's Organisation). He was also a founder member of the Black People's Convention, formed later, in an attempt to broaden the appeal of the philosophy. He died in detention on 12 September 1977 after having been beaten and tortured by the South African security police. He was carried some 750 miles from Port Elizabeth in the Cape province to Pretoria naked in the back of a van the day before he died from a brain injury following blows to the head.
See also **Black Consciousness Movement; Black People's Convention; SASO**.

BIP (*Botswana*) See **Botswana Independence Party**.

'Black' (South Africa)
This is the term used by the South African government to refer to Africans. Black people, however, use 'black' as a term to refer to all those who would identify themselves as black, excluding only those who regard themselves as white. This includes Africans and the groups termed 'Coloured', and 'Indian' by the state. The use of 'black' in this way is part of an explicit rejection of the divisive apartheid categorisation.

In the first stages of colonisation African people in South Africa were referred to as 'Natives'. From 1951, when the Bantu Authorities Act was passed, the National Party Government began using the term 'Bantu' instead in most cases. After 1962 the term 'Native' was no longer used officially. In 1978 the government decided to stop using the term 'Bantu' and instead to use the term 'Black', but to apply it only to the African population. From 1978 onwards, the 'Homelands', for instance, were referred to as 'Black States'. The Department of Bantu Administration and Development became the Department of Plural Relations and Development, while the Bantu Education Department was renamed the Department of Education and Training.
See also **Introduction; apartheid; 'Coloured'; 'Homelands'; 'Indian'**.

Black Consciousness Movement (*South Africa*)
The Black Consciousness Movement (BCM) is a broad term used to describe the ideology of black consciousness and the movements and organisations which developed around it in the late 1960s and early 1970s. It was enormously influential and inspired a generation of young, black South Africans. It was at its most popular during the Soweto uprising of 1976, and all its organisations (some 18), together with many of its leaders, were banned in October the following year. Its most famous leader was Steve Biko who was killed while in police custody in September 1977.

Steve Biko described the emergence of black consciousness in an article he wrote in 1972:

> . . . in South Africa political power has always rested with white society. Not only have whites been guilty of being on the offensive, but, by skillful manoeuvres, they have managed to control the responses of the blacks to the provocation. Not only have they kicked the black but they have also told him how to react to the kick. With painful slowness he is now beginning to show that he realises it is his right and duty to respond to the kick the way he sees fit. . . . The philosophy of Black Consciousness, therefore, expresses group pride and the determination by the blacks to rise and attain the envisaged self.

Whites were to be excluded from all black consciousness activities. The first step, it was argued, was to develop a positive consciousness of being black; psychological liberation from obedient slavery had to take place before political action could result in freedom. Black was defined as African, Indian and 'so-called coloured'. This distanced the movement

b

from the old Africanist tradition which had existed within the ANC and motivated the breakaway and formation of the Pan-Africanist Congress in 1959. Africanists believed that Africans had to free themselves and rejected multiracialism, but they were concerned only with Africans and they were a movement which developed within a political group and which was, moreover, obsessed with strategy and tactics. A more powerful influence on black consciousness was the black power movement from the USA and black theology which was represented in South Africa within the University Christian Movement. Black consciousness drew on many different political currents, but it tended to be against a class perspective. To quote Steve Biko: '(some people) tell us that the situation is a class struggle rather than a race one. Let them go to Van Tonder in the Free State and tell him this. We believe we know what the problem is and will stick by our findings.'

SASO, the South African Students' Organisation, was the first black consciousness organisation to be formed in 1969, and the movement was initially rooted among university students. After SASO many other organisations followed and the idea began to spread beyond the initial group of young intellectuals into the wider community. In 1972 a Black People's Convention (BPC) was established which argued for the implementation of 'Black Communalism' through literacy campaigns, health projects, and a general workers' union, the Black Allied Workers' Union (BAWU). The BPC never developed into a mass organisation and black consciousness itself is now (1987) far less influential. However, the ideas it put forward have had great impact on South African politics.
See also **'Africanism'; AZACTU; AZAPO; Biko; Black Parents' Association; Black People's Convention; CUSA; NACTU; SASO**.

'blackjacks' *(South Africa)*

'Blackjacks' is the name used by township residents to describe the black policemen who patrol the townships. They are the target for reprisals by activists and many have been killed by 'necklacing' (burning to death after a tyre has been placed around the victim's neck) in recent years. They are drawn mainly from people living in the impoverished African 'Homelands'. In 1986 some 300 'blackjacks' went on the rampage in Katlehong township near Johannesburg for more pay.
See also **'necklacing'; townships**.

Black Parents' Association *(South Africa)*

The Black Parents' Association (BPA) was formed in response to the Soweto uprising of 1976. It had links with the black consciousness movement and was drawn from professional and church people. It acted as a support group for the students involved in the unrest, organising funerals and legal and medical help.
See also **Black Consciousness Movement; Soweto uprising, 1976**.

Black People's Convention *(South Africa)*

The Black People's Convention (BPC) was formed in 1972 in order to try and spread the influence of black consciousness from SASO, the black

students' organisation, to the community. The BPC adopted a programme of 'Black Communalism' which sought to promote self-help and self-sufficiency through literacy campaigns, health projects, cultural activities and a general workers' union, the Black Allied Workers' Union (BAWU). It did not develop into a mass organisation although it had some 41 branches and its influence spread far beyond its formal membership. It was banned in 1977 together with the other black consciousness organisations.
See also **Biko; Black Consciousness Movement; SASO**.

Black Sash (*South Africa*)

The Black Sash is a liberal white women's organisation which has opposed apartheid in South Africa. Formed in 1955 as the Women's Defence of the Constitution League, its first campaign was to oppose taking 'Coloured' voters off the common voters' roll. For many years it was best known for its silent 'vigils' of standing women wearing black sashes to signify mourning at the violation of the South African constitution.

After its formation, the Black Sash became more directly involved in the day to day experiences of black people through a network of pass law advice centres established to help them with problems related to the pass laws. More recently, the advice centres have taken on general advice roles for those coming to them for assistance. The Black Sash became more political, expressing support for the Freedom Charter as 'the only viable alternative to the present exploitative and repressive system'.

In 1983, in a departure from its main areas of activity, the organisation was instrumental in establishing the End Conscription Campaign (ECC) to campaign against the compulsory conscription of young white men in South Africa for military service. The president of the Black Sash is Sheena Duncan.
See also **Duncan; End Conscription Campaign; Freedom Charter; pass laws**.

'black spots' (*South Africa*)

'Black spots' are areas of land within 'White' South Africa where 'Black' people live. Under the South African government's apartheid policy this is against the law and the aim is to eliminate what it refers to as 'black spots'. This policy is pursued by 'removing' people – usually against their will – and taking them to the 'Bantustans' or 'Homelands'.
See also **'Homelands'**.

Blank SWA (*Namibia*)

White SWA, a right wing group committed to white supremacy: see **vigilante organisations**.

Blantyre (*Malawi*)

Blantyre is Malawi's largest city, situated in the south of the country. Its population in 1981 was estimated as 295,000. It is named after the Scottish birthplace of Dr Livingstone, the famous missionary and explorer in east and central Africa.

Bloedsap (*South Africa*)
This term, literally 'blood SAP', meant a dyed-in-the-wool member of the United Party, usually very conservative and often a rural-dweller. It originally denoted those members of the United Party who had been supporters of the old South African Party (SAP) led by Jan Smuts (not to be confused with the shortlived South African Party formed in 1977).
See also **National Party; South African Party; United Party**.

b

BNF (*Botswana*) See **Botswana National Front**.

BNP (*Lesotho*) See **Basotho National Party**.

BNP Youth Wing (*Lesotho*)
The BNP (Basotho National Party) Youth Wing was set up by Chief Leabua Jonathan. Chief Jonathan followed a policy of arming the Youth Wing during his last years in power and building it into a politicised unit within the armed forces. Many members of the Youth Wing had close relations with exiled members of the ANC of South Africa and provided the opposition movement with some support within Lesotho.

The growing power of the Youth Wing is cited as one of the reasons for Major General Lekhanya's successful military coup in 1986. The General was a conservative soldier who feared the politicisation of the Youth Wing as well as being concerned that its support for the ANC could bring down the wrath of South Africa on Lesotho.
See also **Basotho National Party; Jonathan; Lekhanya**.

Boesak, Allan (*South Africa*)
Rev. Dr Allan Boesak is the President of the World Alliance of Reformed Churches. He proposed the formation of the United Democratic Front, now South Africa's largest anti-apartheid organisation and has played a prominent role in it ever since. He was elected Patron at its launch in 1983. He is a powerful speaker and a charismatic leader who has been an effective international spokesperson against apartheid.

It is perhaps in the Dutch Reformed Church – the spiritual home of Afrikaners – that Dr Boesak has caused most embarrassment to the government. He is minister of the Bellville congregation of the Dutch Reformed Mission Church (*NG Sendingkerk*) in the Cape, and he has used his position as a church person in the reformed tradition to wage a long campaign against apartheid. In August 1982 at the Ottawa meeting of the World Alliance of Reformed Churches, an organisation representing some 150 Calvinist churches in 76 countries, including the Dutch Reformed Churches and the *Nederduitsch Hervormde Kerk*, from South Africa, Dr Boesak introduced a motion that apartheid be declared a heresy contrary to the gospel and inconsistent with Reformed tradition. The Declaration of Racism was adopted, apartheid declared a sin and the white DRC and the NHK were suspended from the Alliance.
See also **Dutch Reformed Church; UDF**.

BOSS (*South Africa*)
The Bureau of State Security, later renamed the Department of National Security and now the National Intelligence Service: see **NIS**.

Botha, Pieter Willem 'P.W.' (*South Africa*)
P. W. Botha became the Prime Minister of South Africa in 1978 following a bitter battle with the heir apparent, Connie Mulder. Having lost ground to right and left in the 1981 general election, he called a 'Whites' only referendum on a new constitution in 1983 and achieved a personal triumph when it was approved by an overwhelming majority of voters. Under the new constitution the state president became the executive head, Botha becoming, in 1984, the first incumbent.

Despite having built his power-base in the Cape National Party, President Botha was born in the Orange Free State in 1916. He abandoned a university degree in 1935 to become a National Party organiser, and later flirted with the national socialist politics of the *Ossewabrandwag*, eventually resigning from the organisation in 1941. He was elected to parliament as MP for George in 1948, a seat he has held ever since.

Popularly known as 'P. W.', Botha rose to power through the bureaucratic apparatus of the National Party and his long service as Minister of Defence from 1966 to 1980. As premier and later state president he has pursued an aggressive foreign policy, using economic and military pressure to extract agreements from neighbouring countries, bludgeoning them into reducing support for the ANC.

P.W. Botha has made halting attempts to buy off internal opposition to apartheid and to win new friends internationally through the philosophy and practice of 'total strategy' (to counter a purported 'total onslaught' on the country). Such progress as he had made was largely destroyed when, in 1985, he was forced to declare a state of emergency in some areas, and, in 1986, to extend it to the whole of the country. His truculent responses to the efforts of the EPG in 1986, the wide-ranging curbs on both local and international reporting on events in South Africa, and a collapse in international confidence in the economy further isolated his regime. Having started down the 'reformist' road, Botha has been unable to deliver and is currently reaping the harvest of raised but unmet expectations – black South Africa is in a constant state of simmering unrest.
See also **constitution of 1983; EPG; Mulder; National Party; 'reforms'; total strategy**.

Botha, Roelof Frederik 'Pik' (*South Africa*)
'Pik' Botha became Minister of Foreign Affairs in 1977 under Prime Minister Vorster, and has held the post since, adding to it, in 1980, that of Minister of Information. He entered parliament in 1970 as National Party MP for Wonderboom, and became MP for Westdene in 1977. He actively identified himself, from the start, with the *verligte* group in his party.

Born in 1932, 'Pik' Botha completed a law degree before joining the Department of Foreign Affairs and becoming a career diplomat. After

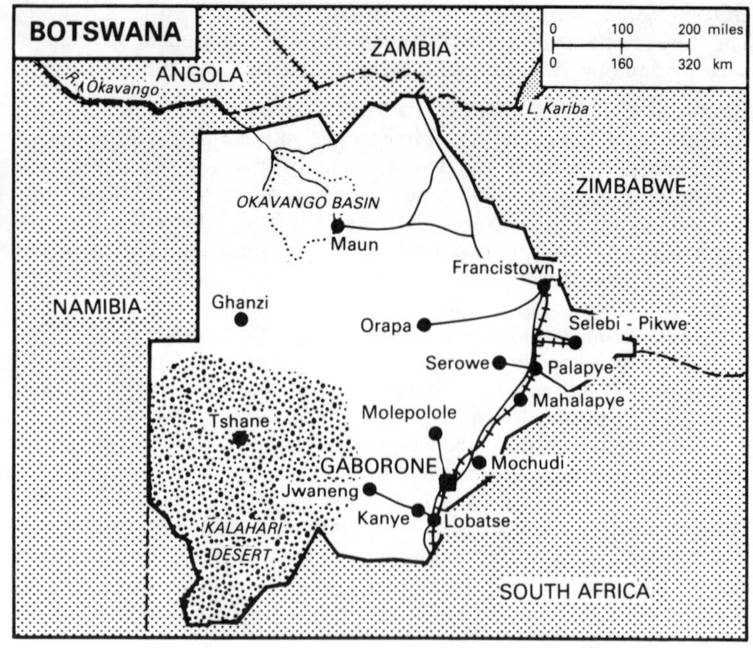

various postings, he represented South Africa at the International Court of Justice hearings on Namibia. He subsequently became South African ambassador to the United Nations for one month, before leaving following the suspension of the country's membership. In 1975 he was appointed ambassador to the USA.

Botha was one of three candidates for National Party leader in 1978, using his candidature to block that of Connie Mulder and assisting P. W. Botha to victory and the premiership. 'Pik' Botha subsequently became a close confidant of the prime minister, playing a central role in the development of his 'reformist' strategy.
See also **Botha, P. W.; International Court of Justice; Mulder; 'reforms'; verligte**.

Botha, Thozamile (*South Africa*)
Thozamile Botha was the leader of PEBCO, the Port Elizabeth Black Civic Organisation. He was also a work study trainee at the Ford factory in the area. In November 1979, after an attempt by management to dismiss him for his political activities there was a successful strike at the Ford plant to get him reinstated. In 1980 he was detained and banned by the state.
See also **PEBCO**.

Botswana

Official title: Republic of Botswana
Head of State: President Dr Quett K. J. Masire
Area: 582,000 sq. km
Population: 1,047,000 (1984)
Capital: Gaberone
Official language: English (Setswana is widely spoken)
GDP per Capita: US$ 589 (1985)
Major exports: diamonds – world's largest producer in 1985 – (72 per cent of total value)
Currency: Pula (P1=100 thebe); US$1=P1.675 (July 1987)
Main political parties: Botswana Democratic Party (ruling party), Botswana National Front, Botswana People's Party

Botswana is a land-locked country buried in the centre of southern Africa. It is surrounded by South Africa to the south and south-east, Namibia to the west and the north (the Caprivi Strip) and Zimbabwe to the north-east. It is a dry country and the Kalahari Desert – the home of the nomadic San people – dominates the south and west. In the north the Okavango River drains into the Okavango delta and Lake Ngami. Botswana is the world's largest diamond producer.

Botswana (then Bechuanaland) became a British protectorate in 1895, and a republic 71 years later, when it gained independence from Britain on 30 September 1966 with Sir Seretse Khama as its first president. The country is a democracy with an executive president who is also commander-in-chief of the armed forces. The president rules through his cabinet (appointed by him) which is responsible to the National Assembly. Legislative power is vested in Parliament which consists of the National Assembly and the president. The National Assembly is made up of the speaker, the attorney-general (who has no vote) 34 elected members and four specially elected members chosen by the assembly. The life of the assembly is five years.

See also **Botswana Democratic Party; Botswana National Front; Botswana People's Party; Masire**.

Botswana Democratic Party (*Botswana*)

The Botswana Democratic Party (BDP) was formed as the Bechuanaland Democratic Party in 1961 by members of the Legislative Council, including Seretse Khama. Khama later became Prime Minister and then President of Botswana. Quett Masire, the current president, was another BDP founder. The party was founded partly in reaction to the militancy of the Botswana People's Party, and was based on an alliance with traditional leaders. It supported independence and its main philosophy was non-racialism. Seretse Khama's personal standing and charisma was responsible for much of the party's success. However, the BDP also gained support from the establishment because of its conservative stance. It was backed by the chiefs, the British administration and many whites living in Botswana as the least threatening nationalist option.

The BDP won an overwhelming victory in the first elections in 1965, securing 28 out of 31 seats and Sir Seretse Khama became Botswana's first prime minister. On independence in 1966 he became president and in the 1969 elections the BDP retained a slightly reduced majority of 24 seats. The party has continued to hold power with no difficulty despite the opposition of the Botswana People's Party and the rise of a new opposition party, the Botswana National Front. In 1980 Seretse Khama died and was succeeded as party leader and president by his old ally and co-founder of the BDP, Dr Quett Masire. The party won 29 of the 34 constituency seats in Parliament in the general election of 8 September 1984 (later reduced to 28).

The BDP has pursued conservative policies at home and abroad, maintaining pragmatic links with white-ruled South Africa while repudiating its racist policies. The military attacks by South Africa on Botswana, allegedly to pursue ANC guerrillas, have increased in recent years.
See also **Botswana National Front; Botswana People's Party; Khama; Masire**.

Botswana Independence Party (*Botswana*)
The Botswana Independence Party (BIP) was founded by Motsamai Mpho, the former secretary-general of the Botswana People's Party (BPP), in 1964. The party grew out of a split within the BPP in which Motsamai Mpho was accused of favouring the South African Communist Party. In fact he retained strong links with the African National Congress of South Africa (ANC) and the other faction in the BPP, led by Philip Matante, supported the South African Pan-Africanist Congress (PAC). The BIP soon faded into relative insignificance, winning only one seat in the 1969 elections.
See also **Botswana People's Party**.

Botswana National Front (*Botswana*)
The Botswana National Front (BNF) was formed in October 1965 by Dr Kenneth Koma. It was a militant nationalist party with socialist policies which sought to unite the opposition to combat the conservative policies of the ruling Botswana Democratic Party. The opposition had hitherto been led by the Botswana People's Party (BPP), which had by now been rendered less effective by splits. Dr Koma brought together the wing of the BPP which supported Kgaleman Motsete (who had been expelled from the party), organised labour groups, civil servants and traditionalists led by Chief Bathoen of the important Ngwaketse ethnic group. Chief Bathoen resigned his hereditary headship of the Bangwaketse and, in the 1969 elections, with two supporters, took all three parliamentary seats in the Bangwaketse Reserve for the BNF, unseating Vice-President Quett Masire.

Over the years the BNF has accused the BDP of being undemocratic and the BDP has sought to brand the BNF as a potentially violent, militant force. The BNF has never succeeded in posing a serious threat to the ruling party. In the 1984 general elections the party won four seats (later five) out of a total of 34.

See also **Botswana Democratic Party; Botswana People's Party**.

Botswana People's Party *(Botswana)*
The Botswana People's Party was founded in 1960 as the Bechuanaland People's Party. It was Botswana's first really modern political party with a nationalist approach to politics. Its founder was Kgaleman Motsete and its secretary-general was Motsamai Mpho who had been tried for treason as a member of the African National Congress of South Africa. He continued to maintain close links with that organisation. The party demanded immediate independence, Africanisation of the administration, an end to rule by the chiefs, an end to the colour bar and the nationalisation of land. Soon internal rivalry between factions influenced by the South African Pan-Africanist Congress and the African National Congress of South Africa destroyed the party. Motsamai Mpho and six other leaders were expelled and formed the Botswana Independence Party in 1964. A further split led to the expulsion of Kgaleman Motsete, and Philip Matante became the party's president.

The splits seriously damaged the BPP's electoral chances and the party won only three seats in the Legislative Assembly in the 1965 elections. However, after independence the party revived and increased its support in the urban centres during the 1969 election campaign. The BPP also did well in local council elections. The ruling Botswana Democratic Party responded by increasing the number of nominated as opposed to elected members of district councils, and this provoked a charge from Philip Matante that the BPP was not only an undemocratic but also, a 'neo-colonialist' party. The party still poses no threat to the ruling BDP; in the 1984 general elections it won only one seat out of a total of 34.
See also **Botswana Democratic Party; Botswana Independence Party; Botswana National Front; Matante**.

BPC *(South Africa)* See **Black People's Convention**.

BPFP *(Botswana)* See **Bechuanaland Protectorate Federal Party**.

British South Africa Company *(Zimbabwe)*
The British South Africa Company was run by Cecil John Rhodes, the famous explorer and entrepreneur who gave his name to the British colonies of Southern and Northern Rhodesia (later Zimbabwe and Zambia). It was the biggest company in southern Africa and effectively administered Southern Rhodesia from 1890 to 1923.

The company was formed in March 1889 by a merger of the Central Search Association and the Exploring Company Ltd. Rhodes became the managing director in Africa and continued in this role for the rest of his life. The British South Africa Company was based in London and obtained a Royal Charter of incorporation on 19 October 1889 to run for 25 years. This meant that the company had rights over a vast, though undefined, territory north of the Limpopo River and west of what is now Malawi and Mozambique. Eventually the company formed the colonies of Northern and Southern Rhodesia.

The Charter allowed the company to make treaties with African rulers, to own and distribute land, to form banks and create its own police in exchange for promises that it would develop its territories economically, respect African law, allow free trade and respect religions. Supervisory powers over administration were retained by the British Secretary of State for Colonies.

In 1915 the British Crown granted the company another ten years on its charter but in 1922 the company surrendered its administration of Southern Rhodesia to the settlers there and in 1923 it handed over administration of Northern Rhodesia to the British Colonial Office. After this it continued to have financial interests in southern Africa. In 1964 the company had to relinquish its mineral rights in what had become Zambia and in 1965 the British South Africa Company merged with other companies to form Charter Consolidated Ltd, which had close associations with the South African-based Anglo American Corporation. See also **Anglo American Corporation; Rhodes**.

Broederbond (*South Africa*) See ***Afrikaner Broederbond***.

Brown Commission (*Zambia*)
Set up in 1966 to report on all aspects of the mining industry in Zambia after a series of strikes in the Copperbelt: see **Mineworkers' Union of Zambia**.

Bulawayo (*Zimbabwe*)
Bulawayo is the second largest city in Zimbabwe. Its population is estimated at 414,000, showing a dramatic growth of some 69 per cent between 1969 and 1982. It is in the south of the country, in Matabeleland, the area where the political leader, Joshua Nkomo, has most of his supporters. Until the 1960s it was the home of most leaders of the modern nationalist movement.
See also **Matabeleland; Ndebele; Nkomo**.

Bureau of State Security (*South Africa*) See **BOSS**.

Burton, Lillian (*Zambia*)
In 1960 there were political disturbances in Northern Rhodesia (later to become Zambia). The main nationalist party, UNIP, was campaigning for changes to the constitution. In particular, it was trying to extract Northern Rhodesia from the Central African Federation. The most serious incident occured on the Copperbelt, in the town of Ndola, on 8 May. The police had dispersed a UNIP meeting and a white housewife, Mrs Lillian Burton, and her two children were driving past the crowd. The crowd attacked the car killing the family. The event made a lasting impression on the people in Northern Rhodesia and led to the banning of UNIP.
See also **Central African Federation; Copperbelt**.

Buthelezi, Mangosuthu Gatsha (*South Africa*)
Chief Buthelezi is one of the most influential black leaders in South Africa. He is prime minister of the KwaZulu 'Homeland' and a traditional chief of the Zulu people. He was born the son of a tribal chief and the

great-grandson of the famous Zulu king, Cetshwayo. He is also the head of the *Inkatha* movement which he founded in 1975 as a political party based on Zulu aspirations.

Chief Buthelezi is the only 'Homeland' leader to have retained any credibility as an opponent of the government. He has consistently refused to accept 'independence' for KwaZulu and has insisted that it remain part of South Africa as a whole. Furthermore, he has in *Inkatha* a loyal and large grassroots following which is claimed to exceed a million people, mainly Zulus. He has tried to broaden his power-base through an alliance with other ethnic groups (the South African Black Alliance) but his real influence continues to be among the Zulu people. He has always taken an anti-apartheid view but has operated very much within the system dictated by the South African government. He has used his control of the KwaZulu administration, and the patronage it provides, to consolidate his power.

Chief Buthelezi is seen by most politically active black people outside *Inkatha* as a collaborator because of his close association with the South African regime. He is bitterly opposed to the ANC (although he was a member of its Youth League from 1948 to 1950) and the anti-apartheid UDF. In Natal, *Inkatha* and the UDF are engaged in an increasingly violent struggle for supremacy. Almost 400 people died in vicious fighting in 1987 alone.

Buthelezi and his *Inkatha* movement are regarded as possible future partners by liberal whites. He has been involved in the so-called *Indaba*, an attempt to set up a multiracial, pro-business, Buthelezi-led provincial government in Natal. Chief Buthelezi is pro-capitalism, arguing that it will increase prosperity which will in turn hasten black liberation. He is also outspoken in his opposition to economic sanctions against South Africa. This makes him a political enemy in the eyes of radical black people in South Africa.

See also **'Homelands'**; ***Indaba; Inkatha***; **SABA; UDF**.

Cabinda (*Angola*)
The enclave of Cabinda forms part of Angola despite the fact that it is wholly surrounded by Zairean territory or the sea. It was separately colonised by the Portuguese in 1885, but was administered as part of their Angolan colony. The Alvor Agreement of 1975 guaranteed its integrity as part of the country.

Cabinda is particularly important to Angola because of its wealth of natural resources – oil, phosphates, potassium, coffee, cocoa and timber are major products. FLEC is engaged in a low-intensity war with Angola to establish Cabinda as an independent state.
See also **Alvor Agreement; FLEC**.

CAF (*Malawi, Zambia, Zimbabwe*) See **Central African Federation**.

Camay, Phirowshaw (*South Africa*)
Phirowshaw Camay became the General Secretary of CUSA, the Council of South African Trade Unions, in November 1980. He is now General Secretary of the new trade union federation, NACTU, a black consciousness grouping set up in 1986.
See also **CUSA; NACTU**.

candonga (*Mozambique*)
Mozambican term for the unofficial and usually illegal market in officially controlled goods.
See also **FRELIMO**.

CANU (*Namibia*)
Caprivi African National Union, a regionalist organisation which merged with SWAPO in 1964: see **SWAPO**. Also the acronym for a subsequent organisation, the Caprivi National Union: see **Caprivi National Union**.

Caprivi African National Union (*Namibia*)
A regionalist organisation which merged with SWAPO in 1964: see **SWAPO**.

Caprivi Alliance Group (*Namibia*)
A constituent party of the Democratic Turnhalle Alliance: see **DTA**.

Caprivi National Union (*Nambia*)
A small, ethinically-based party in the Caprivi Strip, the Caprivi National

Union became part of the 'constitutional council' in 1987, helping to draw up an anti-SWAPO constitution to legitimise the TGNU. It is a different organisation from the Caprivi African National Union (although they share the same acronym, CANU) which joined SWAPO in 1964.
See also **CANU; SWAPO**.

Caprivi Strip (*Namibia*)
This is the long, narrow protrusion of the north-east corner of Namibia which shares borders with four countries, Angola, Zambia, Zimbabwe and Botswana. This geographic curiosity arose from an agreement between Germany and Britain in 1890 on the colonial carve-up of the region. The Caprivi Strip gave Germany access to the Zambezi which the Germans believed to be navigable to the Indian Ocean on the east coast of Africa. Having been relatively sheltered until the 1970s, the lives of Namibians living in the Caprivi were subsequently severly disrupted by the militarisation of this highly strategic area.

South Africa appears to have been making efforts, recently, to find a quasi-legal basis for separation of the Caprivi Strip from the rest of Namibia, possibly in the hope of maintaining a strategic military presence there. This is similar to their efforts to argue that the Walvis Bay enclave is a part of South Africa rather than Nambia.
See also **German colonialism; Namibia; San; Walvis Bay**.

capulana (*Mozambique*)
A *capulana* is a versatile length of cloth, about six feet long and often brightly coloured. Women in Mozambique use them as skirts, slings for babies and for a variety of other purposes.

Carlsson, Bernt (*Namibia*)
Bernt Carlsson was appointed UN Commissioner for Namibia from 1 July 1987, replacing Brajesh Mishra who had held the post since 1982. Carlsson had previously been Swedish Ambassador at Large with responsibility for the Middle East and Africa. He is a former general secretary of the Socialist International and a former international secretary of the Social Democratic Party of Sweden.
See also **Ahtisaari; MacBride; Mishra; UN Commissioner for Namibia**.

Carrington, Peter (*Zimbabwe*)
Lord Carrington was appointed British Foreign Secretary in 1979. On 1 August the Commonwealth Summit Conference opened in Lusaka and an informal meeting of the heads of various countries' delegations took place. At this meeting it was agreed to hold a conference on Zimbabwe, chaired by Lord Carrington. This was the Lancaster House conference held in London in 1979. Lord Carrington was much praised for pushing through a political settlement which led to the independence of the British colony of Southern Rhodesia and majority elections in Zimbabwe in 1980.
See also **Lancaster House; Lusaka**.

Carver, Richard (*Zimbabwe*)
Field Marshall Lord Carver was appointed as the proposed British Resident Commissioner in Southern Rhodesia under the Anglo-American plan for a settlement. He was retired with an outstanding army record. Under the Anglo-American proposals Lord Carver was to be given the powers of a dictator to govern the country for six months during the transition to majority rule. First, however, he had to negotiate a ceasefire. He went to Southern Rhodesia in 1977 together with the United Nations representative General Prem Chand, but the initiative failed and Lord Carver never took up his position.
See also **Anglo-American proposals**.

CCM (*Tanzania*) See ***Chama cha Mapinduzi***.

CDM (*Namibia*)
The Consolidated Diamond Mines company (CDM) is a subsidiary of the South African diamond monopoly De Beers, itself part of the giant Anglo American group. CDM dominates the diamond industry in Namibia and has been accused, in recent years, of asset-stripping and transfer pricing, both at the cost of damage to the long-term interests of the diamond industry in the country.
See also **Anglo American Corporation**.

CEA (*Mozambique*)
Influential *Centro de Estudos Africanos* (Centre of African Studies) at the Eduardo Mondlane University: see **Centre of African Studies**.

Central African Federation (*Malawi, Zambia, Zimbabwe*)
The Central African Federation brought together three British colonial territories: Southern Rhodesia (which was to become Zimbabwe), Northern Rhodesia (which was to become Zambia) and Nyasaland (which was to become Malawi). Southern Rhodesia was a self-governing colony, ruled by the white minority and Northern Rhodesia and Nyasaland were British colonial protectorates. Most Africans of the region rejected the idea of the Federation, arguing that it would serve to benefit the white rulers of Southern Rhodesia more than anyone else. Above all, it was said, they coveted the rich minerals of Northern Rhodesia and its large market place. Africans feared that the Federation would enable the whites to extend their control over the whole region.

In Northern Rhodesia all the nationalist organisations, starting with the Northern Rhodesia African Congress, campaigned against the Federation. In Nyasaland, too, opposition was fierce; this changed the political atmosphere and undermined trust in the British colonial administration.

In 1953, under a Conservative British government and against African wishes, the Central African Federation was formed. As the federal government tried to increase in power, it clashed with emerging African nationalism. At this time the Federation was led by Sir Roy Welensky, who became Federal Prime Minister in 1955. He sought independence from British control and was bitterly resented by African nationalists as he worked to entrench the interests of the whites in the region. In 1959

a state of emergency was declared in Southern Rhodesia and Nyasaland after anti-Federation demonstrations. In 1962 elections in Northern Rhodesia led to a nationalist coalition government and the end of federation. The Central African Federation was formally dissolved in 1963.
See also **Northern Rhodesia African Congress; Welensky**.

Central Committee (*Mozambique*)
Ruling organ of the governing party, FRELIMO: see **FRELIMO**.

Centre Party (*Zimbabwe*)
The Centre Party was formed before the 1969 constitutional proposals were put to the minority white electorate in a referendum. One prominent member was Major-General Sam Putterill, the Rhodesian armed forces commander for three years. The new opposition Centre Party urged a 'no' vote, fearing that the new constitution would end the possibility of a settlement with Britain.

The 1970 white general election was overwhelmingly won by the Rhodesian Front, while the Centre Party won only seven black seats. The new party was fatally weakened by its divided opinion on the 1971 Anglo-Rhodesian agreement; white and black members took different sides. Eventually it abandoned support for the settlement in protest against the discriminatory measures that were included in it.
See also **Anglo-Rhodesian agreement; constitution of 1969; Putterill; Rhodesian Front**.

Centre for Revolutionary Instruction (*Angola*) See **CIR**.

Centre of African Studies (*Mozambique*)
The Centre of African Studies (*Centro de Estudos Africanos* – CEA) at Mozambique's Eduardo Mondlane University in Maputo is a highly influential institution which has concentrated on studying the impact of FRELIMO's policies on the ground, and the problems facing ordinary Mozambicans in the towns and countryside. Its work made a particularly important contribution to the major re-think which took place at FRELIMO's Fourth Congress in 1983. The Centre was headed by Ruth First until her assassination in 1982.
See also **First; Fourth Congress of FRELIMO**.

Centro de Estudos Africanos (*Mozambique*)
The influential Centre of African Studies at the Eduardo Mondlane University: see **Centre of African Studies**.

Chakuamba Phiri, Gwanda (*Malawi*)
Gwanda Chakuamba was regarded for years as President Banda's right-hand man. He was active in the Nyasaland African Congress before independence and in the government afterwards. He was appointed Minister for Youth and Culture in 1977. However, he met a similar fate to another of the president's closest associates, Albert Nqumayo. In 1981 Gwanda Chakuamba was summarily removed from office, found guilty of treason and sentenced to 22 years imprisonment with hard labour.
See also **Banda; Nqumayo; Nyasaland African Congress**.

C

Chama cha Mapinduzi (*Tanzania*)

In February 1977 TANU (the Tanganyika African National Union), Tanzania's sole political party on the mainland, amalgamated with the Afro-Shirazi Party from Zanzibar to form the *Chama cha Mapinduzi* or the Party of the Revolution.

TANU had been the dominant nationalist party in mainland Tanzania and the Afro-Shirazi Party on the islands Zanzibar and Pemba. President Nyerere of Tanzania proposed the merger of the two parties in 1975 because Tanzania has a one-party constitution. He was elected head of the new party and Aboud Jumbe, leader of the Afro-Shirazi Party, became his deputy. In October 1982 the constitution of the *Chama cha Mapinduzi* was changed to allow for the formation of an eighteen-member Central Committee to run the party. The merger coincided with the end of the clove boom in Zanzibar and the consequent depression in the island's economy. It committed Zanzibar's leaders to the leadership conditions of the Arusha Declaration, the guiding principles of Tanganyika's government. The declared aim of the *Chama cha Mapinduzi* is to establish a socialist democratic state by self-help at all levels in Tanzanian society.

See also **Afro-Shirazi Party; Arusha Declaration; Jumbe; Nyerere; TANU**.

Chama cha Umoja wa Watu wa Afrika (*Tanzania*)

The Association of the Unity of the People of Africa, the popular term for the African Association: see **African Association**.

Chamber of Mines (*South Africa*)

The South African Chamber of Mines is the representative organisation of mining capital in the country. Formed in 1887, it rapidly became the dominant organisation of business and was instrumental in persuading governments to enact the first pass laws and to impose taxes to create a pliant black labour force. It established a centralised recruiting organisation (currently The Employment Bureau of Africa or TEBA) to secure labour for the mines. TEBA was highly effective in preventing wage inflation as a result of competition for workers, establishing a wage cartel to keep wages to a minimum, and overseeing the adoption of the closed compound system by all mining companies after its invention by De Beers for its diamond mines in the nineteenth century.

The Chamber of Mines was so successful in controlling miners' wages that they remained virtually static in real terms (and very low) for 75 years. During the past two decades the influence of the Chamber of Mines on both government and member corporations has declined, with the great multinationals prefering direct contacts with government, and taking unilateral action on wages and mechanisation.

See also **ASSOCOM; FCI; NUM; SEIFSA**.

Chaolane, Phoka (*Lesotho*)

Leader of the exiled opposition group, the United Democratic Alliance: see **United Democratic Alliance**.

chibalo (*Mozambique*)
The name given to labour performed under compulsion: see **forced labour**.

Chidzero, Bernard (*Zimbabwe*)
Dr Chidzero has been the minister in charge of Zimbabwe's economic development since independence. Since 1985 he has been Minister of Finance, Economic Planning and Development. He has had a distinguished United Nations career and was appointed Director of Commodities Division of UNCTAD (the United Nations Conference on Trade and Development) in 1968.
See also **ZANU**.

Chikerema, James Robert Dambaza (*Zimbabwe*)
For many years James Chikerema was active in ZAPU, one of Zimbabwe's main nationalist parties led by Joshua Nkomo. He held various offices in the party and was acting president for some time. However, he did not stay in ZAPU, and in 1972 formed the Front for the Liberation of Zimbabwe. He also became a member of the National Executive of the African National Council. Some time later, during the period of the transitional government, he was a minister, and in 1979 he broke away from Bishop Muzorewa's UANC and founded the Democratic Party.
See also **African National Council; FROLIZI; Nkomo; UANC; ZAPU**.

Chilembwe, John (*Malawi*)
John Chilembwe was one of southern Africa's greatest early nationalists. He was influenced by Joseph Booth, the radical British evangelist, who believed that Africa should be run by Africans. In 1897 Joseph Booth took John Chilembwe to the USA to continue his education. He returned in 1900 and started the Providence Industrial Mission as well as several mission schools. By 1914 John Chilembwe had given up his desire to work within the colonial framework. The famine of 1913 underlined the plight of ordinary Malawians and in November 1914 he sent a letter to a local newspaper complaining of the injustice of forced African recruitment for war work. By December he and some two hundred followers had agreed to 'strike a blow and die'.

The revolt began on 23 January 1915. An attack was made on the manager of the Bruce estates, William J. Livingstone, who was known for the cruel way he treated his workers. He was killed together with two other white planters. The revolt went on until February 4th. John Chilembwe was shot while trying to escape on February 2nd, and his followers were killed, jailed or went into exile. The rising was a military failure but it was of great political significance, for it demonstrated that there was another path open to Africans who rejected colonial rule – open rebellion.

Chiluba, Frederick (*Zambia*)
Frederick Chiluba is the leader of the Zambia Congress of Trade Unions. He was arrested, together with other trade unionists, for some months

during 1981. The trade unions, particularly the Mineworkers' Union of Zambia, have often come into conflict with the Zambian government.
See also **Mineworkers' Union of Zambia; Zambia Congress of Trade Unions**.

Chimurenga (*Zimbabwe*)

Chimurenga means 'rebel' or 'fighter'. It has also come to mean any revolution, struggle or resistance. It is derived from the Shona word *Murenga*, and was first applied to the rebellions against white settlers in Southern Rhodesia (later to become Zimbabwe) in 1896–7. It was commonly used during the guerrilla war against white minority rule in the 1960s and 1970s. Indeed the war was sometimes referred to as 'the second war of *Chimurenga*'. One of ZANU's slogans was *Pamberi ne Chimurenga* meaning 'forward with the struggle'. On 28 April 1966 ZANLA, the liberation army of ZANU, penetrated deep into white Rhodesian territory and seven of its guerrillas were killed at the Battle of Sinoia. Now ZANU marks April 28th as Chimurenga Day, the offical start of the war for liberation.
See also **ZANLA; ZANU**.

Chinamano, Josiah Mushore (*Zimbabwe*)

Josiah Chinamano was an active member of ZAPU, one of Zimbabwe's main nationalist organisations, for many years. He was also treasurer and later vice-president of the African National Council. In 1981 he was appointed a minister in the cabinet of Zimbabwe's first prime minister, Robert Mugabe. However, his government career ended as the political differences between ZAPU's leader, Joshua Nkomo, and Robert Mugabe developed. By the 1985 elections there were no ZAPU supporters in the cabinet, but this changed with the union of the parties in 1988.
See also **African National Council; Mugabe; Nkomo; ZAPU**.

Chipembere, Henry (*Malawi*)

Henry Chipembere was a leading Malawian nationalist. In the mid–1950s he was a leader of the radical group which revived the Nyasaland African Congress (the NAC, later to become Malawi's main political party, the MCP or Malawi Congress Party). Together with other young radicals like Kanyama Chiume, he influenced the party to make militant demands for self-government and universal suffrage. They also reorganised and increased the membership of the NAC. Henry Chipembere was responsible for urging Hastings Banda to return and take over the leadership of the NAC.

In 1960 Henry Chipembere was jailed for three years by the colonial government for sedition. On his release he was given government office by Dr Banda. However, in 1964 he was caught up in the cabinet revolt against Hastings Banda, and resigned in sympathy with his colleagues who had been dismissed for demanding more radical policies from their leader. In 1965 Henry Chipembere, supported by some 200 men, led an armed uprising against the government. However, he was forced to retreat and lived in exile until his death in 1975.

See also **Banda; Chiume; MCP; Nyasaland African Congress**.

Chipenda, Daniel (*Angola*)

Daniel Chipenda joined the MPLA in its early stages and rose rapidly within the organisation. He became the head of JMPLA, the youth wing, in 1964 and subsequently became a member of the Steering Committee and the Political and Military Co-ordinating Committee.

Chipenda played an important role as a military leader in the east of Angola and when he split with Neto in 1973 took with him 2,000 fighters in the east, seriously weakening the MPLA. The so-called Eastern Revolt, followed by the Active Revolt, brought the MPLA to its nadir at a crucial time in the struggle against Portuguese colonialism.

Following the Eastern Revolt, and his expulsion from the MPLA in October 1974, Chipenda took his followers into an alliance with Holden Roberto and became Secretary General of the FNLA. Unlike the rest of the FNLA, Chipenda's forces resisted the South African invasion of 1975. They subsequently disintegrated following defeat by FAPLA in 1976.
See also **Active Revolt; FAPLA; JMPLA; MPLA, Neto; Roberto**.

Chirau, Jeremiah (*Zimbabwe*)

Ian Smith, the rebel leader of Rhodesia, regarded Chief Jeremiah Chirau as a valuable African ally in his struggle to maintain white supremacy. In 1973 the chief was appointed President of the Council of Chiefs which he regarded as the only representative body for Africans in Zimbabwe. Chief Chirau remained opposed to the nationalist struggle, and, together with Chief Ndiweni, set up the Zimbabwe United People's Organisation. This was designed to be a new party to rally African opinion to a negotiated settlement with the Smith regime. Its role was to provide a counterpoint to the nationalists and their parties. Chief Chirau also played an important part in the internal settlement attempted by Ian Smith in 1978.
See also **internal settlement; Ndiweni; Smith; Zimbabwe United People's Organisation**.

Chirwa, Orton Edgar Ching'oli (*Malawi*)

Orton Chirwa was the founder of the Malawi Congress Party (MCP) in 1959. He was its acting president until Hastings Banda was released from prison and took over. The two men worked closely together in the party during the campaign for independence, and in government Orton Chirwa was Minister for Justice and Attorney-General. However, their relationship reached a crisis in 1964 during the cabinet revolt when six senior ministers, including Orton Chirwa, were dismissed or resigned from the government. The ministers concerned had failed to persuade President Banda to speed up his Africanisation policy, whereby Africans replaced Europeans in key positions in the country, and to pursue stronger anti-colonial policies.

After his dismissal Orton Chirwa left Malawi and became the leader of the Malawi Freedom Movement (MAFREMO), a left-wing opposition group. It is alleged that he and his wife Vera were kidnapped from Zambia in December 1981. They were then put on trial for treason in

Malawi and sentenced to death in May 1983 of that year. After appeals from the international community President Banda commuted the sentences to life imprisonment in June 1984.
See also **Africanisation; Banda; Malawi Freedom Movement; MCP**.

Chissano, Joaquim (*Mozambique*)
Joaquim Chissano was elected Mozambique's second president on 3 November 1986 following the death, in an aeroplane crash in South Africa, of his predecessor, Samora Machel. During the liberation struggle, Chissano headed the Security Department of FRELIMO, and was also head of the Transitional Government after the Portuguese withdrawal and before full independence. He became Foreign Minister at independence, and remained throughout one of Samora Machel's closest associates. On taking over the presidency, Chissano pledged to continue the war against the MNR and rejected negotiations, arguing that if there were to be negotiations they should be with the South Africans who controlled the MNR.
See also **FRELIMO; Machel; MNR; Transitional Government**.

Chitepo, Herbert Tapfumanei (*Zimbabwe*)
Herbert Chitepo became Southern Rhodesia's first African barrister in 1954. Gradually his work became more political and in 1966 he joined the National Democratic Party (NDP). In 1961, together with other NDP members, he became a member of ZAPU, one of Zimbabwe's main nationalist organisations. However, he left ZAPU after disagreements with its leader Joshua Nkomo and in 1963 helped to found ZANU, the other main nationalist group. He was elected chairperson and kept this position until he was assassinated in 1975 in Lusaka, Zambia. Controversy surrounded his death and a Zambian government commission declared that he was killed as a result of a power struggle within ZANU. Others believed his death was linked to his opposition to a policy of conciliation with South Africa. Herbert Chitepo was responsible for much of the guerrilla war waged by ZANLA, ZANU's army, and maintained an uncompromising and tough political line within his party.
See also **National Democratic Party; Nkomo; ZANLA; ZANU; ZAPU**.

Chiume, Kanyama (*Malawi*)
Kanyama Chiume was one of Malawi's early radical nationalists, and a close associate of the president. He was a prominent member of the Nyasaland African Congress and, together with Henry Chipembere, led a group of young radicals which sought to make the party more militant. After Hastings Banda returned as leader in 1958, Kanyama Chiume was made Publicity Secretary and was active in Pan-African conferences. In 1964 as Minister of External Affairs, he was part of the cabinet revolt against President Banda. He was dismissed and fled to Zambia. Later he became president of the Congress for a Second Republic, one of Malawi's dissident groups, based in Tanzania.
See also **Banda; Chipembere; Congress for a Second Republic; Nyasaland African Congress**.

Chona Commission (*Zambia*)
The Chona Commission was set up by the Zambian government in 1972. It was led by Vice-President Chona, a prominent Zambian leader and activist in UNIP. The commission's task was to recommend what form a 'one-party participatory democracy' in Zambia should take. It reported quickly and the report was highly praised for its thoroughness and balanced preparation. Zambia became a one-party state later in the year. See also **UNIP**.

Christian Democratic Action for Social Justice (*Namibia*)
An insignificant party which replaced the National Democratic Party, the Christian Democratic Action for Social Justice party is the personal political vehicle of Peter Kalangula.

churches (*Angola*)
While about half of the population of Angola follows traditional African religious beliefs, European-introduced Christian churches attract the support of about an equal number of Angolans. Most significant is the Catholic Church with 3,285,000 members in 1980, compared to about 800,000 Protestants.

The Christian churches played a significant role during the colonial period in spreading Portuguese and European influence and language among indigenous Angolans. Missionaries provided what formal schooling was available to people in the rural areas, as well as much of the medical care available.

churches (*Lesotho*) See **Roman Catholic Church**.

churches (*Mozambique*)
Out of the population of 12½ million (1981 official estimate) in Mozambique, the majority follow traditional, mainly animist, beliefs. Of the remainder, about 2 million are Muslims and a similar number Christians. Of the last group, three quarters are Roman Catholic. Mozambique also has a small Hindu population.

The Roman Catholic hierarchy in Mozambique was extremely conservative during the colonial period, reflecting the dominance of fascist politics in Portugal. Church schools aimed to give black Mozambicans only a rudimentary education in order to fit them for lives of servitude rather than independence. Church radio stations broadcast anti-FRELIMO propaganda and the Church disciplined its own priests when they strayed from the official line. Nonetheless, some did speak out and were arrested and tortured by PIDE or expelled from the country.

By contrast, some Protestant churches were more sympathetic to FRELIMO and even gave some convert support. One Protestant cleric, Rev. Zedequias Manganhela, was murdered by PIDE because of his work for FRELIMO.

churches (*Namibia*)
Almost 70 per cent of Namibians are practising Christians, most of them members of autonomous, mainly black, churches which had their roots

in the nineteenth-century European missionary societies. The largest of the present churches is the Evangelical Lutheran Ovambo-Kavango Church (ELOK), with about 280,000 members, mainly in the north of the country. ELOK was formed in 1956 from the Finnish Missionary Society (FMS) and in 1960 became the first of the churches to elect a black leader, Bishop Leonard Auala. By 1972, 93 of its 97 ministers were black.

C

In 1972 ELOK joined the second largest church in Namibia, the Evangelical Lutheran Church (ELK), to form a federal body, the United Evangelical Lutheran Church. ELK was formed in 1957 when the Rhenish Missionary Society yielded to pressure and granted its Namibian churches local autonomy. However, ELK did not become fully independent for another ten years. It now has a membership of about 130,000. The other large churches in Namibia are the Roman Catholic Church, with about 100,000 members, and the Anglican Church with about 60,000.

As a result of the Africanisation of the church hierarchies in Namibia, the churches became increasingly outspoken in their opposition to South African rule. Church people were instrumental in organising appeals and protests to the United Nations and to international church bodies, and compiled dossiers of torture and murder allegations against the police and military. The result of their confrontationist approach was increased harassment by South Africa.

South Africa also began taking punitive action against individual church leaders, subjecting them to deportation, passport and visa refusals, denials of entry and travel permits, and bannings. The Anglican Church, one of the more outspoken in Namibia, has had four successive leaders deported, Bishop Mize, Bishop Colin Winter, Suffragen Bishop Richard Wood and the church's Vicar-General, Rev. Edward Morrow. In August 1979 the church's most senior official in the north of the country, Archdeacon Shilongo, was arrested for allegedly having 'SWAPO propaganda material' on mission premises. In January 1987 the main Windhoek offices of the Namibian Council of Churches were deliberately burnt down. A report by Amnesty International in 1986 documented such action by South Africa and extreme right-wing groups against the churches in Namibia.

One outcome of the church-state conflict was the perception by churches of the need for a single representative organisation, and in October 1978 the Council of Churches of Namibia was formed. Although the council was boycotted from the start by the Afrikaner Dutch Reform Churches, all other significant churches, including the conservative German Lutheran Church (DELK), became members. Since its formation the council has also taken up issues of repression and racism in Namibia. It has been particularly outspoken in its denunciation of South Africa's introduction of conscription for all young men in Namibia.

Despite the polticisation of the churches in Namibia, most have avoided open support for any one political group. Some prominent leaders, however, such as Rev. Kameeta, elected President of the Nambian National Convention in 1976, are open members of SWAPO and accept

the necessity of armed struggle. In 1986 the Council of Churches initiated a campaign, headed by the Ai Gams Action Committee, for the implementation of UN Security Council Resolution 435.
See also **Ai Gams declaration; CCN; NNC**.

churches (*South Africa*)
Christian churches in South Africa are divided into four main groups, not all mutually exclusive. The largest such group operates under the umbrella of the South African Council of Churches (SACC). The SACC brings together the majority of the Protestant churches excluding the white Dutch reformed churches. The 23 denominations of its membership exclude the Roman Catholic Church.

The SACC has been the most vociferous of all church organisations in its criticism of the South African government and the apartheid system. In 1984 its head, then Bishop Desmond Tutu, was awarded the Nobel Peace Prize in recognition of his work and that of the SACC. In the same year, after three years of deliberation, the government-appointed Eloff Commission of Inquiry into the SACC made its report, accusing the Council of using overseas funds to subvert the political system of South Africa. The criticisms in the Eloff Report drew immediate and widespread rejection and expressions of support for the SACC and its work.

The other large Protestant grouping is the family of Dutch reformed churches. Controversial for their role in supporting and attempting to find a theological justification for apartheid, these churches have recently begun to distance themselves from this position. The largest of the churches is the white *Nederduitse Gereformeerde Kerk* (NGK) or Dutch Reformed Church (sometimes called 'the National Party at prayer'). It and the *Nederduitsch Hervormde Kerk* (NHK) or Dutch Reconstituted Church have been the most adamant supporters of apartheid in the past, and continue, on the whole, to exclude black people from membership and worship. One in three of the members of the third largest, the *Gereformeerde Kerk in Suid-Afrika* (GKSA) or Reformed Church in South Africa, are black and it has some of the leading liberal Afrikaner theologians among its ministers.

The Dutch reformed movement includes some almost exclusively black churches, the *Nederduitse Gereformeerde Sendingkerk* or Dutch Reformed Mission Church which caters for 'Coloured' people, the 'Indian' Reformed Church in Africa, and the African *Nederduitse Gereformeerde Kerk in Afrika* (NGKA) or Dutch Reformed Church in Africa. The black churches have been leading the pressure for the white churches to drop their support for apartheid, and were instrumental in persuading the international Reformed Ecumenical Synod to declare apartheid a heresy. The assessor of the *NG Sendingkerk*, Rev. Dr Allan Boesak, is currently president of the World Alliance of Reformed Churches and a senior vice-president of the SACC.

While many predominantly black churches are members of the SACC or the Dutch reformed group, they and other black churches also come together under the Alliance of Black Reformed Christians (ABRECSA).

ABRECSA has given active support to trade unions and the mass struggle against apartheid.

Among independent African churches are the African Congregational Church, the Council of African Independent Churches and the Zion Christian Church. The Reformed Independent Churches Association, a highly conservative body, has been outspokenly critical of Archbishop Desmond Tutu's advocation of sanctions and praised US President Reagan's opposition to them.

The Roman Catholic Church in South Africa, through the Southern African Catholic Bishops Conference, has been consistently although not always outspokenly critical of the government and the apartheid system over a long period. In the late 1970s it confronted the state over the issue of segregation in its schools, and, in defiance of warnings from the government, began to admit small numbers of black pupils to its previously all-white establishments.

Although organised religion in South Africa is dominated by the Christian churches, there are also significant Jewish, Muslim and Hindu communities as well as adherents of traditional African religions. Most Muslims are concentrated among so-called 'Coloured' people in the Western Cape, while the Hindu community is concentrated among people of Indian origin in Natal.

See also **apartheid; Boesak; FAK; National Party; Tutu**.

CIR (*Angola*)

The CIR's, or Centres for Revolutionary Instruction, were institutions set up by FAPLA for the military instruction of its soldiers. In August 1975 Angolan President Neto requested Cuban help with military instruction at the CIRs. The first Cuban instructors arrived in October 1975, by which time South African military activity in the south of Angola had been continuing for the previous month.

See also **Cuban troops; FAPLA**.

'Coloured' (*Namibia, South Africa*)

The South African government describes the group of people of mixed descent as 'Coloured'. Most of those in this group of people live in the Cape province of South Africa and in Namibia and would regard themselves as part of the black community as opposed to part of the white community.

See also **Introduction; 'Black'; 'Indian'**.

Coloured People's Congress (*South Africa*)

The Coloured People's Congress (CPC) was founded in 1953 as the South African Coloured People's Organisation (SACPO). It aimed to form a militant political movement with mass support among South Africa's 'Coloured' population. The organisation grew out of the Franchise Action Committee, an alliance formed in 1951 to oppose the removal of 'Coloured' people from the common voters' roll – the campaign failed and 'Coloureds' were excluded by the government in 1956. Soon it changed its name to the Coloured People's Congress to reflect its membership of the

Congress Alliance, a common front under the leadership of the ANC of all groups opposed to the government. The CPC took part in the 1955 Congress of the People and, as a result, some of its members were arrested the following year on a charge of high treason (these charges were later dropped). Other CPC leaders were detained during the 1960 state of emergency.

The Coloured People's Congress was banned in 1960 together with the other groups in the Congress Alliance (except the provincial branches of the South African Indian Congress). In 1966 many of its exiled members left the ANC and joined the Pan-Africanist Congress. The PAC – unlike the ANC at that time – was prepared to admit 'Coloureds' to full membership. It also did not suffer from the tensions generated by the debate within the ANC between the South African Communist Party and the other nationalists. By 1969, however, most CPC members had left or been expelled from the PAC because of internal feuding within the organisation.

See also **ANC; Congress Alliance; South African Coloured People's Organisation**.

C

Comite Revolucionário de Moçambique (*Mozambique*)

The main opposition party to FRELIMO in Mozambique: see **COREMO**.

Committee of Ten (*South Africa*)

The Committee of Ten was formed in June 1977 by professional people living in Soweto. It was one of the community organisations which sprang up in the aftermath of the Soweto uprising of 1976. The riots led to the collapse of the state-imposed Soweto Urban Bantu Council and the Committee of Ten was formed to organise the administration of Soweto. A committee of some 300 people elected the Committee of Ten which was headed by Dr Nthato Motlana. The government sought to impose community councils in Soweto and the Committee led a campaign against these. In October 1977 the Committee members were detained and elections for the Soweto Community Council announced. The elections were boycotted and the leaders released. In 1979 the Committee of Ten formed the Soweto Civic Association (SCA) in an attempt to try and broaden its support. Events in the townships have made the Committee of Ten less important now (1987) than it once was. However the SCA has become involved in the setting up of the street committees in the townships.

See also **Soweto Civic Association; Soweto uprising, 1976; street committees; Motlana**.

Committee on South African War Resistance (*South Africa*)

The Committee on South African War Resistance (COSAWR) was set up in December 1978 by merging two groups operating in Britain. It publishes a journal, *Resister*, and gives advice to those avoiding conscription, especially those in exile. Resistance to conscription is growing in South Africa as the society becomes more militarised.

See also **End Conscription Campaign; SADF**.

Communist Party of South Africa (*South Africa*) See **South African Communist Party**.

compromised (*Mozambique*)
'The compromised' were those who collaborated willingly with the Portuguese in their war against FRELIMO during the colonial period, and were later identified: see **OPV**.

'comrades' (*South Africa*)
'Comrades' is the term given to the young black radicals who have emerged in recent years in the townships. The comrades are in political sympathy with the anti-apartheid UDF organisation, and are said to work through street committees, groups elected to organise the townships in times of unrest. They are frequently involved in fighting, particularly with supporters of opposed, conservative groups like *Inkatha*. There have also been reports of clashes with black consciousness groups like AZAPO. One widely publicised case of a violent clash involving comrades took place in May and June 1986 at the Crossroads squatter camp near Cape Town. There the comrades were defeated by the so-called 'fathers' or *witdoeke* (allegedly assisted by the police) who flattened the homes of newcomers to the camp and drove an estimated 60,000 people from their houses. At least 69 people were killed in the clashes.

The comrades are part of a new trend towards the establishment of groups of vigilantes in the townships. Opponents of the government argue that conservative, often rural, workers are now being used by the government to attack young township radicals. The impression given is of violence among black people, but often, it is argued, the government is simply employing or encouraging the conservative vigilantes to do their work. In Crossroads the clearing of the new shanty dwellings was highly convenient to government policy. And in Soweto a group of vigilantes called the 'Russians' re-emerged in December 1986 (they had not been heard of for thirty years). They attacked comrades planning a Christmas boycott campaign against the emergency regulations, leading to a number of deaths.

The 'false comrades' are groups of thugs pretending to be comrades, called by the township residents 'comtsotsis' (*tsotsi* is a word meaning thug or 'wide boy').
See also **Crossroads; 'fathers'; 'Russians'; townships; UDF; *witdoeke***

CONCP (*Angola, Mozambique*)
The *Conferência das Organizações Nacionalistas das Colonias Portuguesas*, or Conference of Nationalist Organisations of the Portuguese Colonies was an alliance of socialist liberation movements in the Portuguese colonies. It sought to promote the common aims of the organisations through collective diplomatic activity at the OAU, through joint publicity work and through sharing military experience and intelligence. CONCP was established in April 1961 and included the MPLA, FRELIMO, the PAIGC (*Partido Africano da Independência da Guiné Cabo Verde* – fighting for the liberation of Guinea-Bissau and the Cape

Verde) and the MLSTP (*Movimento de Libertação de São Tomé e Princípe* – fighting for the liberation of Sao Tome and Principe).
See also **FRELIMO; MPLA; OAU**.

Congress Alliance (*South Africa*)
The Congress Alliance was formed during the 1950s as a united front against the white South African state. It consisted of the ANC, the South African Indian Congress, the Coloured People's Congress, the Congress of Democrats (whites) and, after 1955, SACTU, the trade union group. The Congress Alliance was led by the ANC and convened the Congress of the People in 1955 in which the Freedom Charter was adopted.
See also **ANC; Coloured People's Congress; Congress of Democrats; Congress of the People; Freedom Charter; South African Indian Congress**.

C

Congress for the Second Republic (*Malawi*)
The Congress for the Second Republic is the smallest of Malawi's dissident groups. Opposition parties are banned so they have to operate outside the country's borders. The Congress for a Second Republic is based in Dar es Salaam, Tanzania, under the leadership of Kanyama Chiume, formerly one of President Banda's closest associates. None of the opposition groups is particularly effective but they all claim support within Malawi.
See also **Chiume; Congress for the Second Republic; Malawi Freedom Movement; Save Malawi Committee; Socialist League of Malawi**.

Congress of Democrats (*South Africa*)
The Congress of Democrats (COD) was formed in 1953 after the Defiance Campaign of 1952. Its aim was to provide an organisation of whites which was fully in sympathy with the ANC and could participate in the Congress Alliance – a united front of all races against the South African state. Its president, Piet Beyleveld, said to the organisation's 1958 conference:

> The Congress of Democrats is . . . a loose association of like-minded people, bound together by a common belief in the necessity for, and the desirability of, a democratic society based on the equality of all citizens regardless of race and colour. . . . The Congress should see itself as . . . a small, and strictly secondary, wing of the Congress Movement.

COD members were active in the Congress of the People in 1955, and over twenty of them, including Ruth First, Joe Slovo and Helen Joseph, were arrested and charged with treason. They were released with the others when the charges collapsed.
See also **Congress Alliance; Defiance Campaign of 1952; First; Joseph; Slovo; Treason Trial**.

Congress of South African Students (*South Africa*) See **COSAS**.

Congress of South African Trade Unions (*South Africa*) See **COSATU**.

Congress of the People (*South Africa*)
The Congress of the People was called by the Congress Alliance, a common front against apartheid. It took place in June 1955 in an open space near Kliptown, a 'Coloured' township near Johannesburg. 3,000 delegates of all races from all over the country attended and the conference was broken up by police carrying sten guns on the afternoon of the second day. They announced that they believed treason was being planned and took the names and addresses of all delegates. A treason trial duly followed in 1956, but all those accused were eventually acquitted.

The congress itself was the occasion for the launching of the Freedom Charter which laid out the demands of the people of South Africa for a democratic country. Various clauses were read out, speeches made and the Charter adopted by a show of hands.
See also **ANC; Congress Alliance; Freedom Charter; Treason Trial**.

Congress Youth League (*South Africa*)
The Congress Youth League was founded in April 1944. Membership was open to all Africans or those 'who live like and with Africans' who were aged between twelve and forty. At the age of seventeen members of the Youth League became members of the ANC as well. The Youth League had a strong influence on the ANC and served to radicalise it as well as provide some of its most celebrated leaders including Walter Sisulu, Nelson Mandela and Oliver Tambo. It criticised the ANC of the time for being badly organised, elitist and reactive. The Youth League was determined to act as the 'brains trust' of the ANC and infuse it with 'the spirit of African nationalism'. Its Programme of Action – a militant statement of goals and strategies – was adopted by the ANC at its 1949 annual conference and six Youth League members were elected to the ANC executive. The Youth Leaguers, however, were not ideologues, but rather enthusiastic, radical new activists often inspired by popular reactions to poverty in South Africa.

Anton Muziwakhe Lembede, the main initiator of the Youth League, was a self-educated teacher and spoke Afrikaans fluently. This made him acutely aware of growing Afrikaner nationalism and gave impetus to the ideology of 'Africanism', which he formulated. This strand of thought was to permeate the ANC and later clash with a more multi-racial and class-orientated so-called 'Charterist' ideology. In 1959 the Africanist element split from the ANC and formed the Pan-Africanist Congress.
See also **Africanism; ANC; Mandela, Nelson; Pan-Africanist Congress; Sisulu, Walter; Tambo**.

CONSAS (*South Africa*) See **Constellation of Southern African States**.

conscription (*Namibia*)
Compulsory registration for conscription for military service for all male Namibians between the ages of 16 and 25 was introduced by South Africa at the end of 1980. Selective conscription followed, except for those living in the north of the country where support for SWAPO was too strong

for it to be practical. Widespread protests against conscription followed and resistance by some took the form of flight from the country. SWAPO reported increased recruitment to PLAN as a result and the Lutheran World Federation reported 5,000 extra refugees fleeing to Angola by January 1981 to avoid conscription. In addition, many contract workers in Windhoek returned to the north where they hoped to be exempt from conscription.

Despite the opposition, all schools were required to register boys due to reach 16 by 1981, forcing some children out of school prematurely. Those refusing to register were threatened with fines and imprisonment. In October 1984 a further announcement required all Namibian men aged between 17 and 54 to register for possible military service. Although registration was begun, area by area, the high cost and widespread resistance brought it to a halt in 1985. In 1986, however, a military spokesperson confirmed that male Namibians throughout the country, except in 'Ovamboland and Kavango', were liable to military call-up.
See also **PLAN; SWAPO; SWATF**.

conscription (*South Africa*) See **SADF**.

Conservative Alliance of Zimbabwe (*Zimbabwe*)
The party of the rebel white Rhodesians, led by Ian Smith, the Rhodesian Front, was renamed the Republican Front in June 1981. This became the Conservative Alliance of Zimbabwe before the 1985 general elections. In those elections it won 15 of the 20 seats reserved for white Zimbabweans under the constitution.
See also **Independent Zimbabwe Group; Rhodesian Front; Smith**.

Conservative Party (*South Africa*)
In 1982 18 National Party MPs broke away from their party to oppose it from the right. The new party, the Conservative Party or *Konserwatiewe Party*, was led by former Transvaal National Party leader, Andries Treurnicht. It accused the National Party under President P.W. Botha of 'selling out' white South Africans, and of working towards integration in the country.

The expulsion of the Conservative Party MPs from the National Party followed their refusal to support P.W. Botha's interpretation of the 1977 National Party constitutional proposals. They fiercely opposed his 'reformist' policies, even campaigning with the Progressive Federal Party (which opposes the governing party) for a 'No' vote in the 1983 whites-only constitutional referendum. The government's large majority in the referendum demonstrated the limits of Conservative Party support among whites. Nevertheless, the Conservative Party displaced the Progressive Federal Party as the main opposition to the government in parliament, gaining 26.6 per cent of the vote in the 1987 'White' election and 23 seats.

The search for unity on the far right in South Africa has been a long and bitter one. Three such groups came together to form the Conservative Party: Connie Mulder's National Conservative Party, the extra-parlia-

mentary group *Aksie Eie Toekoms*, and the newly-expelled group from the National Party. Despite calls at the time for all three to join the existing HNP, there were clear political differences which resulted in the rejection of this option. The Conservative Party, unlike the HNP, does not advocate a return to pure Verwoerdian apartheid, and instead argues for a system of apartheid evolved from that. It differs from the National Party on the grounds that while it supports an evolved system of apartheid, the National Party, it argues, is bringing about the abolition of apartheid.
See also ***Aksie Eie Toekoms; Herstigte Nasionale Party*; National Conservative Party; National Party; Progressive Federal Party; 'reforms'**.

C

Constellation of Southern African States (*South Africa*)

The idea of a Constellation of Southern African States (CONSAS) was embraced by South African premier P.W. Botha as a way of establishing an economic and political alliance, dominated by South Africa, to counter what he saw as the growing influence of 'communism' in southern Africa. CONSAS was to include the bantustans, Botswana, Lesotho and Swaziland, Muzorewa's 'Zimbabwe-Rhodesia', Malawi, South Africa and Namibia. Zaire and Zambia were to be considered for inclusion at a later stage. CONSAS, it was hoped, would overshadow and dominate socialist countries in the region such as Mozambique and Angola.

At the same time as Botha proposed his CONSAS, however, the frontline states proposed the formation of an alternative alliance to counter South Africa's regional power, the Southern African Development Coordination Conference (SADCC). This proved more attractive to all countries except South Africa, and CONSAS never materialised.
See also **'frontline' states; Muzorewa; SADCC; Zimbabwe-Rhodesia**.

'constitutional crisis' of 1973 (*Swaziland*)

In the elections of 1972 three opposition party members (from the Ngwane National Liberatory Congress) were elected to parliament. This had the effect of unseating the accepted leader of the lineage which expected to provide the successor to the king, Prince Mfanasibili Dlamini. The ruling monarchy regarded the election of members from a different party to their own, the Imbokodvo National Movement, as a challenge and an insult. The ruling 'King's party' refused to accept the new members of parliament and tried to outlaw them. These measures failed and the King declared a state of emergency, dismissed parliament and dissolved all political parties including his own. He argued that a Westminster type of parliament was not suited to Swaziland and simply encouraged disloyalty to the king. King Sobhuza then ruled by decree until 1978 when he introduced a new constitution not based on political parties, but on traditional structures.
See also **Dlamini, Mfanasibili; Imbokodvo National Movement; Ngwane National Liberatory Congress**.

constitution of 1983 (*South Africa*)

The new constitution, the Republic of South Africa Constitution Act of

1983, came into operation on 3 September 1984. It provided for a new three-chamber Parliament: the 178 seat House of Assembly (for 'Whites'), the 85 seat House of Representatives (for 'Coloureds'; elected on 22 August 1984) and the 45 seat House of Delegates (for 'Indians'; elected on 28 August 1984). There is no provision under the new constitution for the majority African population. 'Means will have to be found to enable African communities outside the national states to participate in political decision making in matters affecting their own interests', P.W. Botha told the first joint sitting of 308 MPs. The first session of the three Houses of Parliament began on 4 September 1984.

Under the new constitution P.W. Botha resigned as Prime Minister (consequent upon the abolition of that office) and was sworn in as acting State President following a meeting of the (all-white) cabinet on 3 September 1984. He was unanimously elected as South Africa's first executive State President on 5 September by an 88–member electoral college presided over by the Chief Justice. Mr Botha was inaugurated on 14 September as State President.

The president is given many new powers under the new constitution leading observers to describe it as a further move towards autocratic government. He has the right to appoint members to the President's Council, the body that is supposed to advise Parliament and resolve disputes among the three chambers, and he can intervene if the procedure does not work. He can initiate legislation into all three Houses, choose the Cabinet (which has executive authority), choose the three Ministerial Councils and act as the Head of State.

The constitution was approved by 66 per cent of 'White' voters in a referendum in November 1983, indicating that it drew support from 'White' voters outside the National Party (NP support at the last election was some 53 per cent) as well as from within it.

There were many calls for a boycott of the elections in 1984, particularly by the UDF which was set up specifically to oppose the elections, and also by the Natal Indian Congress, the Transvaal Indian Congress, the ANC and many others. The official percentage poll of registered voters for the House of Representatives was 30.9 per cent and for the House of Delegates 20.29 per cent.

See also **House of Assembly; House of Delegates; House of Representatives; President's Council; 'reforms'**.

constitution of 1969 (*Zimbabwe*)

The 1969 constitution was described by Ian Smith, the rebel Rhodesian leader, as the 'world-beater'. He introduced it after the failure of Anglo-Rhodesian talks on board the ship HMS *Fearless*. In his words, the 1969 constitution 'sounded the death knell of the notion of majority rule', and 'would entrench government in the hands of civilized Rhodesians for all time'.

The constitution replaced the idea of majority rule with parity, a system under which the majority African population would achieve equal representation in parliament with the whites in the distant future. The new voting system decreed separate racial polls; each race could only vote for and be represented by a member of its own race. Under the constitution, the Land Apportionment Act was replaced by the Land Tenure Act which imposed new restrictions on the African population and permanently divided land on a 'racial' basis. The effect of this constitution was to entrench political power in the hands of the whites. The white parliament could pass any law it liked and all protections that there had been under the 1961 constitution were thrown out. Ian Smith justified scrapping the old constitution by claiming that it was 'no longer acceptable to the people of Rhodesia because it contains a number of objectionable features, the principle one being that it provides for African rule, and inevitably the domination of one race by another and that it does not guarantee that government will be retained in responsible hands.'

The 1969 constitutional proposals were put to the white electorate in a referendum, and nearly 70 per cent supported them; 81 per cent voted in the same referendum for Rhodesia to become a republic. The chiefs, who had supported Ian Smith's Unilateral Declaration of Independence, also approved the proposals. In protest, Britain closed its mission in Salisbury (then the capital of Southern Rhodesia) and the governor, Sir Humphrey Gibbs, resigned. On 1 March 1970 the white rebel leaders of Southern Rhodesia declared the country a republic with Clifford Dupont as the first president.

See also ***Fearless*, HMS; Land Tenure Act; Smith; Unilateral Declaration of Independence**.

Contact Group (*Namibia*)

The Contact Group consisted of the five western countries with the greatest economic stake in Namibia. They had been under strong pressure at the United Nations to implement economic sanctions against South Africa and launched their initiative in the face of this pressure. Throughout 1976, while attempting to set up its first puppet government in Namibia, South Africa had refused to include SWAPO. On 7 April 1977 the ambassadors of the five western members of the UN Security Council (the USA, Britain, France, the FRG and Canada) approached the South African Prime Minister with the offer to mediate between South Africa and SWAPO and the UN.

On 27 April secret talks between South Africa and the so-called Western Five began. What came to be called the Contact Group was born. On 7 July South Africa reportedly indicated the acceptability of some aspects of the western formula for UN-supervised elections (the so-called Western settlement proposals or Proposal for a Settlement of the Namibian Situation), but rejected the withdrawal of its troops during the election period. In September 1977 South Africa appointed an Administrator General to run the country, ostensibly as a precursor to the implementation of the independence plan. In practice it simply became a useful mechanism of direct rule, and lasted through a series of puppet governments which collapsed one after the other.

On 25 April 1978, and after SWAPO had accepted a proposal for some

South African troops to remain in Namibia during the elections, South Africa accepted the western proposals. The proposals were later embodied in UN Security Council Resolution 435. This included the withdrawal of all but 1,500 South African troops and the restriction of these and PLAN forces to bases monitored by the United Nations forces. South Africa refused, however, to accept that Walvis Bay was part of Namibia. Ten days later the South African military launched a huge, highly provocative invasion of neighbouring Angola, killing 600 Namibian refugees at Kassinga, 100 miles north of the border. The invasion, apparently designed to make it impossible for SWAPO to accept the proposals (which, in fact, SWAPO accepted on 12 July 1978), set the pattern for events which followed. For the next six years, until the Contact Group finally collapsed, South Africa found excuse after excuse to refuse to implement the the agreement it had accepted.

By April 1981 it had become clear that the new US administration under Ronald Reagan would take a different attitude to that adopted by the Carter Administration. The Contact Group became increasingly marginalised as the US laid greater emphasis on direct bilateral talks with South Africa. The formulation of the US concept of 'constructive engagement' and the related 'linkage' of Namibian independence to South Africa's 'regional security' (interpreted to mean the withdrawal of Cuban personnel from Angola) was the death knell for the 'Group of Five', which virtually ceased to meet.

French frustration with the ineffectiveness of the Contact Group increased until at its last meeting, in Paris on 24 October 1983, French representatives fell out with other members after heated disagreement over US support for the UNITA rebels in Angola. In December France withdrew from the group. External Relations Minister Claude Cheysson was quoted as saying: 'The Group must be put to sleep, being unable to carry out faithfully its mandate.' Subsequent negotiations with South Africa were left to the US Under-Secretary of State for African Affairs, Chester Crocker.

See also **Kassinga; linkage; SWAPO; UN Security Council Resolution 435; Walvis Bay**.

contract labour (*Namibia*)

Over half of all black workers in Namibia are employed on fixed term contracts, migrating to their places of employment for the duration of each contract. These vary in length from six to twelve months for workers from the Police Zone, and may be up to thirty months for workers from the north of Namibia. This system of induced migration is a source of deep resentment among workers. It results in long separations from families and homes and workers are forced to live in single sex and often squalid compounds or hostels subject to frequent raids by police and other paramilitary forces.

The 1971 general strike in Namibia was largely directed against the contract labour system, and reflected the extent of resentment. Although it led to the abolition of the notorious South West Africa Native Labour

Association (SWANLA), it left the contract system little changed, with recruitment transferred to labour bureaux run by the bantustan administrations in Namibia.

All black adult men under the age of 65 are required to register with labour bureaux which then determine both employer and wage rate for each worker. Although the legal requirement to carry passes was waived in 1977, black Namibians still may not seek, accept or be in employment without official permission, and all contracts have to be registered. Failure to register contracts makes employers liable to large fines, and workers subject to forcible return to their homes in the rural areas. AG 63 of November 1979 introduced identity documents which, with residence certificates, have served similar functions to passes.

C

See also **general strike of 1971; labour bureaux; Police Zone; SWANLA**.

contratados (*Angola, Mozambique*)

The *contratados* were the so-called contract labourers, exclusively African workers who, like contract workers in South Africa and Namibia often had no choice but to resort to migrant labour to support their families in the rural areas. *Contratados* were often forced labourers, rather than actual contract workers, and it was not until 1962 that the Portuguese declared forced labour illegal. The contract labour system was introduced in the Portuguese colonies after the abolition of slavery, replacing one system of coercive exploitation with another.

See also **contract labour; forced labour**.

Copperbelt (*Zambia*)

The Zambian economy is overwhelmingly dependent on copper which is the country's major export. This is mined on the Copperbelt, an area in the north of the country adjoining Zaire. There are several mining towns in the Copperbelt, with Kitwe the most important. This is the area where the Mineworkers' Union of Zambia, the country's largest trade union, is most active.

See also **austerity programme; Mineworkers' Union of Zambia**.

COREMO (*Mozambique*)

COREMO, the *Comite Revolucionário de Moçambique* (Mozambique Revolutionary Committee) was formed in 1965 following an attempt by the Zambian government to re-unify the Mozambique nationalist movement after a series of splits within FRELIMO. FRELIMO left the talks when other participants refused to come together under its umbrella, and the remaining groups formed COREMO under the leadership of Adelimo Gwambe who had formerly been one of the leaders of UDENAMO.

COREMO aligned itself politically with UNITA in Angola and the PAC in South Africa, and like these, failed to establish an effective armed wing. The few raids into Tete province which COREMO did launch achieved few tangible results. COREMO's failure to establish itself as a credible nationalist organisation meant that FRELIMO domi-

nated the struggle in Mozambique and was the sole organisation capable of running the country when the Portuguese withdrew.
See also **FRELIMO; PAC; UDENAMO; UNITA**.

COSAS (*South Africa*)
The Congress of South African Students was formed in 1979 and banned by the government in September 1985. 1985 and 1986 were years of very little education for black pupils in South Africa because of political unrest. The situation was summed up by a students' slogan in 1985: 'liberation now; education later'. This has been replaced in early 1987 by the slogan 'education for liberation', and demonstrates that leaders have had some success in encouraging pupils back to school.

COSAS supports the Freedom Charter of the ANC, South Africa's liberation movement, and stands in opposition to black consciousness organisations such as AZASM, the Azanian Students' Movement. It is non-racial but organises mainly among black students because it argues that the problems of black and white students are different. Its general objective as stated in its second annual congress in 1982 is to fight for compulsory, free and democratic education in a democratic society.
See also **AZASM; AZASO**.

COSATU (*South Africa*)
COSATU, the Congress of South African Trade Unions, is the largest trade union federation in South Africa and carries considerable influence in the country. It is an independent, mainly black, grouping, which represents a formidable achievement in union organisation and worker power in South Africa. It is independent from any political grouping but sympathetic to the anti-apartheid unbrella organisation, the UDF.

COSATU was launched in November 1985 after long and difficult negotiations and had a signed-up membership then of 450,000 and 34 affiliated unions. Its main principles are: non-racialism; a commitment to establishing one union in each major industry; worker control of the unions; representation on the basis of paid-up membership; and national cooperation. FOSATU, the Federation of South African Trade Unions, was one of the main groups which came together to form COSATU, and it has carried its commitment to industrial unionism into the new federation. The other group within COSATU is one which was affiliated to the UDF political grouping, and this includes the Natal-based National Federation of Workers. CUSA, which has leanings towards black consciousness, felt unable to accept the principle of non-racialism and did not join COSATU, establishing instead an alliance with the strongly black-consciousness group, AZACTU, to form NACTU. In the process however, the NUM, previously the largest CUSA affiliate, left and joined COSATU.

Trade unions organising black workers have been politically active in South Africa since SACTU joined the Congress Alliance in the 1950s, and COSATU has continued this trend, although its vice-president (FOSATU's former president), Chris Dlamini, has clearly stated that COSATU will lay down its own conditions very carefully before political

alliances are formed. COSATU's leadership includes Elijah Barayi (president), Jay Naidoo (general secretary), Chris Dlamini (vice-president) and Cyril Ramaphosa (head of the National Union of Mineworkers).
See also **Congress Alliance; CUSA; Dlamini; FOSATU; NACTU; NUM; Ramaphosa**.

Council of Churches of Namibia (*Namibia*)
Representative organisation of churches in Namibia, formed in October 1978: see **churches**.

C

Council of Ministers (*Lesotho*)
The Council of Ministers was appointed by Major-General Justin Lekhanya, Lesotho's ruler after a coup d'état in January 1986. The council was chosen by him as a 'non-political' cabinet to run the affairs of the country under the direction of the Military Council. Major-General Lekhanya is the Chairman of both the Military Council and the Council of Ministers. He is also the Minister of Defence and Internal Security in his Council of Ministers.
See also **Lekhanya; Military Council**.

Council of Ministers (*Mozambique*)
This is the government of the RPM, answerable to the People's Assembly and headed by, and subject to, the authority of the president. Its members are appointed by the president to oversee the ministries of the state.
See also **People's Assembly; RPA**.

Council of Unions of South Africa (*South Africa*) See **CUSA**.

coup attempt, 1974 (*Mozambique*)
On 7 September 1974, after the signing of the Lusaka Agreement between FRELIMO and Portugal, white settlers in Mozambique took over the radio station, the telephone exchange and the newspaper in Maputo (or Lorenço Marques as it then was) and called for UDI to prevent the establishment of a FRELIMO government. Two hundred PIDE agents arrested previously were freed from prison. Some members of the settler community drove into the black neighbourhoods of Maputo, firing automatic weapons indiscriminately into the homes of black residents and at the people themselves. Their action sparked a defensive movement by black residents to take over control of their own areas, and young people in particular were instrumental in building barricades to prevent settlers driving into black areas.

In Beira, Mozambique's second largest port, settlers were also involved in violent action attempting to prevent a FRELIMO take-over. Neither revolt was successful, and calm was restored after a broadcast from Dar es Salaam by Samora Machel calling for an end to the violence. The settlers were ousted from the radio station by Portuguese troops, and FRELIMO soldiers were flown in from Tanzania to police the black suburbs.

There was another outbreak of violence precipitated by whites, on 21 October. Several FRELIMO fighters were murdered in Maputo by a group of Portuguese commandos. There was an immediate reaction from black city dwellers who set up road-blocks and pulled white drivers and passengers out of their vehicles, killing several dozen. Once again the violence was stopped when FRELIMO went into the black areas and called for calm.

Ironically, however, having provoked the violence by its actions, the settler community then began leaving *en masse* out of fear that black Mozambicans would turn on whites to extract revenge for the decades of brutality and exploitation. By independence day over half the white population had fled, and eventually over 90 per cent left.
See also **FRELIMO; PIDE**.

coup attempt, 1977 (*Angola*)
On 27 May 1977 an abortive coup d'état was launched against the Angolan Government. In the course of the coup several leaders of the MPLA were murdered, including the Finance Minister, Saydi Mingas. The coup attempt was led by the former Interior Minister, Nito Alves, who had previously been sacked for factionalism within the MPLA.

Popular discontent at the food shortages and other hardships had followed the massive economic sabotage associated with the Portuguese withdrawal. Shortages had been exacerbated by the civil war which followed independence, and were exploited by the Nitistas (as Alves supporters were known). In the event there was little public support for the coup plotters. Until the discovery of the bodies of Mingas and other leaders, the MPLA leadership had shown a marked leniency towards those involved. This was replaced by anger, however, when the killings were discovered. Several of the coup leaders were eventually executed.
See also **Alves; Mingas; MPLA; Nitistas**.

coup of 1986 (*Lesotho*)
Military coup of 1986 which overthrew government of Leabua Jonathan: see **Lekhanya**.

CP (*South Africa*) See **Conservative Party**.

CPSA (*South Africa*) See **South African Communist Party**.

Crossroads (*South Africa*)
Crossroads is a squatters' settlement just outside Cape Town. It has been the focus of the government's 'removals' policy, which seeks to relocate illegal or inconvenient residents from the urban areas and then sends them back to remote bantustans. The campaign to save Crossroads began in 1978 when the government announced it intended to 'clear the Western Cape of illegal squatters'. It demolished houses in the Unibell camp and announced it would do the same to Crossroads. However the Crossroads Committee was formed under Johnson Ngxobongwana and a long battle began to save the settlement. The publicity surrounding the campaign embarrassed the government into an attempt at a negotiated settlement. Some families, it said, could be moved to a new township, Khayelitsha, some 15km away (35km from Cape Town). Later it agreed to upgrade a

smaller, newly organised settlement at Crossroads to the status of a legal black township.

The issue has not yet been resolved, for Crossroads has grown. Newer camps have sprung up around the edges of Crossroads. They are known as Nyanga Extension, Nyanga Bush and Portland Cement, and house between 100,000 and 200,000 people, mainly 'illegal' residents from the 'Homelands' of the Transkei and Ciskei. These were razed to the ground in violent clashes in May and June 1986. At least 69 people were killed and some 60,000 left homeless in brutal fighting. The clashes developed between older residents known as 'fathers' or *witdoeke* (an Afrikaans word describing the white strips of cloth they wear for identification) and the 'comrades' or militant youth often associated with the UDF. The comrades, who were supporting the new residents, were defeated and thousands of people have had to find new places to live. The 'fathers' were reported to be under the leadership of Johnson Ngxobongwana, who headed the Crossroads Committee.

Opponents of the government have accused it of using the 'fathers' in the way they argue the government is using other vigilante groups in the townships – to root out and destroy the young radical comrades in what seems like violence confined to the black community. 'The government wants the world to think it was black-on-black violence in Crossroads', Allan Boesak is reported to have said, 'But black forces were being used to do the state's dirty work'.

See also **'comrades'; 'fathers'; removals; *witdoeke*.**

Cuban troops (*Angola, Namibia*)

The presence of Cuban personnel in Angola, including troops, has been a major factor cited by Western countries and the USA in particular, for the tension between their governments and that of the RPA. The stated aim of the Reagan Administration in the USA is the removal of Cuban personnel from Angola and to this end it has supported both South African intransigence on Namibia and UNITA rebellion. The 'linkage' of the Cuban presence in Angola to the Namibian struggle has been used by South Africa and the US as an excuse for lack of progress towards independence in Namibia.

Friendly relations between the MPLA and the Cuban Government date from a visit to Cuba by Agostinho Neto in 1966 when, despite cordial exchanges, Neto refused offers by Cuban President Fidel Castro of substantial aid. The MPLA did, however, accept the offer of training opportunities for its cadres in Cuba.

In May 1975 Neto requested shipments of light arms from Cuba for the MPLA, in the face of the substantial arming of the FNLA and UNITA then in progress. In August he requested the help of Cuban instructors to train FAPLA soldiers in four new CIRs then being established. The first of these arrived in October and a month later found themselves joining their new trainees in desperate resistance to the South African invasion of Angola. Ten days after the start of that invasion, when its extent became clear, the MPLA appealed to Cuba for urgent help and Cuban combat troops were sent into the country to defend it against the twin invasions from north and south. The first Cuban volunteers arrived in Luanda on 10 November, by which time the South Africans were poised to advance on Luanda while the FNLA/Zairean force was a few miles north of the city.

While Cuba continues to provide military support to Angola, other forms of aid have also become important. Cuban personnel have contributed in such areas as engineering and construction, communications, agriculture and fishing, public health and education generally.
See also **CIR; FNLA; linkage; MPLA; Neto; PRA; South African invasion; UNITA**.

C

cuca shop (*Namibia*)
Cuca shops are small local stores which serve rural communities, especially in the north of Namibia. They are often social focal points for the communities.
See also **shebeen**.

Cunene River Hydro-Electric Scheme (*Angola, Namibia, South Africa*)
This is a major joint project on the Cunene River (which forms part of the border between Angola and Namibia). Conceived in 1926 between the Portuguese and the South Africans, work on the scheme did not begin until 1969.

When complete, the scheme is planned to encompass 27 dams and power stations and to provide water to irrigate large areas of southern Angola and power for use in Angola, Namibia and South Africa. Work has been completed on three out of five generators at the Ruacana Falls, but the building of the dam at Calueque has been suspended. Work on dams planned for Gove, Matala and Matunto has yet to begin.

Since Angolan independence, the Cunene scheme has become the subject of political contention between the governments of Angola and South Africa. Angola has expressed doubts over whether the planned irrigation schemes would benefit the peasant farmers of southern Angola and is reluctant to negotiate the sale of hydro-electric power to the illegal South African administration in Namibia.

CUSA (*South Africa*)
CUSA, the Council of Unions of South Africa, joined with AZACTU, the Azanian Confederation of Trade Unions, to form CUSA-AZACTU in 1986 (renamed NACTU in 1987). CUSA had a policy of black leadership and had criticised other unions, like FOSATU, for their non-racial policy. This was the reason given for its failure to join COSATU, the largest trade union federation in South Africa. CUSA was black consciousness orientated, but not to the same degree as AZACTU, which admitted only blacks to its unions. CUSA insisted only on black leadership positions and provided for membership of all races; its position was 'anti-racist' as opposed to COSATU's 'non-racial' position.

CUSA was formed in 1980 and generally promoted a non-political approach to trade unionism, although this became less of an issue as most unions came to accept a more political role. Its unions were not particularly radical, giving more influence to officials than to members in negotiations. Most CUSA unions accepted the government's registration regulations following the Wiehahn legislation without significant involvement in the debate about the particular wisdom of this.

Its general secretary was Phirowshaw Camay and in 1986 it had thirteen unions affiliated to it and claimed a membership of some 180,000. It successfully organised miners setting up the National Union of Mineworkers, now the largest single union in the country. However, the NUM is no longer affiliated to CUSA, having joined COSATU in 1986. One of CUSA's largest members was the South African Chemical Workers' Union, now affiliated to NACTU.

See also **Camay; COSATU; FOSATU; NACTU; Wiehahn Commission**.

CUSA-AZACTU (*South Africa*)

CUSA-AZACTU was renamed NACTU (the National Confederation of Trade Unions) in 1987. It was formed in 1986 by merging CUSA and AZACTU, two union groupings rooted in the black consciousness tradition.

See also **AZACTU; CUSA; NACTU**.

Dadoo, Yusuf Mahomed (*South Africa*)
Dr Dadoo was the president of the South African Indian Congress from 1948 to 1952 when he was banned by the government. He is the current chairperson of the South African Communist Party. He was also a founder and the first secretary of the Non-European United Front, formed in 1937. He played a central role in promoting cooperation among all races opposed to white minority rule in South Africa, and signed the Xuma-Dadoo-Naicker Pact in 1947 which formed the basis for cooperation between the ANC and the South African Indian Congress.
See also **Natal Indian Congress; South African Communist Party; South African Indian Congress; Xuma-Dadoo-Naidoo Pact**.

Damara Chiefs' Council (*Namibia*)
Better known as Damara Council, it was a constituent party of the Namibia National Front and later informally allied to SWAPO: see **National Convention of Namibia**.

Damara Council (*Namibia*) See **Damara Chiefs' Council**.

Dar es Salamm (*Tanzania*)
Dar es Salaam is the capital city of Tanzania and the country's main port. It is also the centre of commerce and government, despite the fact that the administrative functions of the city are in the process of being transferred to Dodoma. Dar es Salaam has been growing at an annual rate of 10 per cent, and attempts are being made to decentralise industrial development to other towns. In the 1978 census its population was 757,346.
See also **Dodoma**.

death squads (*Lesotho*)
There have been allegations that death squads are operating in Lesotho. Father Michael Worsnip, the General Secretary of the Lesotho Christian Council, was deported in October 1986 after a report appeared in a Lesotho Sunday paper in which he had spoken of the alleged activities of South African hit squads. The report claimed that one squad had been responsible for the death of at least three members of the anti-apartheid ANC. Father Worsnip was reported to have said that the Christian Council had heard 'consistent allegations' that 'an anti-ANC vigilante squad' was operating in Lesotho.

Between June and August 1986 three ANC members were shot dead in Lesotho and two others, including ANC representative Simon Makhetha, were abducted and have not been heard of since. In November 1986 Desmond Sixishe and Vincent Makhele, the former information minister and the former foreign minister respectively, were murdered together with their wives and a friend. Both were former members of Chief Jonathan's government. Makhele had been connected with the BNP Youth League and Sixishe was one of the most pro-ANC members of the government.
See also **Jonathan**.

Decree No. 1 *(Namibia)*

In 1970 and 1971 the UN Security Council agreed resolutions 276 and 301 respectively, outlawing economic links with South African-occupied Namibia. The validity of the resolutions was upheld in a 1971 opinion of the International Court of Justice, but the major western governments refused to act in accordance with them. In September 1979 the United Nations Council for Namibia enacted its Decree No. 1 for the Protection of the Natural Resources of Namibia. The decree was approved by the UN General Assembly in December 1974. According to its provisions, the exploitation, processing or export of Namibia's natural resources is illegal; any concessions or licences emanating from the South African administration are null and void; any illegally exported resources are liable to seizure; and anyone contravening the decree may be held liable in damages by the future government of independent Namibia. The provisions apply unless permission to undertake such activities has previously been granted by the Council. The main provisions of the decree are contained in the following extracts:

'1. No person or entity, whether a body corporate or unincorporated, may search for, prospect, for, explore for, take, extract, mine, process, refine, use, sell, export, or distribute any natural resource, whether animal or mineral, situated or found to be situated within the territorial limits of Namibia'.

'2. Any permission, concession or licence for all or any of the purposes specified in paragraph 1 above whensoever granted by any person or entity, including any body purporting to act under the authority of the Government of the Republic of South Africa or the "Administration of South West Africa" or their predecessors, is null, void and of no force or effect'.

'3. No animal resource, mineral or other natural resource produced in or emanating from the Territory of Namibia may be taken from the said Territory by any means whatsoever to any place whatsoever outside the territorial limits of Namibia'.

'4. Any animal, mineral or other natural resource produced in or emanating from the Territory of Namibia which shall be taken from the said Territory . . . may be seized and shall be forfeited'.

'5. Any vehicle, ship or container found to be carrying animal, mineral or other natural resources produced in or emanating from the Territory of Namibia shall also be subject to seizure and forfeiture'.

'6. Any person, entity or corporation which contravenes the present decree in respect of Namibia may be held liable in damages by the future government of an independent Namibia'.

On 2 May 1985 the United Nations Council for Namibia decided to begin sample prosecutions of violators of Decree No. 1 commencing with legal action in the Netherlands.
See also **UN Council for Namibia**.

Defiance Campaign of 1952 (*South Africa*)
The 1952 'Defiance Campaign Against Unjust Laws' was launched on 26 June that year by the ANC, the South African Indian Congress and the Franchise Action Committee (a loose alliance formed to oppose the removal of 'Coloureds' from the voters' roll). The campaign was enormously successful – one of the most successful campaigns ever organised by the ANC, resulting in a membership increase from 7,000 to nearly 100,000 in a few months. It also had the effect of building up mass support for the ANC as well as paving the way for joint organisation with other groups. This soon led to the formation of the Congress Alliance.

The Defiance Campaign was implemented as part of the Programme of Action which called for massive civil disobedience, and which had been adopted by the ANC in 1949. The campaign began in the main urban centres around Johannesburg and Port Elizabeth and led to many arrests following the breaking of racist laws such as those prohibiting the use by black people of 'White' facilities at post offices, railway stations and on trains; and those laying down curfew regulations and pass laws which applied to Africans only. The campaign was accompanied by 'days of prayer', and volunteers overwhelmingly chose to serve prison sentences rather than pay fines. The peak of the campaign was reached in September with some 2,500 arrests that month in twenty-four centres. In October rioting broke out in the Cape province and the resulting bans and curfews made political activity very difficult. The campaign had ground to a halt by 1953 as a result of bans on the leaders and associated organisational problems.
See also **ANC; Congress Alliance; Congress Youth League**.

DELK (*Namibia*)
The German Lutheran Church: see **churches**.

Democratic Turnhalle Alliance (*Namibia*) See **DTA**.

Department of National Security (*South Africa*) See **DONS**.

detention centres (Namibia)
Prisons in Namibia are overcrowded and inadequate for the large number of people held by the South African administration for political or related offences. As a result, a series of open air detention centres have been set up around the country. In 1981 eighteen such camps were known to exist, most of them in the north of the country. All major military bases

have detention centres. Some of the most notorious are the Hardap Dam centre near Mariental, where the ex-Kassinga refugees abducted by the South Africans in 1978 were held, the Onuno camp, where people were held in a fenced enclosure without shelter or blankets, and the military detention centres at Grootfontein, Ondangua, Oshikango and Oshakati.

Under proclamation AG9, people detained without trial may be held in any 'suitable place'. Reports, including one by Amnesty International in 1986, indicate that detentions of civilians occur in the 'war zone' without public announcement, that detainees are held in unknown places, and that their families are not notified of their detention or whereabouts. They simply 'disappear'.

The churches and other agencies in Namibia have documented the consistent use of torture, flogging and maltreatment of detainees at all the detention centres over the past ten years. Despite South African denials of torture, the reports and allegations from ex-detainees and independent observers indicate that the practice is a normal part of interrogation procedures in Namibia.

See also **AG9; banning order; detention without trial; Kassinga massacre; Robben Island**.

detention without trial (*Namibia, South Africa*)

This is a common practice in both Namibia and South Africa and is used by the South African authorities to silence political opponents, to intimidate, to extract information from detainees and to imprison those it would be politically embarrassing or difficult to bring to trial. The main legislative measure which provides for such detention in Namibia is proclamation AG9. In South Africa it has been used to hold thousands of people, including young children, since the declaration of a state of emergency. Detentions in South Africa have been mainly under the Internal Security Act.

See also **AG9; banning order; detention centres; Internal Security Act**.

DGS (*Angola, Mozambique*)

Direcçã de Seguranca, previously PIDE, the Portuguese security police: see **PIDE**.

DIAMANG (*Angola*)

DIAMANG, or the *Companhia de Diamantes de Angola*, had monopoly control of diamond production in Angola until 1986. Its activities were concentrated in the north of the country where it continued to mine the bulk of Angolan diamonds despite attacks by UNITA forces and the abduction of foreign workers on the mines by UNITA. DIAMANG was established in 1917 by a consortium of Anglo-American, Belgian and South African companies. Despite the nationalisation of a majority shareholding after Angolan independence, DIAMANG continues to have links with the South African-based diamond conglomerate De Beers. In 1986, in a reorganisation of diamond production, DIAMANG was dissolved and replaced by a number of smaller companies. These retained the same shareholders as DIAMANG.

Direcção de Investigação e Segurança de Angola
(*Angola*) See **DISA**.

Direcção de Segurança (*Angola, Mozambique*)
DGS, the new name, in 1969, for the Portuguese security police PIDE: see **PIDE**.

Directorate of Information and Security of Angola
(*Angola*) See **DISA**.

DISA (*Angola*)
The *Direcção de Investigação e Sergurança de Angola* (DISA) or Directorate of Information and Security of Angola, maintained state security in the country until its increasing unpopularity led to its disbandment in 1979. A commission was subsequently appointed to investigate allegations of abuses by its officials.

disinvestment (*South Africa*)
The withdrawal of a company from South Africa, often confused with divestment, the sale of shares in a company because of its South African links.
See also **sanctions**.

Dlamini (*Swaziland*)
Dlamini or Dhlamini is the ruling clan of Swaziland. Those bearing its name are members of this ruling clan. The prefix *Nkosi* is used as an acknowledgement that this clan produces royalty.

Dlamini, Bhekimpi (*Swaziland*)
Prince Bhekimpi Dlamini became prime minister in 1983 after the dismissal of his predecessor, Prime Minister Mabandla Dlamini. Prince Bhekimpi himself was summarily dismissed by King Mswati III on 6 October 1986, just a month after the coronation, and replaced with Sotja Dlamini. This followed the dissolution of the *Liqoqo* (Supreme Ruling Council) and was widely interpreted as the new king's attempt to consolidate his power. At the time of Prince Bhekimpi's appointment he was described as a 'conservative traditionalist'. He was appointed as part of the *Liqoqo's* ousting of Queen Regent Dzeliwe and her supporters. Prince Bhekimpi supported the *Liqoqo* in its bid for greater power, whereas Prince Mabandla had been engaged in an attempt to break the power of the *Liqoqo* politicians.
See also **Dlamini, Mabandla; Dlamini, Sotja; Dzeliwe; *Liqoqo*; Mswati**.

Dlamini, Chris (*South Africa*)
Chris Dlamini was the President of FOSATU and is now the Vice-President of COSATU, the largest trade union federation in South Africa. He is also President of FAWU (the Food and Allied Workers' Union) which is a 50,000 strong union formed after a merger between the Food and Canning Workers' Union and the Sweet, Food and Allied Workers' Union in 1986. He is a full-time shop steward at the Kelloggs plant in Springs, near Johannesburg.
See also **COSATU; FOSATU**.

Dlamini, Clement Dumisa (*Swaziland*)
Prince Dumisa Dlamini was a charismatic and eloquent leader of the 1963 strikes in Swaziland. He received a prison sentence at the time for disturbing the peace. His political loyalty was to the opposition Ngwane National Liberatory Congress (NNLC) of which he was secretary-general. In 1965 he fled the country and worked with the Organisation of African Unity Liberation Committee. Later he returned to Swaziland and was again imprisoned. After his release he said he was no longer associated with the NNLC and joined the ruling party of the monarchy, the Imbokodvo National Movement.
See also **Ngwane National Liberatory Congress**.

Dlamini, Mabandla (*Swaziland*)
Prince Mabandla Dlamini was prime minister of Swaziland from 1979 to 1983 when he was dismissed by Queen Regent Dzeliwe under pressure from the *Liqoqo* or Supreme Ruling Council; his replacement was Prince Bhekimpi Dlamini. He had the reputation of being a modernist and was a popular prime minister, who sought to break the power of the established *Liqoqo* politicians. He also opposed the land deal with South Africa concerning the transfer of South African-controlled land, including the *KaNgwane* bantustan, to Swaziland. He was noted for his attacks on corruption and set up a commission of inquiry into corruption soon after taking office, prosecuting offenders rigorously. However, once he began to find evidence against members of the royal family, the King closed down Prince Mabandla's commission.
See also **Dzeliwe; KaNgwane-Ingwavuma land deal; *Liqoqo***.

Dlamini, Makhosini (*Swaziland*)
Prince Makhosini Dlamini was Prime Minister of Swaziland from independence in 1967 until 1976 when he was retired by King Sobhuza. He was also president of the ruling political party, the Imbokodvo National Movement, known as the 'King's party'.
See also **Imbokodvo National Movement**.

Dlamini, Mfanasibili (*Swaziland*)
Prince Mfanasibili Dlamini was one of the most powerful members of the *Liqoqo* or Swaziland's Supreme Ruling Council. He was dismissed in 1985 by the Queen Regent as part of a campaign to downgrade the *Liqoqo*. On 28 May 1986 the Supreme Court found him guilty of attempting to 'defeat the ends of justice' and sentenced him to two seven year terms in prison to run concurrently.

Prince Mfanasibili was a senior member of the cabinet for many years and he is also the accepted leader of the lineage which expected to provide the successor to King Sobhuza. Together with supporters like Chief Maseko and a commoner, George Msibi (also dismissed in 1985), he had succeeded in elevating the status of the *Liqoqo* and deposing Queen Regent Dzeliwe. He also won a faction struggle against Sishayi Nxumalo, a powerful member of the government and once leader of the monarchy's investment organisation, *Tibiyo Taka Ngwane*.

See also **Dzeliwe; *Liqoqo*; Ntombi; Nxumalo; *Tibiyo Taka Ngwane*.**

Dlamini, Polycarp Ka-Lazarus (*Swaziland*)
Polycarp Dlamini is the managing director of the *Tibiyo Taka Ngwane* fund, the most powerful economic organisation in the country. He is a former justice minister and became managing director of the fund in 1983. Polycarp Dlamini is one of the most influential people in Swaziland and has held high office in government ever since independence.
See also ***Tibiyo Taka Ngwane*.**

Dlamini, Sozisa (*Swaziland*)
Prince Sozisa Dlamini held the powerful hereditary office of the 'Authorised Person' in Swaziland during the crucial period after the death of King Sobhuza. This is the leading position among the Elders of the Nation and in recent years the power of the office has been extended. After a political battle with the queen regent, Prince Sozisa effectively became the regent himself. However he was involved in a further power struggle and the *Liqoqo* or Supreme Ruling Council moved to limit his power. In September 1984 the *Liqoqo* suspended him for allegedly attempting to stage a coup in June of that year. The prince was the senior royal prince and head of the *Liqoqo*.
See also **'Authorised Person'; Elders of the Nation; *Liqoqo*; Sobhuza.**

Dlamini, Sotja (*Swaziland*)
King Mswati III appointed Sotja Dlamini the new Prime Minister on 6 October 1986, and dismissed Prince Bhekimpi Dlamini. The new appointment took place soon after the coronation and was widely interpreted as an attempt by the new king to consolidate his power and oppose the growing power of the *Liqoqo* (Supreme Ruling Council) – which he dissolved – and its politicians. Sotja Dlamini is the first Prime Minister of Swaziland who is not a member of the royal family. Before his appointment he served as assistant police commissioner until 1974 when he was summarily dismissed with no explanation by the *Liqoqo*. He was then employed as head of security on a sugar estate in the south of the country.
See also **Dlamini; Dlamini, Bhekimpi; *Liqoqo*; Mswati.**

Dodoma (*Tanzania*)
At the 1973 TANU conference it was announced that Dodoma was to be Tanzania's new capital city. Dodoma was chosen to create a new central city in the country since the present capital Dar es Salaam is on the coast. However, the cost of the move is very high and there are problems with Dodoma itself. It is in a semi-desert environment with water shortages. A plan was produced for the move and some building work begun. The prime minister's office and TANU headquarters moved, but since 1975 the idea has lost its momentum.
See also **Dar es Salaam.**

Do Nascimento, Lopo (*Angola*)
A veteran MPLA organiser, Lopo do Nascimento was first arrested by PIDE for his nationalist activities when he was only 17 years old. After

working as a trade unionist, he was appointed to the MPLA's Political Bureau and Central Committee in 1974. He became the first prime minister after independence in 1975, but was dismissed in 1978.
See also **MPLA**.

DONS (*South Africa*)
Department of National Security, formerly the Bureau for State Security (BOSS) and subsequently renamed the National Intelligence Service: see **NIS**.

Dos Santos, José Eduardo (*Angola*)
José Eduardo dos Santos became the second President of the People's Republic of Angola when he assumed office on 21 September 1979 to take the place of the recently deceased Agostinho Neto. Dos Santos joined the JMPLA at the age of nineteen and subsequently saw guerrilla service. He joined the Central Committee and Political Bureau in 1974, was the first post-independence foreign minister and later First Vice Prime Minister. Despite being dismissed in the 1978 reshuffle, he subsequently became Minister of National Planning. Like the man he succeeded, dos Santos is a pragmatic politician and in the political centre of the party.
See also **JMPLA; MPLA; Neto**.

Dos Santos, Marcelino (*Mozambique*)
A veteran of the struggle in Mozambique, and one of the few of FRELIMO's original leaders to have maintained his position until independence, Marcelino dos Santos is the respected elder of the revolution and one of the party's chief ideologists. Dos Santos was elected FRELIMO's first Secretary for External Relations at its inaugural congress in 1962. With Samora Machel and Uria Simango, he was a member of the ruling triumverate elected by the Central Committee after the assassination of FRELIMO president Eduardo Mondlane in 1969. He was Vice-President of FRELIMO from 1970–78 and Secretary for Economic Affairs from 1980.

Dos Santos, although arguably the most senior member of the party after Machel, reputedly refused to put himself forward for the presidency after Machel's death, and it was he who made the announcement in November 1986 on behalf of the Central Committee, that Joaquim Chissano would succeed Machel as president.
See also **FRELIMO; Machel; Mondlane; Simango**.

DTA (*Namibia*)
In 1977, following encouragement from South Africa, a group of ethnically-based political parties met in the so-called 'Turnhalle Constitutional Conference' to discuss a possible 'internal' political settlement excluding SWAPO. The parties held their meetings in the Turnhalle, a building from the German colonial period in Namibia, and used its name for the alliance which resulted. On 7 November, they agreed to form a joint organisation to campaign for internal self-government, the Democratic Turnhalle Alliance. The DTA is dominated by the white Republican

Party and its leader, Dirk Mudge, is the chairperson. In addition it includes small ethnically-based parties from each of the ten other apartheid groups. Most important of these is the National Unity Democratic Organisation led by Herero chief Kuaima Riruako, DTA President. Others include the Caprivi Alliance Group, the Kavango Alliance Group, the Namibia Democratic Turnhalle Party, the Namibia People's Liberation Front, the *Rehoboth Baster Vereniging*, the Seoposengwe Party, the SWA People's Democratic United Front and the Tsumkwe Group.
See also **interim government; Mudge; NUDO; Republican Party**.

Dube, John Langalibalele (*South Africa*)
Rev. John L. Dube was the first President of the ANC of South Africa (or the South African Native National Congress as it was then named). He was the editor and founder of the first Zulu language newspaper in South Africa, *Llanga lase Natal* (The Natal Sun), in which he published many articles critical of the white government and in favour of Zulu rights. He led the ANC campaign against the Land Act of 1913 which formally gave most of South Africa's land to the whites. As part of this protest he and other leaders went to London to protest to the British government; they were not favourably received by the Colonial Office.
See also **ANC; Plaatje**.

Duncan, Sheena (*South Africa*)
Sheena Duncan is the current president of the Black Sash, a post she held from 1975 to 1978 and from 1982. Born in Johannesburg she later moved to what was then Southern Rhodesia (now Zimbabwe) and lived there for some years. On her return she became involved in voluntary work in a Black Sash advice office, work she has continued ever since. A liberal and a practising Christian, she has also been involved in the work of the Anglican Church in South Africa and of the SACC.
See also **Black Sash; churches (South Africa)**.

Dupont, Clifford (*Zimbabwe*)
On 1 March 1970 the white rebel leaders of Southern Rhodesia declared the country a republic with Clifford Dupont as the first president.
See also **constitution of 1969**.

Dutch reformed churches (*South Africa*) See **churches**.

dynamising groups (*Mozambique*)
The GDs or *Grupos Dinamizadores* were political action groups of FRELIMO militants, usually of eight to ten people, established by FRELIMO during the period of the Transitional Government to explain, promote and implement FRELIMO policies. In the factories of Mozambique the GDs worked with the factory production councils and management to run the factories. They were largely replaced by other organisations in 1976 and 1977, but survived in urban areas as neighbourhood councils. They were an important part of the early stages of 'People's Power'.
See also **factory production councils; FRELIMO; *Poder Popular***.

Dzeliwe (*Swaziland*)
Princess Dzeliwe took over as queen regent or *Indlovukazi* ('She Elephant') on the death of King Sobhuza in August 1982. The king had appointed her specifically because she was childless. He believed this would help avoid a situation in which the Queen Regent would involve herself in the factional struggle to promote the chances of her own son as the successor to the king.

However, soon after she took power, faction fighting erupted. On 2 August 1983 the *Liqoqo*, or Supreme Ruling Council, presented her with a document transferring many of her powers to the office of the 'Authorised Person', an extremely powerful traditional political position. Queen Regent Dzeliwe refused to sign the document on the advice of Douglas Lukhele, a prominent lawyer, and Arthur Khoza, King Sobhuza's former interpreter and the principal Secretary for Agriculture. A week later an issue of the *Government Gazette* announced Dzeliwe's dismissal. She challenged the ruling but failed to get it overturned; her advisers and supporters were placed in detention. She refused to accept her dismissal but was soon forcibly removed from her residence and a new queen regent, Ntombi Latfwala, installed. Queen Regent Ntombi is the mother of the present king, Mswati III.

See also **'Authorised Person'; *Liqoqo*; Ntombi**.

EAC (*Tanzania*) See **East African Community**.

East African Community (*Tanzania*)
The 1967 Treaty for East African Cooperation set up the East African Community (EAC). The treaty attempted to redress the balance between Kenya – a relatively wealthy country – and the other two poorer members, Tanzania and Uganda. For instance the headquarters of the Community was moved from Kenya to Arusha, in Tanzania. However, relations with Kenya were not good and in 1976 President Nyerere closed his country's border with Kenya. It was not reopened until 1983. In 1971 there was a *coup d'état* in Uganda in which Milton Obote was deposed by Idi Amin. President Nyerere of Tanzania supported Milton Obote in his attempt to regain control of his country and gave him refuge in Tanzania. Eventually the Tanzanian army helped expel Amin and Milton Obote was returned to power in 1980. By then, however, the EAC had finally collapsed in 1977. The Community, it was widely felt, had been of very little benefit to Tanzania, its poorest member.

Eastern Revolt (*Angola*)
The *Revolta do Leste* or Eastern Revolt was the result of a 1973 split in the MPLA between its president, Agostinho Neto, and the eastern commander of the MPLA's forces, Daniel Chipenda. Chipenda took about 2,000 fighters out of the MPLA and into an alliance with the FNLA, undermining the military strength of the MPLA.
See also **Chipenda; FNLA; MPLA; Neto**.

EEC Code of Conduct (*South Africa*)
The EEC Voluntary Code of Conduct for Companies operating in South Africa was agreed by the Foreign Ministers of the EEC countries on 20 September 1977. The code was proposed by the UK Foreign and Commonwealth Secretary at the time, Dr David Owen. It was prompted by the development of black independent trade unions in South Africa and revelations of poor wages paid by many foreign companies and bad conditions of employment practised by them.

The central provisions of the code of conduct were as follows: trade unions should be actively encouraged in EEC companies or their subsidiaries; migrant labour should be opposed wherever possible and the minimum wage should exceed by 50 per cent the minimum level required

to satisfy the basic needs of an employee and a family; job reservation should be opposed and there should be equal pay for equal work; training schemes should be provided for black workers; places of work should be desegregated; there should be public annual reports on the implementation of the code of conduct by parent companies.
See also **Sullivan Code**.

Eglin, Colin Wells (*South Africa*)

Colin Eglin is the leader of the Progressive Federal Party. He was born in 1925 and trained as a Quantity Surveyor. He was active in the Torch Commando, which opposed the removal of 'Coloureds' from the common voters' roll, in the early 1950s, and later in the Liberal Association (a forerunner of the Liberal Party) before becoming United Party MP for Pinelands in 1958. The following year Eglin, together with other liberals, left the United Party to form the Progressive Party. Along with all but one of his colleagues, he lost his parliamentary seat in 1961.

In 1971 Eglin succeeded Jan Steytler as leader of the Progressive Party, and under his leadership it won six parliamentary seats in 1974 (including Eglin's Sea Point seat) and, as the Progressive Federal Party (after two mergers with other groups), became the official parliamentary opposition in 1977. The party lost this position in the 1987 general election.

Eglin was replaced in 1979 as party head by Frederik van Zyl Slabbert, following growing dissatisfaction with his leadership. He remained active, however, becoming chair of the party, and its deputy leader. Following the controversial resignation of Slabbert in 1986, Eglin once again assumed the leadership.
See also **Progressive Federal Party; Slabbert**.

Elders of the Nation (*Swaziland*)

Under traditional law the Elders of the Nation, together with the 'Authorised Person', oversee Swazi politics. They are extremely powerful within the traditionalist power structure and are made up of the male heads of family of direct descendants of the nineteenth-century King Mswati, or the hereditary heads of the main clans of the country. Their spokesperson is known as the 'Authorised Person', an extremely powerful office in its own right. The elders collectively have the right to check, and even sometimes reverse, the actions of the king. They operate as a crucial part of the system of checks and balances within the Swazi monarchy.
See also **'Authorised Person'; monarchy**.

election of 1973 (*Namibia*)

In August 1973 South Africa called elections for the bantustan or 'Homeland' 'administrations' in Namibia. The 50,000 registered electors were only one third of those eligible for the vote. Of those registered, only 2.5 per cent actually voted, an overall poll of less than one per cent of those enfranchised. The derisory turnout dealt a serious blow to the 'Homelands' policy in Namibia, depriving it of all credibility.
See also **AG8; 'Homelands'**.

ELK (*Namibia*)

Evangelical Lutheran Church: see **churches**.

ELNA (*Angola*)
ELNA, the *Exército de Libertação Nacional de Angola* was the military wing of the FNLA. Based in Zaire, its main area of operations was the north of Angola. It later established a presence in the east, but collapsed following the defeat of the joint FNLA and Zairean forces at Kifangondo Marshes north of Luanda in 1975.
See also **FNLA; Kifangondo Marshes**.

ELOK (*Namibia*)
Evangelical Lutheran Ovambo-Kavango Church: see **churches**.

Eminent Persons Group (*South Africa*)
The Eminent Persons Group (EPG) was announced by the Commonwealth Secretary-General, Shridath ('Sonny') Ramphal, on 25 November 1985. It had been agreed at the Nassau Summit in 1985 to entrust the setting up of this group to seven heads of government, namely, the President of Zambia, the Prime Ministers of Australia, the Bahamas, Canada, India, the UK and Zimbabwe. The two EPG co-chairpersons were Malcolm Fraser, former Australian Prime Minister and General Olusegun Obasanjo, former Nigerian head of state. The EPG was to report on progress towards ending apartheid in South Africa and the Commonwealth would decide whether to pursue sanctions against that country on the basis of that report.

On the EPG's second visit to South Africa from 13 to 19 May 1986, South Africa's President Botha said that his government would not tolerate 'unsolicited interference' in its affairs. Confidential proposals made by the EPG were also believed to have been rejected by the ANC at a meeting on 14 and 15 May, and the idea collapsed when South Africa launched military raids against ANC targets in Zimbabwe, Botswana and Zambia on 19 May.

The EPG's report was entitled *Mission to South Africa: the Commonwealth Report*. It was completed at a meeting in London from 4 to 7 June 1986 and later published by Penguin. The report was negative on every count: apartheid continued as did the state of emergency; political prisoners remained in prison; the ANC remained banned and political freedom was limited; there was no prospect of dialogue. The EPG concluded that 'it is not sanctions which will destroy the country but the persistence of apartheid and the government's failure to engage in fundamental reform'.

> The question in front of heads of government is in our view clear. It is not whether such measures (sanctions) will compel change; it is already the case that their absence, and Pretoria's view that they need not be feared, defers change. . . . Such action may offer the last opportunity to avert what could be the worst bloodbath since the Second World War.

The EPG held its final meeting on 30 July 1986 in London.

See also **sanctions**.

End Conscription Campaign (*South Africa*)
The End Conscription Campaign or ECC was set up in 1983 to coordinate the various groups within South Africa campaigning against conscription. According to official figures 7,589 'White' South African men failed to answer the call to arms in 1985. 6,000 were said to be studying, but the ECC believe that as many as 3,000 were draft-dodgers. The state of emergency has led to a situation where conscripts may be sent into the townships to fight anti-apartheid protesters. This is now presenting many young white South Africans with an acute moral dilemma. In 1985 a 'Peace Festival' to popularise the call to 'cancel the call-up' was organised, and in 1986 the ECC initiated a 'Troops out of the Townships' campaign.

The Committee on South African War Resistance (COSAWR) is another influential body opposing conscription. It was founded in 1978 and operates from Britain providing practical advice to resisters. It publishes a journal, *Resister*.
See also **Committee on South African War Resistance; SADF**.

e

EPG (*South Africa*) See **Eminent Persons Group**.

EPLA (*Angola*)
The EPLA, or *Exército Popular de Libertação de Angola*, was, until 1974, the name of the military wing of the MPLA. In that year FAPLA was formed to replace the EPLA. During the period between 1961, at the start of the armed struggle, and the formation of FAPLA, the EPLA established a presence in six 'Political-Military Regions' in Cabinda and the north, centre, east and south of the country.
See also **FAPLA; MPLA**.

Etango (*Namibia*)
A propaganda organisation linked to the military in Namibia: see **SWATF**.

European Advisory Council (*Botswana*)
The European Advisory Council was created in 1921 to provide a forum for discussing matters directly affecting European interests in Bechuanaland (later to become Botswana). It excercised considerable influence on the administration and effectively promoted European interests, lobbying for closer ties with South Africa. In 1950 the colonial administration decided to set up a Joint Advisory Council to supplement the African and European councils. However, all these bodies ceased to exist with the creation of the Legislative Council in 1961.
See also **African Advisory Council**.

European Advisory Council (*Swaziland*)
The European Advisory Council was set up in 1921 in order to deal with affairs affecting whites in Swaziland. It also served to express white political interests and consisted of nine elected members. In 1949 the Reconstituted European Advisory Council was created by a government proclamation replacing the European Advisory Council. The body was

conservative by nature and represented the pro-South African sympathies felt by many whites in Swaziland. Its members eventually formed the United Swaziland Association as a counter-force to the emerging nationalist organisations.
See also **Reconstituted European Advisory Council; United Swaziland Association**.

Evangelical Lutheran Church (*Namibia*) See **churches**.

Evangelical Lutheran Ovambo-Kavango Church (*Namibia*) See **churches**.

Evangeliese Lutherse Kerk (*Namibia*) Evangelical Lutheran Church: see **churches**.

Evangeliese Lutherse Owambokavangokerk (*Namibia*) Evangelical Lutheran Ovambo-Kavango Church: see **churches**.

Exército de Libertação Nacional de Angola (*Angola*) See **ELNA**.

Exército Popular de Libertação de Angola (*Angola*) See **EPLA**.

Ezuva (*Namibia*)
A propaganda organisation linked to the military in Namibia: see **SWATF**.

F

factory production councils (*Mozambique*)

Introduced in October 1976 to provide for worker participation in factory management and planning, these formed part of a tripartite system in factories: the councils, management and the 'dynamising groups' of FRELIMO militants working in the factories. In April 1983 a conference of the production councils decided that the stage had been reached for the establishment of trade unions in the factories.
See also **dynamising groups; trade unions**.

FAK (*South Africa*)

The *Federasie van Afrikaanse Kultuurverenigingе* (FAK) or Federation of Afrikaans Cultural Associations is the cultural offshoot of the *Afrikaner Broederbond* (AB), and its public face. Formed by the AB in 1929, its membership includes almost 3,000 organisations broadly concerned with Afrikaans culture. The FAK was the main originator of the concept of Christian National Education which is now at the centre of the education system in South Africa, and played a central role in mobilising Afrikaner capital. It was also one of the most important bodies in welding together the political forces of Afrikaner nationalism to capture state power in 1948.
See also ***Afrikaner Broederbond***.

FAPLA (*Angola*)

FAPLA, *Forças Armadas Populares para a Libertação de Angola* or the People's Armed Forces for the Liberation of Angola, was established in August 1974 to replace the EPLA as the armed wing of the MPLA liberation movement. It is now Angola's national army. The Commander-in-Chief is the President of the People's Republic of Angola.

As the government struggles to defend Angolan sovereignty against South Africa and UNITA, all Angolans between 18 and 35 have been subjected to a military draft. Those called up serve two years before joining the military reserve. 1987 estimates of armed forces at Angola's disposal were: FAPLA – 50,000 in the field and a reserve militia of roughly the same size; Cuban troops – 27,000; other foreign advisors and instructors – 2,000.
See also **Dos Santos; EPLA; MPLA**.

'fathers' (*South Africa*)

The 'fathers' is the term used to describe the established, conservative

residents of the Crossroads squatter camp near Cape Town. They have clashed, sometimes violently, with newer, more radical residents known as the 'comrades'. The 'fathers' have also been called *witdoeke*, an Afrikaans word referring to the white strips of cloth they wear to identify themselves. The leader of the 'fathers' in Crossroads is said to be Johnson Ngxobongwana who headed the Crossroads Committee. He is likely to have influence with the government in deciding who will live in the smaller, rebuilt Crossroads community.
See also **'comrades'; Crossroads; *witdoeke***.

FCI (*South Africa*)
The FCI or South African Federated Chamber of Industries, as it is correctly known, is the primary representative organisation of manufacturing capital in South Africa. Its support is strongest among indigenous companies of non-Afrikaner origins. While seen as slightly more progressive than other organisations representing capital in South Africa, the FCI has been a staunch supporter of President P.W. Botha's 'total strategy' and has been among those organisations which, in recent years, have advocated strategies of co-option and incorporation of middle-class black people.

The FCI was historically at odds with the Chamber of Mines over the issue of protectionism and labour allocation, but with the integration and interpenetration of capital in recent years has found itself having more in common with other representative bodies of capital. It has been consistently more liberal than the other main representative organisation of industry, SEIFSA.
See also **ASSOCOM; Chamber of Mines; SEIFSA**.

f

FDLA (*Angola*)
The *Frente Democrãtica de Libertação de Angola* (FDLA) was an alliance between the MPLA and several small organisations active in the north of Angola. Negotiated in 1963, the FDLA disintegrated two years later following opposition from within the MPLA. The groups party to the alliance were the MPLA, UNTA (*União Nacional dos Trabalhadores de Angola* – the trade union wing of the MPLA), MNA (*Movimento Nacional Angolano*), MDIA (*Movimento de Defesa dos Interesses de Angola*) and *Ngwizako* (*Ngwizako Ngwizani a Kongo*).
See also **MPLA; UNTA**.

***Fearless,* HMS** (*Zimbabwe*)
A second attempt to reach a settlement on the UDI dispute was made at a meeting between the Rhodesian rebel leader, Ian Smith, and the British prime minister, Harold Wilson, in 1968, and took place on board the British cruiser HMS *Fearless*. The proposals emerging from the meeting provided for a long-term transition to majority rule. Harold Wilson abandoned most preconditions for the return to legality, agreeing to leave Ian Smith in control. The major obstacle to agreement this time was the safeguards to prevent the specifically entrenched clauses in the

constitution from being amended by the Rhodesian government. Britain also insisted on the right of appeal to the Privy Council.

The proposals were rejected by the white Rhodesians and provoked Ian Smith to make his notorious declaration: 'There will be no majority rule in my lifetime – or in my children's'. After the failure of these talks, Ian Smith pressed ahead with the constitution of 1969, which he nicknamed the 'world-beater'.
See also **constitution of 1969; *Tiger*, HMS**.

Federal Party (*Namibia*)
In October 1975 the United National Party of South West Africa (UPSWA) was re-constituted as the Federal Party. The UPSWA had been affiliated to the United South African National Party (UP). When the UP was dissolved in 1975, the UPSAW became the Federal Party, under the leadership of a white lawyer, Brian O'Linn. In 1977 the party declared itself in favour of a non-racial system in Namibia and of national reconciliation. It opposed the Odendaal report which laid out the basis for the introduction of apartheid in Namibia. It called for an internationally acceptable settlement based on direct negotiations involving South Africa, SWAPO and the 'frontline' states, and for the immediate implementation of the UN Plan as the only route to independence. It rejected participation in whites only elections and from March 1978 to June 1979 was part of the largely black Namibia National Front, an anti-apartheid but also anti-SWAPO alliance. The party advocates a federal and non-racial constitution for Namibia.
See also **Namibia National Convention; Odendaal Commission; SWAPO; UPSWA; United Party**.

Federasie van Afrikaanse Kultuurvereniginge (*South Africa*)
See **FAK**.

Federation of African Societies (*Zambia*)
The colonial government in Northern Rhodesia would not allow trade unions so African workers formed welfare societies throughout the country. In 1946 a Federation of African Societies was formed with the aim of bringing various African groups together. In 1948 the Society renamed itself the Northern Rhodesia African Congress and became the country's first nationalist political party.
See also **Northern Rhodesia African Congress; welfare societies**.

Federation of Afrikaans Cultural Associations (*South Africa*) See **FAK**.

Federation of Rhodesia and Nyasaland (*Malawi, Zambia, Zimbabwe*)
The Federation of Rhodesia and Nyasaland is another name for the Central African Federation, joining together Nyasaland (later Malawi), Zambia (Northern Rhodesia) and Zimbabwe (Southern Rhodesia) under a federation government, set up in 1953.
See also **Central African Federation**.

Federation of South African Trade Unions (*South Africa*) See **FOSATU**.

Federation of South African Women (*South Africa*)
The Federation of South African Women (FSAW) was founded as a non-racial group in 1954 by women like Helen Joseph and Lillian Ngoyi. It was part of the Congress Alliance of groups opposed to apartheid. Its most famous campaign was that launched against the extension of passes to African women during the 1950s. On 9 August 1956 some 20,000 women marched on Prime Minister Strijdom's offices at the Union Buildings in Pretoria, rejecting the extension of the pass laws to women. They chanted the slogan 'Strijdom when you struck the women you struck a rock'. The 9th August is now celebrated as South African Women's Day in South Africa.

FSAW was never formally banned although after 1960, when the ANC and other groups were banned, state harassment made it very difficult for it to continue with its activities. In recent years it has been revived, and continues to fight against the particular injustices experienced by black women at the hands of a white male state, campaigning on issues such as maternity leave and equal pay. It is also trying to combat sexism within the liberation movements. Albertina Sisulu has been active in FSAW's revival and is its president.
See also **Congress Alliance; Joseph; Ngoyi; Sisulu, Albertina**.

Fifth Brigade (*Zimbabwe*)
The Fifth Brigade is a specialist army brigade which was set up after independence by the Zimbabwean prime minister, Robert Mugabe. It has been widely criticised outside ZANU, the ruling party, as a partisan force operating often in a brutal way to quell all those who oppose ZANU. It is composed of ZANU sympathisers trained by personnel from the Democratic People's Republic of Korea. Early in 1983 and in 1984 serious allegations of atrocities and indiscipline were made against the Fifth Brigade as it sought to crush dissidents in the country. These were supported by many church people including the Roman Catholic authorities who had previously helped to expose atrocities committed by the illegal white Rhodesian regime led by Ian Smith.
See also **Mugabe; ZANU**.

Finnish Missionary Society (*Namibia*) See **churches**.

First, Ruth (*Mozambique, South Africa*)
Ruth First was a South African radical journalist, writer and campaigner against injustice in South Africa. She helped to found the Congress of Democrats, the white branch of the Congress Alliance. In 1956 she was one of those arrested after the Congress of the People and tried for treason, only to be released with the others when the charges collapsed. She was also an active member of the South African Communist Party.

After the declaration of the 1960 state of emergency she escaped to Swaziland, later returning to work in South Africa. She was constantly harassed by the South African state and forced into exile in 1964. She was killed in 1982 in her office at Mozambique's Eduardo Mondlane University in Maputo by a parcel bomb allegedly sent by the South African secret services. At the time of her death she was the head of the Centre for African Studies at the university. A committed member of the ANC, she put all her considerable energies into studying and supporting the revolution in Mozambique. She argued that it was misguided to see the struggles for freedom in South Africa and for socialism and self-determination in Mozambique as separate. Many of the ideas which contributed to the success of the Fourth Congress of FRELIMO in 1983 derived from the work of the Centre for African Studies.

In the course of her career Ruth First did path-breaking work on, among other things, migrant labour and Namibia. One of her best known books on the struggle for freedom in Africa is *The Barrel of a Gun: Political Power in Africa and the Coup d'Etat* (Penguin, 1972).
See also **ANC; Centre of African Studies; Congress Alliance; Congress of Democrats; Treason Trial**.

Fischer, Abram 'Bram' (*South Africa*)

Bram Fischer was born in 1908, son of the Judge President of the Orange Free State. He became an advocate (barrister), QC, and was a leading defence lawyer in the political trials of the 1950s and 1960s. He was also a prominent member of the underground South African Communist Party. He was part of the defence team in the Treason Trial, and led the defence in the Rivonia Trial.

In September 1964 Bram Fischer was arrested and charged with membership of the South African Communist Party (SACP) and activities against the state. After a period on bail to pursue a case in London, he returned in November for the start of his trial. In January 1965, however, after it became clear that he was unlikely to escape conviction, he jumped bail and went underground in the country to continue his political work. He was rearrested ten months later, and, in 1966, was sentenced to life imprisonment. In a speech from the dock before sentence, Fischer made his now famous statement on the SACP. Explaining why he had joined the party, he spoke of 'a gradual realization, as I became more and more deeply involved in the Congress movement of those years, . . . that it was always members of the Communist Party who seemed prepared, regardless of cost, to sacrifice most: to give their best, to face the greatest dangers in the struggle against poverty and discrimination.'

Life imprisonment for political activitists in South Africa means imprisonment until death. Bram Fischer, already suffering from a number of physical complaints when sentenced to life, received no quarter. He eventually died of cancer in 1975, being allowed out of prison for only the last six weeks of his life.
See also **Rivonia Trial; South African Communist Party; Treason Trial**.

Five principles (*Zimbabwe*)

The 'five principles' were the conditions the British government stated were essential to any Southern Rhodesian constitutional settlement. They

were part of the philosophy of NIBMAR (No Independence Before Majority Rule) which the British government claimed to adopt in its relations towards the white Rhodesian rebels.
See also **NIBMAR**.

FLEC (*Angola*)
In 1963 the *Frente para a Libertação do Enclave de Cabinda* (FLEC – Front for the Liberation of the Enclave of Cabinda) was established to campaign for Cabindan secession from Angola. It launched a military struggle in 1974 after it had been banned by the Portuguese, and has continued its military campaign despite Angolan independence. The FLEC has received external support from Zaire, France and Gabon, all attracted by the oil wealth of the enclave, but has been ineffective owing to internal divisions.
See also **Cabinda**.

flechas (*Mozambique*)
The *flechas* were an elite group set up by PIDE in Mozambique to combat FRELIMO at a time when the liberation movement had made significant advances against the demoralised conscripts of the Portuguese colonial army. The *flechas* were a particularly ruthless group of irregular soldiers who countered FRELIMO's successes with the killing of members of the peasant populations in areas where the liberation movement enjoyed the support of local people. They were responsible for some of the worst atrocities of the liberation war.
See also **FRELIMO; PIDE; Wiriamu massacre**.

FNLA (*Angola*)
The *Frente Nacional de Libertação de Angola*, or the National Front for the Liberation of Angola (FNLA), operated primarily in the north of the country under the leadership of Holden Roberto. Based in Zaire with the support of Zairean President Mobutu, the FNLA received covert support from the CIA and other Western sources as well as China. The FNLA had its roots in a peasant revolt in the north of Angola which began in March 1961, instigated by a precursor organisation, the Union of the Peoples of Angola (UPA). Holden Roberto had founded the UPA in 1958, and headed the FNLA when it was formed in March 1962 out of a merger between the UPA and the smaller but more radical Angolan Democratic Party or PDA, another northern group.

During the 1960s and 1970s the FNLA received escalating military and financial aid from the USA through the CIA in Zaire. It also received some military aid from China in the early 1970s. Shortly after the Portuguese coup the FNLA signed a ceasefire and in September 1974, just before he too was ousted, General Spinola reached a secret agreement, the Sal plan, to bolster the FNLA and UNITA against the MPLA. By March 1975 the FNLA was attacking MPLA personnel in the north of Angola with the assistance of Western mercenaries and of Zairean troops dispatched by President Mobutu. By July it had been displaced from Luanda by the MPLA and the civil war had begun in earnest. The

FNLA disappeared as a significant force after its military defeat at the Kifangondo Marshes.
See also **FNLA; GRAE; Kifangondo Marshes; MFA; Roberto; Transitional Government; UNITA**.

Food and Canning Workers' Unions (*South Africa*)
The Food and Canning Workers' Unions were a combination of the Food and Canning Workers' Union and the African Food and Canning Workers' Union. They were long-standing militant unions and were active in SACTU during the 1950s and 1960s. Dr Neil Aggett, who died in custody in 1982, was the Transvaal regional secretary of the unions. In 1986 the Food and Canning Workers' Union joined with the Sweet, Food and Allied Workers' Union to form the 50,000 strong Food and Allied Workers' Union.
See also **Aggett; Dlamini, Chris; SACTU**.

food riots (*Zambia*)
Occurring in December 1986, these were the worst riots in 22 years of independence in Zambia: see **austerity measures**.

Forças Armadas Populares para a Libertação de Angola (*Angola*) See **FAPLA**.

Forças Populares de Libertação de Moçambique (*Mozambique*)
Mozambique's armed forces: see **FPLM**.

forced labour (*Angola, Mozambique*)
In all of Portugal's African colonies black people who had not attained *assimilado* status were subject to forced labour or *chibalo*, a way of ensuring a labour supply without having to pay workers a living wage. In Lourenço Marques (renamed Maputo at independence), for instance, the black-white wage differential widened over a fifty year period from 1:3 to a 1960 figure of 1:13.

Corporal punishment (using the painful *palmatória*), imprisonment and other punishments were meted out to those who resisted the requirement to perform forced labour.
See also ***assimilado*; *chibalo*; *palmatória***.

FOSATU (*South Africa*)
FOSATU, the Federation of South African Trade Unions, was one of the main groups that joined to form COSATU. It was an independent, largely black trade union grouping which had a policy of non-racialism. It was the most highly organised of all South African trade union groupings and made the trade union movement a force to be dealt with during its growth in the 1980s.

FOSATU was formed in April 1979 with about ten affiliates. By 1985, when it joined with other unions to form COSATU, FOSATU had 140,000 members, and a strong tradition of organising in industries throughout the country. FOSATU supported the registration of unions under government legislation, arguing that this would allow its members as much freedom as possible to organise and grow. The federation took the view that unions should not get involved prematurely in political campaigns but concentrate instead on building up their membership on the shop floor. FOSATU never opposed political activity but sought to build a sound base and to engage politically at a more grassroots level. Its general secretary, Joe Foster, speaking at the 1982 congress, argued that building an organised working class base was itself a political activity.

In recent years there has been a general move towards greater participation in political campaigns by unions, and FOSATU has supported this. The change was marked by its decision to support the 1984 work stay-away – a success because of FOSATU's backing. However, FOSATU always put workers' issues before political campaigns and the debate about union involvement in specifically political issues continues in COSATU.

See also **COSATU; Dlamini, Chris; trade unions; Wiehahn Commission**.

FP (*Namibia*) See **Federal Party**.

FPLM (*Mozambique*)

The FPLM or *Forças Populares de Libertação do Moçambique* (Popular Forces for the Liberation of Mozambique) is Mozambique's army employed in both the defence and the reconstruction of the country. Since 1978, all men and women in Mozambique over the age of 18 have been subject to two years compulsory military service. In 1982 the strength of the FPLM stood at 21,600, but it was supported by an additional paramilitary force of 6,000, and by the people's militia. Internal security is maintained by SNASP, the National Service of Popular Security (*Serviço Nacional de Segurança Popular*), which has the duty of detecting and preventing subversion, sabotage and acts against People's Power, the economy, or the state.

See also **SNASP**.

Freedom Charter (*South Africa*)

The Freedom Charter was adopted by a show of hands at the Congress of the People on 26 June 1955. It outlined the basic demands of the people of South Africa for a democratic state. In 1956 the ANC formally adopted the Freedom Charter and its programme for a liberated South Africa.

The suggestion of a national congress to 'draw up a Freedom Charter for the Democratic South Africa of the future' was made by Professor Z.K. Matthews at the ANC Cape provincial conference in 1953. The idea was adopted by the ANC and the Congress Alliance planned to set up a National Action Council with provincial branches to plan the Congress of the People and the drawing up of the Charter. Freedom volunteers were to be asked to collect demands for the Charter from people throughout the country. This process did not take place exactly in the form proposed, but the campaign for the Charter succeeded in evoking a popular response. In early 1955 a drafting committee produced the Charter drawing on material prepared by the National Action Council sub-committees. It was presented to the ANC executive just before the Congress of the People.

The Charter began its preamble thus:

'We, the people of South Africa, declare for all our country and the world to know:
That South Africa belongs to all who live in it, black and white, and that no government can justly claim authority unless it is based on the will of the people;
That our people have been robbed of their birthright to land, liberty and peace by a form of government founded on injustice and inequality; . . . And therefore, we, the people of South Africa, black and white together – equals, countrymen and brothers – adopt this FREEDOM CHARTER. And we pledge ourselves to strive together, sparing nothing of our strength, and courage, until the democratic changes here set out have been won.'

The main demands of the Charter are as follows:

the people shall govern;
all national groups shall have equal rights;
the people shall share in the country's wealth;
the land shall be shared among those who work it;
all shall be equal before the law;
all shall enjoy equal human rights;
there shall be work and security;
the doors of learning and of culture shall be opened;
there shall be houses, security and comfort;
there shall be peace and friendship.

See also **ANC; Congress of the People**.

Free Market Foundation (*South Africa*)

A much smaller organisation than the Urban Foundation or the South African Foundation, the Free Market Foundation has the more limited objective of propagandising within the country on the benefits of free enterprise and of persuading black people that such a system is in their best interests. It is supported by organisations such as the *Afrikaanse Handelsinstituut*, ASSOCOM, the Chamber of Mines, the Confederation of Labour, the FCI, NAFCOC and TUCSA.

See also ***Afrikaanse Handelsinstituut*; ASSOCOM; Chamber of Mines; Confederation of Labour; FCI; NAFCOC; South African Foundation; TUCSA; Urban Foundation**.

FRELIMO (*Mozambique*)

FRELIMO, the *Frente de Libertação de Moçambique* (Front for the Liberation of Mozambique), was founded on 25 June 1962 in Dar es Salaam, through the merger of three nationalist organisations, the *União Democrática Nacional de Moçambique* (UDENAMO), the Mozambique African National Union (MANU) and the *União Africana de Moçambique Independente* (UNAMI). The three constituent parties had been formed in Southern Rhodesia, East Africa and Nyasaland, respectively, but had all moved their headquarters to Dar es Salaam after Tanganyikan independence in December 1961. Following pressure from Tanganyikan prime minister Julius Nyerere, they came together to form FRELIMO.

At FRELIMO's first congress, held in September 1962, Eduardo Mondlane was elected president, Rev. Uria Simango vice-president, and Marcelino dos Santos secretary for external relations. FRELIMO was established as a broad nationalist front, and like other similar organisations spent its first years plagued by divisions and infighting as different political factions struggled for control. By 1965 many of its founders had left the organisation, and in mid–1965 many of these formed a rival group, COREMO. COREMO's subsequent failure to establish itself as a viable organisation allowed FRELIMO to dominate the liberation struggle in Mozambique which followed.

The first shots in the war of liberation were fired on 25 September 1964, and over subsequent years FRELIMO received training and equipment from Algeria, Egypt, Israel, Zaire, the USSR, Eastern Europe and China. While an increasingly successful armed struggle was being waged, the internecine struggles continued within the leadership of FRELIMO. Between 1967 and 1969 it was virtually paralysed by divisions. So many of its early leaders left the organisation, that only three members of its first central committee remain on the present one. Despite the ideological struggles at the top, however, the armed struggle proved effective and itself attracted recruits to the organisation.

FRELIMO's second congress was held in Niassa province in Mozambique in July 1968, at a time when the leadership struggles had reached a peak. The congress included representatives of FRELIMO's military personnel and was boycotted by some of its more conservative leaders opposed to their involvement. It approved the establishment of liberated zones and re-elected Mondlane and Simango as president and vice-president respectively.

The assassination of Mondlane, allegedly by PIDE, in February 1969 led to a further crisis in FRELIMO's leadership with vice-president Simango attempting to take over from Mondlane. He was thwarted by the Central Committee which appointed a triumverate of Simango, Samora Machel and Marcelino dos Santos instead, precipitating Simango's resignation and denunciation of FRELIMO later that year. Samora Machel was subsequently elected president.

Following the resolution of the leadership crisis, FRELIMO was able to concentrate on the armed struggle, and the early 1970s saw the war spreading to more and more of the country. Eventually, following the MFA coup and the overruling of the conservative General Spinola, Portuguese Foreign Minister Mario Soares signed an agreement with Samora Machel on 7 September 1974, the so-called Lusaka Agreement, which established a ceasefire in the war, a Transitional Government to take power in Mozambique from 25 September, and full independence from 25 June 1975.

On independence day FRELIMO nationalised the land, and did likewise to rented property on 3 February 1976. Although those who could afford to do so were allowed to retain their main home and a holiday home, the loss of investments in property affected a significant proportion of the settler population which had invested in the high rise apartment blocks of Maputo. Nationalisation of the social services meant improvements in health and education for the majority, but declining quality of services for the privileged. Together with the settler-instigated violence at the time of the Portuguese hand-over, the loss of their privileges persuaded over 90 per cent of the settlers to leave by the end of the first year of independence. Most took with them whatever they could transport and destroyed everything else, including machinery, buildings and livestock.

At its third (and first post-independence) congress, in February 1977, FRELIMO was established as a 'Marxist-Leninist Vanguard Party' with mass organisations, such as those for women and the youth, to connect the party to the people. Spelling out the democratic centralist nature of the party, the Central Committee described it as 'a centre expressing the will of all'.

A fourth congress took place in April 1983 and implemented a number of reforms including the expansion of the Central Committee of FRELIMO from 54 people drawn largely from the existing leadership to 128, the majority of whom were elected from outside central government. The build-up to the congress was used to remotivate party members and build the party among Mozambicans. Announcements at the congress set membership at 110,000, about two per cent of the country's population. The membership included 54 per cent peasants and 19 per cent paid workers.

While in some ways the Fourth Congress of FRELIMO marked a return to radical roots and a crack-down on *candonga*, it also accepted the role of the private sector in the Mozambican economy. The revitalisation of the party which followed the congress, however, was undermined by the escalation of problems in the country as a whole following increased activity by the MNR and a series of natural disasters. In 1984 Mozambique was forced to reschedule its debts and turn to the IMF for help. The final blow was the death of Samora Machel in an aeroplane crash on 19 October 1986. By mid–1987 emergency appeals to feed the starving population of Mozambique had been launched throughout the world.

See also ***candonga*****; Chissano; COREMO; Dos Santos; Machel; MANU; MFA; Mondlane; Nkavandame; Operation Production; People's Assembly; Simango; UDENAMO; UNAMI**.

FRELIMO-Partido (*Mozambique*)

The Marxist-Leninist ruling party in Mozambique which was formed out of FRELIMO, the Mozambican liberation movement, in 1977.
See also **FRELIMO**.

Frente de Libertação de Moçambique (*Mozambique*)
Full title of FRELIMO until it was changed into a Marxist-Leninist party in 1977: see **FRELIMO**.

Frente Democrática de Libertação de Angola (*Angola*) See **FDLA**.

Frente Nacional de Libertação de Angola (*Angola*) See **FNLA**.

Frente para a Libertação do Enclave de Cabinda(*Angola*) See **FLEC**.

FROLIZI (*Zimbabwe*)
FROLIZI or the Front for the Liberation of Zimbabwe, was a nationalist organisation formed in Lusaka, Zambia in 1971. It was set up after disagreements between the two main nationalist parties, ZANU and ZAPU, and was an attempt to bring them together. President Kaunda of Zambia urged the groups to unite and warned of his reluctance to continue allowing them to organise in Zambia if so much time was spent in rivalry. The leader of FROLIZI was Shelton Siwela, who represented the younger generation of nationalists, ambitious to replace the old-style party leaders. However, large sections of ZAPU and ZANU refused to join FROLIZI and it soon became irrelevant, and dominated by the older politicians – James Chikerema became the leader and Shelton Siwela left. It did, however, launch one successful guerrilla raid into Zimbabwe in 1973.
See also **Chikerema; ZANU; ZAPU**.

Front for the Liberation of Mozambique (*Mozambique*)
English translation of the full title of FRELIMO until it was reconstituted as a Marxist-Leninist party in 1977: see **FRELIMO**.

Front for the Liberation of Zimbabwe (*Zimbabwe*) See **FROLIZI**.

'frontline' states (*southern Africa*)
The 'frontline' states are those states in southern Africa which regard themselves as in the 'frontline' of the struggle against apartheid and white minority rule in South Africa. They are formally constituted as a group led by a president who is elected periodically and they meet regularly to coordinate their approach towards South Africa. The member states are Zimbabwe, Zambia, Mozambique, Botswana, Swaziland, Lesotho and Tanzania.
See also **CONSAS; SADCC**.

FSAW (*South Africa*) See **Federation of South African Women**.

f

Gaberone (*Botswana*)
Gaberone is the capital of Botswana and the seat of government. Its population in 1981 was estimated as 59,657.

GDs (*Mozambique*)
Grupos Dinamizadores or dynamising groups established to explain and promote FRELIMO policies: see **dynamising groups**.

Geingob, Hage (*Namibia*)
Director of the United Nations Institute for Namibia since 1975, he was born at Otjiwarongo, Namibia. He acted as SWAPO representative to the United Nations from 1964 to 1971 and subsequently became associate political affairs officer in the UN Secretariat from 1972 to 1975 before being appointed to his current position. He has been a member of SWAPO since 1962 and a member of the SWAPO Central Committee since 1969.
See also **SWAPO; UNIN**.

general strike of 1971 (*Namibia*)
In December 1971, following a call from workers at Walvis Bay, 6,000 workers in Windhoek and 2,000 in Walvis Bay went on strike in protest against the contract labour system. The resultant general strike lasted two months and involved up to 25,000 migrant workers in Windhoek, Walvis Bay and other smaller towns, mines and farms throughout the country. The mining industry ground to a halt and business generally was seriously affected, with over 13,000 workers being transported back to the rural north.

While the strike enjoyed wide support among black Namibians, the white community regarded it as a political assault on white minority rule. South Africa sent police reinforcements to Windhoek, surrounded Katutura township (the African residential area of Windhoek), assaulted the inhabitants, destroyed property and forcibly removed thousands of strikers from the city. As the strike continued, units of the SADF were sent to northern Namibia, gatherings were fired on and thousands of people were arrested. In the Ovambo area of northern Namibia a state of emergency was declared, all meetings were banned and wide powers of arrest were instituted.

The strike successfully brought about the dissolution of SWANLA and

its replacement with a system of regional labour bureaux administered by bantustan officials. It also led to increased militancy and political consciousness among black Namibians, and increased support for SWAPO which had been instrumental in organising the protests.
See also **contract labour; labour bureaux; NUNW; SWANLA; trade unions**.

Geneva Conference (*Zimbabwe*)
Talks opened in Geneva, Switzerland, in October 1976 between the Smith government and the nationalist parties. The nationalists were divided despite efforts by the presidents of the 'frontline' states to unite them. The main two nationalist leaders, Joshua Nkomo and Robert Mugabe had, however, formed the Patriotic Front, a political alliance, earlier in the month. Ndabaningi Sithole and Abel Muzorewa, leaders of other parties, attended the conference separately. Ian Smith, the white rebel leader of Rhodesia, insisted that the purpose of the conference was to implement the Kissinger proposals, which included white control of defence and law and order. The nationalists rejected these proposals out of hand. Ivor Richard, Britain's ambassador to the United Nations, presided over the conference which lasted for seven weeks. The talks were adjourned in December and never started again.
See also **'frontline' states; Patriotic Front**.

German colonialism (*Namibia*)
Germany colonised Namibia through a series of land 'purchases' and annexations in the 1880s, ratified at the Berlin Conference. The year 1884 is generally accepted as the start of German colonial rule in Namibia. It was characterised by the progressive alienation of the land and cattle of black people, genocide against the Nama and Herero ethnic groups during the 1904–7 colonial wars (when 80 per cent of the Herero and 50 per cent of the Nama were killed), the creation of a dispossessed African wage-labour force and a programme of white settlement. In 1915, at Britain's request, a South African expeditionary force invaded German South West Africa and ended German rule. In the course of the period of German hegemony, the Herero and Nama lost all their land and their social structures were completely destroyed. The Germans introduced passes, differentiated between Africans and whites at all levels and forbade 'mixed' marriages.
See also **Caprivi Strip; Namibia; Police Zone; Walvis Bay**.

German Lutheran Church (*Namibia*) See **churches**.

German South West Africa (*Namibia*)
The name of Namibia after German colonisation: see **German colonialism**.

Gesuiwerde Nasionale Party (*South Africa*)
The 'Purified National Party', an extreme right-wing Afrikaner nationalist party, later subsumed into the National Party: see **National Party**.

Govêrno Revolucionário de Angola no Exílio (*Angola*) See **GRAE**.

GRAE (*Angola*)
The Revolutionary Government of Angola in Exile (*Govêrno Revolucionário de Angola no Exílio*, or GRAE) was the creation of Holden Roberto, leader of the FNLA. Based in Zaire and established on 5 April 1962, it was an alliance of the UPA, the PDA and the da Cruz breakaway from the MPLA. It was recognised by the OAU in 1963 as the sole viable representative organisation of the Angolan people and recommended as the sole recipient of military aid.

GRAE was weakened by the breakaway of Jonas Savimbi and the formation of UNITA, and by the growth in influence and effectiveness of its ideological rival, the MPLA. In 1971 it lost its OAU recognition, and eventually disappeared from the scene. During its strongest period, GRAE set up subsidiary organisations for women, youth and workers. See also **FNLA; MPLA; PDA; Roberto; Savimbi; UNITA; UPA**.

Great Zimbabwe (*Zimbabwe*)
Great Zimbabwe is the most spectacular of Zimbabwe's many stone ruins, and it is after Great Zimbabwe that the country takes its name. The dry stone walling and tower at Great Zimbabwe date from the thirteenth to the fifteenth centuries and are a major architectural feat reflecting the complexity and achievements of past African societies in Zimbabwe. They act as a reminder of the former prosperity and stability of the societies of the majority Shona-speaking inhabitants. During the period of white rule there was an attempt to deny this achievement and suggest that some non-African community built Great Zimbabwe. This fitted in with the ruling whites' theory that Africans were not capable of such an architectural feat.

Group Areas Act (*South Africa*)
The Group Areas policy is one of the cornerstones of apartheid. The original act, Number 41 of 1950, was among the first introduced by the National Party government after its 1948 election victory. It is a complicated act and has been amended many times; it was re-enacted in Consolidation Acts, Number 77 of 1957 and Number 36 of 1966. The act was imposed throughout the entire country with the exception of African townships, African Reserves and 'Coloured' mission stations and reserves.

Its aim is to divide the country into separate 'Group Areas' where 'Whites', 'Coloureds' and 'Indians' may live and work; Africans do not fall under the Group Areas Act. Group Areas are designated areas of land – usually in urban areas – which can only be owned and occupied legally by people of a particular 'racial' group. In 1950 the government made the reason for the act very clear. The then Minister of the Interior said it was needed to reduce points of contact among the races to a minimum. 'The paramountcy of the white man and of Western civilisation in South Africa must be ensured in the interests of the material, cultural and spiritual development of all races', he was reported as having said in Parliament.

The Group Areas Act has resulted in many people being 'removed' from their homes against their wishes. In February 1985 the government stated that 126,176 families had been moved since the implementation of the act until the end of August 1984. By the end of 1984, 899 'group areas' had been proclaimed, 86.6 per cent for 'Whites', 10.6 per cent for 'Coloureds' and 5.6 per cent for 'Indians'. The Native Trust and Land Act of 1936 (incorporating the 1913 Natives Land Act) already ensured that Africans might not acquire further land outside the 'Reserves'.

In 1981, following a recommendation of the President's Council the government appointed a technical committee of inquiry (the Strydom Committee) to investigate the Group Areas Act and related laws. It recommended later that year that the Group Areas and Separate Amenities Acts be repealed and replaced by a Land Affairs Act. In 1985 President Botha said he had referred the report to the President's Council for consideration, and that he was 'not prepared to undertake anything more in connection with the act'. While he favoured 'reform', he was not prepared to deviate from the constitutional principle that each group should retain control of its own areas and social welfare. The Group Areas Act still remains as part of the government's apartheid policy (1987). However, the Minister of Foreign Affairs, 'Pik' Botha, said in April 1987 that the government had never argued that the act could not be changed but it was in favour of the principle of maintaining 'own community life' and wanted to remove the racial sting in such laws.
See also **apartheid; Natives Land Act of 1913; Native Trust and Land Act of 1936; 'reforms'**.

g

Grupos Dinamizadores (*Mozambique*)

GDs or dynamising groups, these were part of People's Power or *poder popular:* see **dynamising groups**.

Gumede, Josiah Tshangana (*South Africa*)

Josiah Gumede was president-general of the ANC in 1927. He tried to broaden the base of the organisation and involve more grass-roots participation in it. He also sought to align the organisation more closely with the Communist Party. In 1927 he led the ANC delegation to a conference in Europe held by the League Against Imperialism and visited Moscow soon afterwards. In 1929 he joined the Communist Party's League of African Rights and became the president. This association caused his downfall within the ANC as many members felt that the independence of their organisation was threatened. As a result he was voted out of the presidency in 1930 at the annual conference.
See also **ANC; South African Communist Party**.

Gumo (*Mozambique*)

Following the easing of controls on political parties in Mozambique after the 1974 coup in Portugal, one of the new political parties which emerged was Gumo. It had actually been formed in 1973, before the restrictions

had been lifted, but had been tolerated by PIDE because the white politicians and business people who were behind it envisaged it as a party to provide an alternative to FRELIMO for black Mozambicans. It became the main opposition to FRELIMO, and organised mass rallies after the fall of the Caetano regime. Divisions among its leaders led to its collapse, however, and it was replaced by the National Coalition Party.
See also **FRELIMO; MFA; National Coalition Party; PIDE**.

Harare (*Zimbabwe*)

Harare is the capital city of Zimbabwe. Between 1969 and 1982 its population expanded by about 70 per cent to some 656,000 people. During the period of white rule the name Harare was applied only to the oldest and largest African township just south of what was then called Salisbury. After independence it became the name of the whole capital city. Harare is situated in the north of the country in the heart of Mashonaland, where Zimbabwe's Shona-speaking people come from.
See also **Mashonaland; Shona**.

Hendrikse, H.J. 'Allan' (*South Africa*)

The Rev. Allan Hendrikse is South Africa's only 'Coloured' cabinet minister. He has been leader of the Labour Party of South Africa for many years and was elected to Parliament in the 1984 elections under the new constitution which has a three-chamber parliament, one for 'Whites', one for 'Coloureds' and one for 'Indians'. He has always taken the line that his party will use the institutions of apartheid to try and destroy the system. However, the political difficulties of pursuing this path are substantial. One illustration of the kind of situation that arises concerns a public anti-apartheid demonstration that Rev. Hendrikse made in January 1987 by swimming on a 'Whites only' beach (King's Beach) in Port Elizabeth. He was later forced into a humiliating apology by President P.W. Botha who threatened him with a 'Coloured' general election. The polls for the 'Coloured' House of Representatives were very low and many 'Coloured' MPs did not relish the prospect of another election so soon.
See also **constitution of 1983; House of Representatives; Labour Party of South Africa**.

Herenigde Nasionale Party of Volksparty (*South Africa*)

The HNP/V or Reunited National Party or People's Party, the chief Afrikaner nationalist party during the 1940s and the senior party in the 1948 government: see **National Party**.

Herero Chiefs' Council (*Namibia*)

Initially part of SWANU, the council left in 1964 to form NUDO. It joined SWAPO and other parties in the National Convention of Namibia

in 1971, but subsequently became part of the DTA, taking a militantly anti-SWAPO line under Chief Clemens Kapuuo.
See also **National Convention of Namibia; NUDO; SWANU**.

Herero Royal House (*Namibia*)
Merged with SWAPO in 1977 after participating in Namibia National Convention: see **National Convention of Namibia**.

Herstigte Nasionale Party (*Namibia*) See **HNP (Namibia)**.

Herstigte Nasionale Party (*South Africa*)
The *Herstigte Nasionale Party* (Reconstituted National Party) or HNP is the longest-standing opposition party to the right of the National Party. It was formed in 1969 when four National Party MPs, led by the former Minister for Posts and Telegraph, Albert Hertzog, broke away after years of strife between so-called *verligtes* and *verkramptes* following the death of the high priest of apartheid, Prime Minister Hendrik Verwoerd.

The HNP was effectively forced out by Verwoerd's successor, John Vorster, in an attempt to discipline the right of his party and allow for the emergence of a consensus within the National Party. It remained in the political wilderness, with only marginal support, until the onset of the 'reformist' policies of P.W. Botha. It then gained rapid support among disaffected Afrikaners, and in the 1981 general election won over 14 per cent of the vote, narrowly missing winning its first parliamentary seat.

The formation of the Conservative Party in 1982 removed from the HNP its previous status as the only party opposed to the National Party from the right, and it lost some support. The failure of the HNP and Conservative Party to reach an electoral agreement has meant that the splitting of the hard right vote between them has greatly reduced its potential representation in the 'White' parliament. In the 1987 election the HNP lost much of its support, gaining only 3 per cent of the vote and losing its single seat in parliament. The current leader of the HNP is Jaap Marais.
See also **Botha, P.W.; Conservative Party; Marais; National Party; Verwoerd; Vorster**.

Highland Water Project (*Lesotho*)
The Highland Water Project is a projected scheme which would involve a vast complex of dams, tunnels and pumps designed to carry some of Lesotho's plentiful water supplies into the dry industrial centres of Johannesburg and Pretoria. The project is one of the most dramatic civil engineering schemes being discussed in Africa. It would provide important benefits for both countries – foreign exchange for Lesotho (South Africa's payments would almost double Lesotho's foreign exchange income) and water for the dry industrial heartland of South Africa. There is also a chance that the water might generate electricity for Lesotho which at present has to buy its supplies from South Africa. The project will cost over US$2 thousand million, with finance coming from the World Bank, the EEC and the USA. The Highland Water scheme was signed by Lesotho and South Africa in October 1986, marking a high

point in the new friendly relations between the two countries under Lesotho's military government.
See also **relations with South Africa**.

HNP (*Namibia*)
The *Herstigte Nasionale Party* (HNP) in Namibia is a branch of the extreme right-wing party of the same name in South Africa. It seeks Namibia's incorporation into South Africa and is opposed to the black client (or 'puppet') governments, to the devolution of power to 'ethnic administrations' and to independence for the country.
See also ***Herstigte Nasionale Party***.

HNP (*South Africa*) See ***Herstigte Nasionale Party***.

'Homelands' (*South Africa*)
The 'Homelands' or bantustans are based on what used to be called the 'native reserves' – formally constituted under the 1913 Natives Land Act. These amounted to 13 per cent of the land (in 1936) and were the only areas in which Africans could acquire land legally. In 1951 the Bantu Authorities Act provided for tribal, regional and territorial authorities to control matters such as roads and hospitals. The first bantustan proposals were passed in the 1959 Promotion of Bantu Self-Government Act. This removed from Africans their right to elect four 'Whites' to the Senate through a system of electoral colleges, and was passed in the face of strong protests by black people. The Bantu Homelands Constitution Act of 1971 empowers the state to promulgate constitutions for any 'Bantu' area for which a territorial authority has been set up.

The new plan (under the 1959 act) provided for the restructuring of the Bantu Authorities system on the basis of eight Territorial Authorities which were later expanded to ten and renamed 'Bantu Homelands' (after 1978 the government called these 'Black States'). It also envisaged their eventual 'independence' from South Africa. The idea of 'Homelands' amounted to a reformulation of white domination or *basskap* policies and caused some dissension within the ruling National Party. In 1961 the then prime minister, Dr Verwoerd, admitted that 'in the light of the pressure being exerted on South Africa' the government had decided on the policy of 'bantustans', later renamed 'Homelands'. These had been conceived as 'a form of fragmentation which we would not have liked if we were able to avoid it thereby buying the white man his freedom and right to retain domination in what is his country'. It was a way of appearing to give some political rights to the majority African population while not interfering with the political reality of white control in South Africa.

The apartheid theory behind the ten bantustans or 'Homelands' is that the African population is not South African but belongs to ten different 'ethnic nationalities' with different languages, customs and cultures. Hence these separate 'nations' will be given the right to 'self-determination' in their 'traditional areas'. The aim is that each 'Homeland' should become 'independent' leaving 'White' citizens only in 'White' South

Africa. The 1970 Bantu Homelands Citizenship Act provided that all 'Bantu' in South Africa should be made citizens of one of the 'Homelands' even where they had always lived in the 'White' areas. Such urban dwellers were regarded as citizens of one of the 'Homelands' and supposed to exercise their political rights there, not in 'White' South Africa where they in fact lived and worked. The question of urban 'Blacks' was the subject of a special cabinet committee set up in 1983 and its recommendations have resulted in some minor changes.

The 'Homelands' which have accepted 'independence' and are set up with the administrative apparatus of government (they are still heavily dependent on South Africa) are: Transkei (became 'independent' in 1976 under Chief Kaiser Matanzima), Bophutatswana ('independent' in 1977 under Chief Lucas Manyane Mangope), Ciskei ('independent' in 1981 under Chief Lennox L. Sebe), Venda ('independent' in 1979 under Chief Patrick Mphephu). None are recognised as independent states outside South Africa.

The other 'Homelands' which have not accepted 'independence' are KwaNdebele, Lebowa, KaNgwane, Qwaqwa, Gazankulu and KwaZulu. KwaZulu, under Chief Gatsha Buthelezi has always refused to accept a nominal 'independence', arguing that this would give support to the government's apartheid policies.

Many of the 'Homelands' are not economically viable (the Tomlinson Commission was set up to plan them in 1950 and urged massive expenditure by the government to make them so – it never occurred) and some consist of several areas of scattered territory. The land tends to be overcrowded and most of the people live in poverty. The policy has resulted in many forced 'removals' of people from 'White' South Africa over the years.

h

See also **apartheid; 'black spots'; Buthelezi; migrant labour; 'removals'; Tomlinson Committee; urban 'Blacks'**.

household subsistence level (*southern Africa*)
The household subsistence level (HSL) is an estimate of the cost of the basic essentials necessary for a family to survive. A variety of such measures of the costs of subsistence in southern Africa have been calculated. They have been attacked, however, by those who argue that such measures are used to legitimise the payment of subsistence wages to black workers, rather than the rate for the job for all, irrespective of colour.

House of Assembly (*South Africa*)
The House of Assembly is the 'White' chamber in South Africa's three-chamber parliament set up under the 1983 constitution. Before this it was the only House of Parliament. Members must be 'White' citizens, must be registered voters and resident for at least five years in the Republic. Elections to the House of Assembly occur every five years unless called early. It consists of 178 'White' members who are directly elected by 'White' voters over the age of 18, as well as twenty members nominated in proportion to the position of parties in the Assembly.

The National Party has been the governing party since the 1948 general

election. In the May 1987 election the results were as follows: National Party, 133 seats; Conservative Party, 23 seats; Progressive Federal Party, 20 seats; New Republic Party, 1 seat.
See also **constitution of 1983; House of Delegates; House of Representatives; South Africa**.

House of Delegates (*South Africa*)

The House of Delegates is the chamber of parliament under the new South African constitution reserved for 'Indians'. Elections were held for this House on 28 August 1984. It consists of 40 elected members, two members appointed by the State President, and three others elected by the members themselves. The line-up of parties within the House of Delegates is as follows: National People's Party – 18 seats, 29,930 votes; Solidarity Party – 17 seats, 30,039 votes (in December 1984 four members of this party who had won seats left it and remained in the chamber as independents); Progressive Independent Party – 1 seat, 1,322 votes; independents – 4 seats, 20,825 votes.

The election of the House of Delegates was widely boycotted – the turnout was about 20 per cent of registered voters – and opposed by the Natal Indian Congress and the Transvaal Indian Congress. The UDF was formed in order to campaign against the implementation of the new constitution which does not include any representation for the African majority.
See also **constitution of 1983; House of Assembly; House of Representatives; Natal Indian Congress; UDF**.

House of Representatives (*South Africa*)

The House of Representatives is the chamber of parliament reserved for 'Coloured' voters under South Africa's new constitution. Elections were held for this House on 22 August 1984. The results were as follows: Labour Party of South Africa – 76 seats, 201,111 votes; People's Congress Party – 1 seat, 31,681 votes; Freedom Party – no seats, 15,080 votes; Reformed Freedom Party – no seats, 2,058 votes; independents – 2 seats, 19,209 votes. In September 1984 an independent joined the Labour Party giving it 77 seats and another independent joined the People's Congress Party giving that party 2 seats. As well as the 80 seats in the House two members are appointed by the State President and three more elected by the members.

The elections under the new constitution were widely opposed by many organisations including the UDF, formed in 1983 specifically to campaign against the constitution which ignores the country's majority African population. The elections to the House of Representatives were widely boycotted, with a turnout of thirty per cent of registered voters, varying considerably in different areas.
See also **constitution of 1983; Hendrikse; House of Assembly; House of Delegates; Labour Party of South Africa; UDF**.

Hove, Byron Reuben Mtouliodzi (*Zimbabwe*)

Byron Hove is a well-known Zimbabwean lawyer and politician. He was

active in ZAPU, one of the main nationalist organisations, and later in the African National Council. He became a minister in Bishop Muzorewa's transitional government of 1978, and was the only one to speak out on the need for reform. He called for positive discrimination in favour of black policemen and argued that law enforcement agencies, including the judiciary would have to reflect the change of political leadership in the country. The white establishment was outraged and Byron Hove was dismissed. The resulting row finally discredited the transitional government and fully demonstrated how little power was exercised by its members.
See also **African National Council; Muzorewa; transitional government; ZAPU**.

'humanism' (*Zambia*)
President Kaunda's philosophy of 'humanism' is well known in Zambia. He pronounced it to be the national philosophy in 1967, soon after the President of Tanzania, Julius Nyerere, published his Arusha Declaration. Humanism was designed to give some direction to the development of Zambia and it is a simple idea based on the importance of people. As Kenneth Kaunda expresses it: 'This high valuation of MAN and respect for human dignity which is a legacy of our tradition should not be lost in the new Africa.' Humanism has been widely preached in Zambia and in 1969 a Ministry of National Guidance was set up specifically to promote the idea.
See also **Arusha Declaration; Kaunda; Nyerere**.

I

IG (*Namibia*)
The *Interessengemeinschaft Deutschsprachiger Sudwester* (IG) is an organisation of German-speaking settlers which previously only supported the DTA. More recently, however, it has had talks with SWAPO's leadership in exile and has agreed that the UN Plan is the best way forward for Namibia.
See also **DTA; German colonialism; UN Security Council Resolution 435**.

Imbokodvo National Movement (*Swaziland*)
The Imbokodvo National Movement was known in Swaziland as the 'King's party'. It was the ruling political party in the country from 1964 until political parties were banned in 1973. It was controlled by the monarchy, and led by Prince Makhosini Dlamini. It was formed by the *Libandla* or Swazi National Council in 1964 at the suggestion of South African advisers when the British had made it clear there would be independence elections. At first the King was reluctant to form a political party since this was not part of the traditional ruling pattern. However, the South African lawyer, Van Wyk de Vries, convinced him of the need for an electoral base.

The Imbokodvo National Movement was always very successful at the polls. During the 1964 elections the party cooperated with the white conservative organisation, the United Swaziland Association. After winning these elections, however, it tried to broaden its support and abandoned some of its earlier positions such as very close ties with South Africa and restricted voting rolls. The party had considerable support in the country partly because the King was seen to be personally incorruptible and fighting to win back the land for the Swazi people, and because of popular support for traditionalism. The Imbokodvo National Movement had an organisational structure, but its support derived mainly from its ties to the *Libandla*.

No political parties are now permitted to operate in Swaziland, including 'the King's party'. In 1973 after the election of three opposition Ngwane National Liberatory Congress members of parliament and the unseating of Prince Mfanasibili Dlamini, the King suspended political activities and ruled by decree. In 1978 a new constitution was introduced which provided for parliament to be 'elected' through traditional struc-

tures known as the *tinkundla* or local councils in such a way that it is in effect entirely nominated by the monarch.
See also **'constitutional crisis' of 1973; Dlamini, Mfanasibili; Libandla; Mswati; Ngwane National Liberatory Congress; *nkundla*; United Swaziland Association**.

Immorality Act (*South Africa*)

In June 1985 the South African government abolished the Immorality Amendment Act, number 23 of 1957 and the Prohibition of Mixed Marriages Act, number 55 of 1949. The Immorality Amendment Act of 1957 forbade 'unlawful carnal intercourse' or 'any immoral or indecent act' between a 'White' person and a black person. The Prohibition of Mixed Marriages Act made marriages between 'Whites' and members of any other 'racial' group illegal. It was one of the first laws to be passed by the National Party government when it came to power in 1948.

Between 1950 and the end of 1980 over 11,500 people were convicted under the Immorality Act and more than twice that number charged. In 1984, 171 people were charged and 114 convicted. The penalty for 'unlawful carnal intercourse' or 'any immoral or indecent act' was up to seven years with hard labour in prison and a maximum of ten lashes when the male was under fifty years of age.
See also **apartheid; 'reforms'**.

Indaba (*South Africa*)

The *Indaba* (Zulu for meeting or discussion) is the term given to the attempt to set up a multi-racial, pro-business provincial government in Natal province. This would administer the affairs of Natal and KwaZulu but would not have any control over foreign or defence policy. It would be headed by Chief Gatsha Buthelezi, at present leader of the KwaZulu 'Homeland'. The *Indaba* is supported by many white liberals and businessmen as well as the white opposition Progressive Federal Party. It is opposed by most black organisations including the ANC, the UDF and AZAPO. They argue that it is irrelevant and detracts from the issue of one-person-one-vote in a unitary South Africa. The National Party government too, is opposed so far to the Natal Option, as it is also called.
See also **Buthelezi; Inkatha; Progressive Federal Party**.

Independent Zimbabwe Group (*Zimbabwe*)

In 1983 a faction of the main white party in Zimbabwe, the Republican Front, broke away to form the Independent Zimbabwe Group, led by Bill Irvine. This party was sympathetic to the government of Robert Mugabe and won four seats in the 1985 elections. The Republican Front had formerly been known as the Rhodesian Front, and was the ruling white party led by Ian Smith during the days of Southern Rhodesia's rebellion against Britain.
See also **Republican Front; Rhodesian Front; Smith**.

'Indian' (*South Africa*)

'Indian' is an apartheid term used by the South African government to describe South Africans of Indian or Asian descent. Many of this group

live in Natal province and would regard themselves as part of the black community as opposed to the white community.
See also **'Black'; 'Coloured'; Introduction**.

indigena (*Angola, Mozambique*)
A Portuguese word meaning, literally, native, *indigena* was applied as a legal term to the majority of Africans in the Portuguese colonies to distinguish them from Portuguese citizens. *Indigenas* had none of the rights which went with Portuguese citizenship and were subject to forced labour. Acquisition of citizenship rights and privileges necessitated achieving the status of *assimilado*. *Indigena* was abolished as a legal category in 1961.
See also ***assimilado***.

'indirect rule' (*Tanzania*)
The British ruled most of their colonies under the system of 'indirect rule'. In this system conservative local leaders, usually chiefs, exercised authority over the people as the agents of the colonial government. This made British rule more acceptable, and meant that most unpopular measures were implemented by local chiefs rather than the colonial authorities.

indvuna (*Swaziland*) See ***ndvuna***.

Information Scandal (*South Africa*)
This is the name generally given, along with 'Muldergate', to the controversy which rocked white South Africa in the late 1970s when details of dirty dealing by the Department of Information under Connie Mulder began to emerge. The reputations of both Mulder and, to a certain extent, Premier John Vorster, were severely damaged by the revelations and allegations of embezzlement, non-accountability to Parliament, illegal activities and jet-setting by department civil servants. Mulder was subsequently defeated in the National Party leadership elections by P.W. Botha, largely as a result of the scandal.
See also **Botha, P.W.; Vorster**.

Inkatha yeNkululeko ye Sizwe (*South Africa*)
Inkatha was formed in 1975 by the Zulu leader Chief Gatsha Buthelezi as a political party based on Zulu aspirations. The party allows Chief Buthelezi to have his own power-base outside the KwaZulu administrative structures and thus a claim to power in a future black-ruled South Africa. *Inkatha* claims over a million members, but many do not join out of conviction – an *Inkatha* party card makes life a lot easier if you live in KwaZulu. *Inkatha* is the dominant party in the South African Black Alliance, a group set up by Chief Buthelezi, and including the 'Coloured' Labour Party of South Africa and the 'Indian' Reform Party as well as *Inkatha*.

Inkatha yeNkululeko ye Sizwe, the National Cultural Liberation Movement, is Zulu for 'Mystical coil'. The term refers to the coil worn by Zulu women to help carry weights upon their heads and conveys the idea of the lightening of burdens. Chief Buthelezi argues that the party is more than simply a Zulu power-base since it includes non-Zulus amongst its members. But his opponents point to the extent to which the movement emphasises Zulu history. *Inkatha* undeniably has such roots for an earlier version, *Inkatha ka Zulu*, was founded in 1928 by King Solomon ka Dinizulu, the late uncle of Chief Buthelezi. This movement was short-lived but designed to preserve the Zulu heritage.

Inkatha is pro-business and has an investment company, Khulani Holdings, which is in partnership with private firms and has interests in trading and insurance. The party also controls the economic life of KwaZulu through the bantustan's development company. It has its own trade union, the United Workers' Union of South Africa (UWUSA) which was established in 1986 in Durban. The party has been involved in negotiations for the so-called 'Natal option' or *Indaba*, an attempt to set up in Natal a pro-capitalist, multiracial provincial government led by Chief Buthelezi.

Inkatha is denounced by the ANC and the largest umbrella anti-apartheid group in South Africa, the UDF, for its collaboration with the South African regime by taking part in the KwaZulu 'government', its anti-sanctions stand and its tribal base. There have been frequent, violent clashes between supporters of *Inkatha* and those of rival groups, such as the UDF, leading to many deaths. One such clash in March 1987 near Durban left seven high school pupils dead from knife and bullet wounds. See also **Buthelezi; *Indaba*; SABA; UDF**.

inkundla (*Swaziland*) See **nkundla**.

Interessengemeinschaft Deutschsprachiger Sudwester (*Namibia*) See **IG**.

Interim Constitution Amendment Bill (*Tanzania*)
In June 1975 the Tanzanian National Assembly passed the Interim Constitution Amendment Bill. It officially incorporated into the constitution the basic principles of self-reliance and socialism as spelled out by the president, Julius Nyerere, in the Arusha Declaration. It also gave legal supremacy to TANU as the national political party.
See also **Arusha Declaration; TANU**.

Internal Security Act (*South Africa*)
The Internal Security Act No. 74 was passed in 1982. It is the act most commonly used against critics of the South African regime and governs most detentions, including detentions without trial and the issuing of banning orders. Section 29 deals with the detention of people for interrogation and is substantially the same as Section 10(6) of the Internal Security Act No.44 of 1950 and Section 6 of the Terrorism Act.

The new act consolidated the Internal Security Act of 1950, the Suppression of Communism Act 1953, Riotous Assemblies and Suppression of Communism Amendment Act 1954, Riotous Assemblies Act 1956 and sections of the General Laws Amendment Act 1964. It seems at first sight to provide more protection for the accused than

before. Civil rights groups and others, however, argue that abuse and torture of detainees is still prevalent and people continue to be banned and detained without trial. The number of such detainees increased dramatically after the declaration of the state of emergency in 1986 and included children of all ages.
See also **banning order; state of emergency**.

Internal settlement (*Namibia, Zimbabwe*)

'Internal settlement' is the term used for an accommodation reached between white minority regimes and internal conservative opposition leaders giving the latter token power in 'joint' governments. Such settlements are designed to undermine and bypass liberation movements. Their main aim is to confer some legitimacy on the ruling regimes by including local black leaders – often accused of being puppets by nationalist supporters – and to undercut support for the guerrilla wars.

The main examples of internal settlements in southern Africa are those attempted in Zimbabwe and Namibia. In Zimbabwe Ian Smith, the white rebel leader, reached an internal settlement with Bishop Abel Muzorewa which led to the Bishop being installed as prime minister of Zimbabwe-Rhodesia on 1 June 1979. In Namibia the illegal South African occupying regime has contracted a series of internal settlements establishing so-called 'governments' of the territory; for example, the Democratic Turnhalle Alliance (DTA) and the Multi-Party Conference (MPC).
See also **Democratic Turnhalle Alliance; Multi-Party Conference; Muzorewa; Zimbabwe-Rhodesia**.

International Court of Justice (*Namibia*)

Opinions by the International Court of Justice (ICJ) on the South African presence in Namibia have been central to arguments about jurisdiction. After the establishment of the United Nations in 1945, countries administering League of Nations mandates entered into trusteeship agreements with the UN for the mandated territories. These agreements were intended to lead to full independence for the territories. South Africa refused to enter into a trusteeship agreement for Namibia and in 1949 ceased its annual reports to the UN.

In its first opinion on the legality of South Africa's position, the ICJ held, in 1950, that the mandate was still in force, that South Africa could not unilaterally change the status of the country, but that it was under no legal obligation to enter into a trusteeship agreement. When Ethiopia and Liberia instituted proceedings against South Africa in 1966 for violating the mandate by introducing apartheid in Namibia, the ICJ found that they had no legal standing on Namibia and refused, therefore, to give an opinion on the substance of their petition. Within South Africa the result was hailed as a vindication of its policies on Namibia.

In response to the position taken by the ICJ and international pressure for a resolution of the dispute, the United Nations General Assembly adopted Resolution 2145 (October 1966) terminating South Africa's mandate and placing Namibia under direct UN control. UN Security

Council Resolution 264 of 1969 confirmed the General Assembly resolution and called for immediate South African withdrawal.

Finally, on 21 June 1971, the ICJ gave an opinion that South Africa's continued presence in Namibia was illegal and that it had a legal obligation to withdraw immediately and transfer control to the United Nations. South Africa rejected the opinion and continued its illegal occupation of the country. Three other countries, the UK, France and the FRG have explicitly rejected paragraphs 2 and 3 of the opinion.

The final ICJ opinion read, in part:

(1) that, the continued presence of South Africa in Namibia being illegal, South Africa is under obligation to withdraw its administration from Namibia immediately and thus put an end to its occupation.

(2) that State Members of the United Nations are under obligation to recognize the illegality of South Africa's presence in Namibia and the invalidity of its acts on behalf of or concerning Namibia, and to refrain from any acts and in particular any dealings with the Government of South Africa implying recognition of the legality of, or lending support or assistance to, such presence and administration.

(3) that it is incumbent upon States which are not members of the United Nations to give assistance, within the scope of subparagraph (2) above, in the action which has been taken by the United Nations with regard to Namibia.

See also **League of Nations Mandate on Namibia; UNCN**.

International Police for the Defence of the State (*Angola*) See **PIDE**.

i

J

Jehovah's Witnesses (*Malawi*)
Jehovah's Witnesses reject any allegiance to a political party or a government. As such they are often considered subversive and sometimes persecuted. This has happened across southern Africa, and particularly in Malawi. As the nationalist party, the MPC, prepared for elections, the Jehovah's Witnesses refused to register as voters and tried to dissuade others. The party retaliated and there were violent episodes. The conflict continued and in 1967 the Jehovah's Witnesses were banned. In the early 1970s some 20,000 members of the religious group fled to neighbouring Zambia and Mozambique. In 1975 some 2,000 were arrested – most were later released.

JMPLA (*Angola*)
The JMPLA or *Juventude do MPLA* is the youth organisation of the MPLA.
See also **MPLA**.

job reservation (*South Africa*)
For many years black South Africans were restricted in the jobs they could do under 'job reservation' laws. These were strongly supported by reactionary white trade unions. For instance, in terms of proclamations R3, R4 and R5 of 1968 issued under the Group Areas Act of 1966, anyone, other than a 'White', was barred from being employed as a 'chargehand, executive, professional, technical or administrative employee, manager, or supervisor' in a trading or business concern in a 'White' area. The Riekert Commission of 1979 recommended that this restriction should be abolished and the government accepted this recommendation. Job reservation was formally abolished in 1980.

Job reservation remained in one part of the mines, however, until 1986 when it too was abolished. Until then, Section 12 of the Mines and Works Act of 1911 granted 'blasting certificates' for certain scheduled positions (such as managers and surveyors) only to 'Whites', persons born in the Republic and ordinarily resident there who are 'Coloured' and 'Cape Malayan', and people known as 'Mauritius Creoles' or 'St Helena' persons and their descendants.
See also **apartheid; International Confederation of Labour; 'reforms'**.

Joint Monitoring Commission (*Angola, Namibia, South Africa*)
A joint Angolan and South African body set up in 1984 to monitor the South African withdrawal from Angola: see **Lusaka Agreement between South Africa and Angola, 1984**.

Joint Security Commission (*Malawi, Mozambique*)
Set up in 1986 to deal with MNR attacks from Malawi: see **accord between Malawi and Mozambique, 1986**.

Jonathan, Leabua (*Lesotho*)
Chief Leabua Jonathan led Lesotho from independence in 1966 until he was deposed in a military coup in 1986. In 1959 he founded the Basutoland National Party (Basotho National Party) which offered for the first time a challenge to the Basutoland Congress Party. The new party was conservative and traditional and had the support of the strongly anti-communist Roman Catholic church. Nevertheless it did badly in the 1960 elections. By 1965, however, Chief Jonathan had financial and organisational support from South Africa and his party won the crucial pre-independence election. He was able to portray himself as the man likely to achieve the best arrangement with South Africa while at the same time maintaining Lesotho's independence; he described his policy as 'peaceful coexistence'. As a result of this approach he was dismissed by his enemies as South Africa's puppet.

Gradually Chief Jonathan became more and more estranged from South Africa, partly because he perceived how much international support he lost because of his previous close relationship. This process of estrangement continued in the late 1970s and 1980s until Lesotho gained a brave reputation (and much international assistance) for resisting South Africa's demands, adding considerably to Chief Jonathan's popularity and stature.

Chief Jonathan became increasingly authoritarian as Prime Minister of Lesotho. In 1970 the country's last (abortive) elections were held. Their results were never officially announced and instead the prime minister declared a state of emergency, suspended the constitution and governed by decree. It was widely believed that the elections had in fact been won by the opposition Basutoland Congress Party. Opposition leaders were arbitrarily imprisoned, their publications and parties banned and the courts in effect temporarily abolished. The King had to go into exile leaving his wife as regent and his brother in prison; he returned in December having accepted a proclamation forbidding him from taking part in politics.

Chief Jonathan took advantage of the state of emergency to pursue a nationwide campaign of purges in an attempt to crush opposition to his rule. During this time, according to reports, he relied on South African military support to help crush some 1,000 men who took up arms against him. Fred Roach, the commander of the police paramilitary unit and an English expatriate, was an important support to the prime minister. At the end of the year Chief Jonathan declared a five year 'holiday from politics', and in 1971 he released political prisoners and opened discussions with the opposition. In 1973 he established a 93–member Interim National Assembly to draw up a new constitution compatible with 'traditional institutions'. The opposition did not accept this, however, and there was more violence the following year which was brutally suppressed. In 1975 a cabinet reshuffle admitted some opposition members.

There has always been the possibility of violent political change in Lesotho because of the repressive regime; and in 1986 there was a military coup led by Major General Lekhanya, a conservative soldier who feared the growing disaffection of South Africa and was concerned about the militarisation of the Youth Wing of Chief Jonathan's party, the Basotho National Party. Chief Jonathan was deposed and a Military Council took over the country. Chief Jonathan died on 5 April 1987, aged 73.
See also **Basotho National Party; Basutoland Congress Party; Lekhanya; relations with South Africa**.

Joseph, Helen Beatrice May (*South Africa*)

A veteran of the anti-apartheid struggle, Helen Joseph was born and educated in the UK. After teaching in India, she travelled to South Africa in 1931, where she subsequently settled. She was politicised by the ANC's Defiance Campaign, and took an active part in the formation of the Congress of Democrats in 1953, and in the formation of the Federation of South African Women the following year. As national secretary of the latter, she was one of the leaders of the mass anti-pass demonstrations and the protest march by women on the Union Buildings in Pretoria.

From 1956 to 1961 Helen Joseph was one of the defendants in the Treason Trial facing, in addition, a banning order and detention. The banning order was reimposed and intensified in 1962, after she had resumed political work, and continued uninterrupted until 1971. She was again banned in 1980 for a further two years.

As a 'listed person' (a former member of a banned organisation, in her case the Congress of Democrats) she was prohibited from membership of political organisations, and was therefore prevented from accepting repeated elections as Honorary President and Honorary Vice-President of NUSAS. In 1975 she was elected an Honorary Fellow of Kings College, London, and in 1983, a patron of the United Democratic Front.
See also **Congress of Democrats; Defiance Campaign of 1952; Federation of South African Women; NUSAS; Treason Trial; UDF**.

Jumbe, Aboud (*Tanzania*)

Aboud Jumbe became President of Zanzibar in 1972 and thus Tanzania's First Vice-President. He was a schoolteacher and had been Minister for Health in the union cabinet. In 1984 there was a political crisis in Zanzibar and there were signs among the islanders of dissatisfaction with the union with Tanganyika. As a result, amid some controversy, Aboud Jumbe resigned. The new president was Ali Hassan Mwinyi.
See also **Mwinyi; Zanzibar**.

Juventude do MPLA (*Angola*) See **JMPLA**.

JUWATA (*Tanzania*) See **Union of Tanzanian Workers**.

j

K

Kalahari Desert (*Botswana, Namibia, South Africa*)
The Kalahari is a vast inland desert covering much of the east of Namibia, the bulk of west and central Botswana and stretching to the north of the Cape Province in South Africa. Although not as arid as the Namib, it receives little rain and when it does fall it is confined to three months of the year. The Kalahari has been a final refuge for persecuted groups, historically, including the San and the Herero.
See also **Namib Desert; San**.

Kalangula, Peter (*Namibia*)
Leader of Christian Democratic Action for Social Justice: see **Christian Democratic Action for Social Justice**.

Kangueehi, Kuzeeko (*Namibia*)
Elected President of SWANU in 1984: see **SWANU**.

KaNgwane (*South Africa, Swaziland*)
KaNgwane is a Swazi word used to mean Swaziland. It is 'the place or country of *Ngwane*', one of the country's most famous early rulers.
See also **KaNgwane-Ingwavuma land deal; Ngwane**.

KaNgwane-Ingwavuma land deal (*South Africa, Swaziland*)
Negotiations on a possible land transfer from South Africa have been a significant issue in Swazi foreign relations over the past few years. What is at issue is the transfer to Swaziland of the KaNgwane bantustan – what the South African government calls the 'Homeland' of the Swazi ethnic group in South Africa – and the Ingwavuma region of KwaZulu (known as the Zulu 'Homeland'). This would give landlocked Swaziland vital access to the sea. The issue is being considered by South Africa as part of an attempt to persuade Swaziland to support its regional plans and refuse help to ANC guerrillas fighting to overthrow the white minority regime. In fact Swaziland has proved very compliant and in 1982 signed a security pact with South Africa; since then the Swazi government has stepped up its harassment of ANC members considerably. The deal is opposed by Chief Gatsha Buthelezi, the head of KwaZulu in South Africa. To date (1987) there has been no sign of the land transfer taking place.
See also **ANC (South Africa); Buthelezi; CONSAS; 'Homelands'; *KaNgwane*; total strategy**.

Kappie Kommando (*South Africa*)
Formed in the late 1970s, the *Kappie Kommando* (Bonnet Commando) is made up of women campaigning for a return to the values and politics of *Voortrekker* days, as they see them. Conspicuous for their wearing of *Voortrekker* style clothing for their events, they are part of the resistance to P.W. Botha's style of politics and supported the formation of the Conservative Party in 1982.
See also ***Aksie Red Blank Suid-Afrika*; Conservative Party; *Voortrekker***.

Kapuuo, Clemens (*Namibia*)
A conservative political leader and traditional Herero chief, Clemens Kapuuo was the leader of the National Unity Democratic Organisation (NUDO). He was instrumental in breaking up the National Convention of Namibia (NCN). Having previously supported the NCN's call for internationally recognised independence for Namibia, Kapuuo threw his weight behind the Turnhalle Conference in 1975 and later became President of the DTA and a leading light in the puppet government it headed.

Kapuuo was assassinated on 27 March 1978, an event which was used by South Africa as an excuse to unleash a massive campaign of repression and arrests against SWAPO. The assassin has never been identified.
See also **DTA; Herero Chiefs' Council; NCN; NUDO; SWAPO; Turnhalle Constitutional Conference**.

Kapwepwe, Simon (*Zambia*)
Simon Kapwepwe was one of Zambia's leading nationalists. He was a contemporary of Kenneth Kaunda, Zambia's future president, and active in the African National Congress (ANC) in its early days. Both he and Kenneth Kaunda left the ANC after a row with its president, Harry Nkumbula, and set up the Zambia African National Congress. This was a more radical party and later became Zambia's main party, UNIP.

Simon Kapwepwe became a high office holder in UNIP and one of the key members in the party organisation. At one point he was vice-president of the country. He was a strong character and an orator who had loyal followers – most from the Bemba-speaking people in the Copperbelt. In 1971, there was a crisis in UNIP affecting its Bemba-speaking faction which led to Simon Kapwepwe's resignation. He left UNIP and set up a new party, the United Progressive Party (UPP). The UPP was soon banned and Simon Kapwepwe, along with other leaders, detained without trial. In 1978, however, he rejoined UNIP with some of his followers. He died in 1980, still an important member of the party. His support was seen as vital for President Kaunda as Zambia faced serious economic troubles.
See also **African National Congress (Zambia); Nkumbula; UNIP; United Progressive Party; Zambia African National Congress**.

Karakul farming (*Namibia*)
The farming of karakul sheep for the skins of the new-born lambs is one of Namibia's major agricultural foreign exchange earners. The curly-woolled skins of the lambs are sold to produce expensive fur fashion

garments, chiefly for Western Europe and North America. The furs are auctioned regularly, but illegally in terms of UNCN Decree No. 1, in London. Karakul farming is mainly the prerogative of large white-owned ranches in the semi-desert areas of Namibia where the hardy sheep are able to thrive in otherwise agriculturally unproductive areas. The industry is vulnerable to changes in fashion preferences. The wholesale marketing of karakul is monopolised by Eastwood and Holt Ltd, a UK subsidiary of the Hudson's Bay Co. of Canada.
See also **Decree No. 1**.

Kariba (*Zambia, Zimbabwe*)
Lake Kariba is on the Zambezi River which forms the boundary between Zimbabwe and Zambia. Soon after the Central African Federation was formed in 1953 Kariba Dam was built. The dam provided hydro-electric power for Southern Rhodesia (later Zimbabwe) and Zambia on Kariba North Bank.
See also **Zambezi River**.

Karume, Abeid (*Tanzania*)
Abeid Karume was leader of the Afro-Shirazi Party in Zanzibar. He maintained close links with TANU, the Tanganyika African National Union, and its president, Julius Nyerere. In 1964 he became President of Zanzibar after ousting John Okello and soon was negotiating a union with the mainland. In April that year Abeid Karume became the United Republic of Tanzania's First Vice-President. His rule became increasingly arbitrary and his regime survived two plots against it in 1969 and 1971. He was assassinated in April 1972.
See also **Afro-Shirazi Party; Okello; TANU**.

Kassinga massacre (*Angola, Namibia*)
In May 1978 South African forces invaded Angola and attacked SWAPO's Kassinga settlement for Namibian refugees, massacring over 600 people and abducting hundreds more. The massacre put a stop to negotiations then in progress for Namibian independence, but despite the production of detailed evidence and photographs by the Angolan Government, failed to attract much attention in the international press. SWAPO mounted an international campaign for the release of those abducted from Kassinga. Despite denials by South Africa, over a five-year period, that it was holding prisoners from Kassinga, SWAPO was able to show that over 100 such prisoners were being held at the Hardap Dam detention centre near Mariental. 137 people were eventually released by South Africa from the camp, the final group in September 1984. The fourth of May is now known as 'Kassinga Day' and is marked by SWAPO supporters as a reminder of South African brutality and the suffering of Namibians in the struggle for independence.
See also **Contact Group; detention centres; SWAPO**.

Katilungu, Lawrence (*Zambia*)
Lawrence Katilungu was one of Zambia's most important trade union leaders. In 1949 he helped found and was president of the African Mine-

workers' Union – later to become the Mineworkers' Union of Zambia, the most powerful trade union in the country. A year later he started the Trades Union Congress and became its first president, retaining this office until December 1960.

He started life as a teacher at Roman Catholic missions, soon becoming a headmaster. However, he left this career and began work in 1936 at the age of twenty-two as a 'spanner boy' at the Nkhana copper mines for 9d a day. By 1940 he was leading strikes demanding a minimum wage and union rights. Lawrence Katilungu was also involved in politics and supported the African National Congress of Zambia all his life, remaining loyal to Harry Nkumbula, the president, when a more radical breakaway group under Kenneth Kaunda formed their new party, later to become UNIP. Nevertheless, he opposed the politicisation of the TUC and the congress split over this issue and his lack of radical leadership in 1959. He died in November 1961.

See also **African National Congress (Zambia); Mineworkers' Union of Zambia**.

Katiusha rocket (*Angola*)

USSR-made Katiusha mobile rocket-launchers, BM 21s, are credited with having provided the FAPLA forces with the only available effective counter to South African and Zairean artillery in the 1975 invasions, and were critical to the successful defeat of these. The vehicles are 40–pod launchers of 122mm rockets and were particularly effective against the South African 140mm Howitzers.

See also **FAPLA; Kifangondo marshes; South African invasions**.

Katjiuongua, Moses (*Namibia*)

Former head of SWANU and now leader of a breakaway group which is part of the MPC: see **SWANU**.

Kaunda, Kenneth David (*Zambia*)

Kenneth Kaunda is the President of Zambia. He is one of Africa's leading statespeople and has had considerable influence in forming post-colonial policy on the continent. He is an ascetic, deeply religious man who has spent time in colonial prisons. His philosophy is one of 'humanism' which seeks to put the needs of people above those of profit or anything else.

He started and led Zambia's main nationalist party, UNIP, through the independence struggle to independence in 1964. In 1972 he made Zambia a one-party state acting strongly to suppress his opponents. He has tried to develop his country into a nation with a broader economic base than simply the Copperbelt mines, and he has led Zambia in the direction of a mixed economy. The country has no strong socialist ideology, although, like his friend and fellow president, Julius Nyerere of Tanzania, he supports and helped to formulate the idea of 'African Socialism'.

On the wider front, Kenneth Kaunda consistently supported the liberation struggle in Zimbabwe at considerable economic cost to his country, helping, in particular, Joshua Nkomo and his party, ZAPU. For his part in this and for his refusal to participate in open dialogue with the white South African president, B. J. Vorster, in the early 1970s during the South African government's attempt to spread its influence further north, Zambia has been repeatedly castigated by the white south. The minority Rhodesian regime launched several damaging attacks into the country during the liberation war in a vain attempt to end President Kaunda's support for ZAPU. Similar attacks have continued from South Africa because of Zambian support for SWAPO and the ANC of South Africa. See also **'African Socialism'; 'humanism'; Nyerere; UNIP**.

Kavango Alliance Group (*Namibia*)
Constituent party of the Democratic Turnhalle Alliance: see **DTA**.

Kawawa, Rashidi (*Tanzania*)
Rashidi Kawawa has for years been a close collaborator with President Nyerere. He was prime minister and defence minister and a leading contender to succeed President Nyerere after his retirement. However, he is a controversial figure with a tough style of leadership. He is known to be strongly committed to Tanzania's original socialist concept and unlikely to continue President Nyerere's moves towards greater pragmatism. He lost the presidency to Ali Hassan Mwinyi.

Rashidi Kawawa's career began in the trade union movement. He was General Secretary of the Tanganyika Federation of Labour when it was first formed in October 1955. Before this he had been President of the Tanganyika African Government Servants Association. By 1962 many of the union leaders had been drawn into government and Rashidi Kawawa was a cabinet minister, working to limit the power of the trade unions. In January 1962 when Julius Nyerere resigned as Prime Minister to build up his party, TANU, Rashidi Kawawa took over and passed a new constitution replacing the British monarch by a president as head of state. In December 1962 Julius Nyerere was elected president under the new constitution.
See also **Mwinyi; Nyerere; Tanganyika Federation of Labour**.

Khaketla, Bennet Makalo (*Lesotho*)
Bennet Khaketla founded and led the conservative Basutoland Freedom Party in 1960 after resigning from the Basutoland Congress Party. In 1962 the new party merged with the Marema-Tlou Party to form the Marema-Tlou Freedom Party.
See also **Basutoland Freedom Party; Marema-Tlou Freedom Party**.

Khama, Seretse (*Botswana*)
Sir Seretse Khama became Prime Minister of Botswana in March 1965 when his party, the Botswana Democratic Party, won the territory's first direct election leading to internal self-government before independence. On 30 September 1966 Botswana became independent and Sir Seretse Khama was sworn in as the country's first president. On the same day he was appointed Knight Commander of the Order of the British Empire (KBE). He governed the country until his death on 13 July 1980 when

he was succeeded as president by his old friend and ally, the vice-president, Dr Quett Masire.

Seretse Khama's political career took a very different course from the original plan. He was the hereditary chief of the important Bamangwato ethnic group and studied in England (at Balliol College, Oxford) before taking up his traditional role. However, he met and married an English woman, Ruth Williams, in September 1948, while pursuing his legal studies in London at the Inner Temple. This caused a political crisis in Botswana where his uncle, the acting regent, Tshekedi Khama, opposed the marriage. Heavy pressures to bar Seretse Khama from taking up the chieftainship came from white-ruled South Africa and Rhodesia (later Zimbabwe) which feared the political consequences of a mixed race marriage on their borders. In 1950 the British Secretary for Commonwealth Relations, Patrick Gordon Walker, persuaded Seretse Khama to visit Britain and then stopped him returning home. Eventually, Seretse Khama returned in 1956 under the condition that he renounce his chieftainship. However, the Bamangwato people accepted his marriage and never appointed another chief. This position gave Seretse Khama all the traditional power of the chieftainship while at the same time allowing him to act freely within the modern political system, where he was remarkably successful.

In 1961 Khama founded what was to become the ruling party, the Botswana Democratic Party, as a conservative group to oppose the more militant Botswana People's Party (BPP). He was also, in the same year, elected to the Legislative Council and recruited ten of its twelve African members to his party. In 1965 his party won the first direct elections with support from the country's establishment: liberal whites, the colonial administration, traditionalists and conservatives who opposed the BPP. Khama subsequently ruled Botswana for fifteen years.

Sir Seretse Khama pursued conservative policies, dominated by the geographical necessity of seeking a *modus vivendi* with South Africa. He once told his party conference 'Botswana cannot allow itself to be used as a springboard for violence against the minority regimes [South Africa and Rhodesia]. Our task is to insulate ourselves from the instabilities their policies provoke'. Nevertheless he gave increasing support to African resistance movements in neighbouring territories over the years particularly as the black states became more prominent and had more influence in their opposition to South Africa. In 1979 the Southern African Development Coordination Conference (SADCC) came into being at a meeting in Botswana when the leaders of the 'frontline' states met to discuss concrete ways of escaping South African economic control.

See also **Botswana Democratic Party: Botswana People's Party; Masire; SADCC**.

Khayelitsha (*South Africa*)

Khayelitsha is a new African township set up by the government in 1975. It is situated 35 km from Cape Town on the sand dunes of False Bay and is designed to house some 400,000 people when it is completed. It

has been described by black radicals as 'apartheid's dream town' and by officials as 'orderly urbanisation'. The government has been trying to remove many of the inhabitants of the Crossroads squatter settlement to Khayelitsha but its policy has not yet succeeded. By February 1987 some 150,000 people were living in the new township.
See also **Crossroads; 'removals'; townships**.

Khoi (*southern Africa*)
The Khoi people, like the San, were among the original inhabitants of southern Africa. By the start of European settlement they were largely concentrated in the Cape (in South Africa). Known by early Dutch settlers as 'Hottentots', the Khoi were decimated by small-pox introduced by the settlers. Those who survived became integrated into the settler community as part of the 'Coloured' group, many of them moving, eventually, to other parts of the Cape and into south and central Namibia. Unlike the San, there are no surviving communities of wholly Khoi descent.
See also **San**.

Khoisan (*southern Africa*)
The collective name for the Khoi and San people of southern Africa: see **Khoi** and **San**.

Kifangondo marshes (*Angola*)
Popularly credited with saving Luanda from the FNLA/Zairean invasion in November 1975, the marshes lie on either side of the northern route into Luanda, forming a natural barrier bridged by the road. The defending FAPLA and MPLA forces chose this point to make their stand against the invasion force, positioning what artillery they had, and a Katiusha rocket battery only available to them from the night before the crucial battle, on the heights above the marshes. When the invasion force attacked in strength with armoured vehicles, infantry and artillery, and attempted to negotiate the narrow road through the marshes, it was routed by the concentration of the defenders' firepower on the road.

Kissinger plan (*Zimbabwe*)
In September 1976 the then US Secretary of State, Henry Kissinger, met Ian Smith, the rebel leader of the Rhodesian regime, in Pretoria, South Africa. He announced the existence of a plan for a Rhodesian settlement which was supported by South Africa as well as by other countries. The plan consisted of an agreement to reach majority rule in two years. It also proposed an interim government giving the whites in Rhodesia an effective veto. Defence and law and order were to be in white hands during the transition, sanctions would be lifted, a ceasefire would operate and a trust fund would help the country's economic development. The 'frontline' states rejected these proposals and called on Britain to convene an international conference on its colony. Their view was echoed by the nationalist movements fighting for independence.

The Kissinger plan was most remarkable for the psychological blow it delivered to Ian Smith's regime. For the first time the white leader

admitted the possibility of majority rule in the forseeable future. 'Rhodesia agrees to majority rule within two years', Ian Smith announced on Rhodesian radio. He was forced to go along with the plan mainly because South Africa's prime minister, B. J. Vorster, supported it, and South Africa was Rhodesia's only economic and political support during the rebellion.
See also **'frontline' states; Smith; Vorster**.

Koevoet (*Namibia*)
80 per cent of the deaths of alleged PLAN soldiers in the war in Namibia are attributed to *Koevoet*. It is a paramilitary police 'counter insurgency' unit with mainly black members and is officially part of the SWA Police in Namibia. *Koevoet* was set up by South Africa to take over the brunt of the fighting inside Namibia when the toll of white South African conscripts had reached politically embarrassing levels. Its name is an Afrikaans word meaning 'crowbar'. It is the most notorious of the various military and para-military organisations operating in Namibia and its members have been convicted of incidents of torture, murder and rape, especially in the north of the country. *Koevoet* has itself become something of an embarrassment to the South Africans and has more recently been renamed the South West Africa Police Counter Insurgency Unit or COIS.
See also **Namibia; PLAN; SWA Police**.

Koma, Kenneth (*Botswana*)
Dr Kenneth Koma, founder of the Botswana National Front (BNF) in 1965: see **Botswana National Front**.

Konservatiewe Party (*South Africa*) See **Conservative Party**.

KP (*South Africa*)
Konserwatiewe Party: see **Conservative Party**.

Kwacha (*Malawi*)
Kwacha means 'the dawn'. It was a slogan introduced in the 1950s by the Nyasaland African Congress, the country's main nationalist organisation, calling for freedom and independence from colonial rule. It is now the name of the national currency.
See also **Nyasaland African Congress**.

L

Labour Party (*Namibia*)

A small party drawing its support mainly from people designated 'Coloured' in Namibia, the Labour Party was formerly part of the DTA. See also **DTA**.

Labour Party of South Africa (*South Africa*)

The Labour Party of South Africa is a political party for what the South African state calls 'Coloured' people. It was formed in 1965 to contest elections for the state-organised Coloured Representative Council. The Labour Party's political position has always been to use the institutions of apartheid for the purpose of getting rid of apartheid. This has caused heated arguments within the party and over the years has led to considerable loss of support. The last by-election for the Coloured Representative Council in 1978, for instance, had the lowest poll hitherto, 13 per cent. The party is the majority party in the 'Coloured' chamber of South Africa's three-chamber parliament.

Despite gaining most seats in the 1969 election for the Coloured Representative Council the state prevented the Labour Party from becoming the majority party in the CRC. This it did not achieve until 1975 when it provoked a crisis in the institution and gained the majority. Thereafter the party negotiated with the state until it clashed over proposals for the new constitution put forward by the then prime minister, P. W. Botha. Botha promptly abolished the Coloured Representative Council and threatened to install a nominated Coloured People's Council. The party then turned to other popular resistance organisations such as AZAPO. However, the Prime Minister managed to persuade the Labour Party to participate in the new constitution, and it agreed to contest seats for the 'Coloured' House of Representatives. This decision led to some resignations and the suspension of the party from the South African Black Alliance (in April 1983 the Labour Party resigned from SABA). The party's leader, Rev. Allen Hendrikse, is the only 'Coloured' cabinet minister in South Africa.
See also **constitution of 1983; Hendrikse; House of Representatives**.

Labour Relations Act 1985 (*Zimbabwe*)

The old Industrial Conciliation Act of 1959 was replaced by the Labour Relations Act in 1985. All the racially-discriminatory clauses, such as those allowing white trade unions to dominate unions by weighting skilled votes against unskilled votes, were eliminated. The new act also added new clauses to allow for the statutory rights of workers' committees in relation to trade unions and it set up a Labour Relations Board to deal with unfair labour practices. Furthermore, the rights of employers and employees have been defined by the new act, which incorporated the 1980 Minimum Wage Act and the Employment Act.

The act has been criticised by some trade unionists and the International Labour Organisation for giving excessive power to the Minister of Labour who can interfere in union affairs under the act, for instance, by deciding what the level of union dues shall be.
See also **Zimbabwe Congress of Trade Unions**.

Lake Malawi (*Malawi*)
Lake Malawi (formerly Lake Nyasa) is the third largest lake in Africa. It takes up most of the northern and central part of Malawi. There are two other, smaller lakes in the south and south-east, Lake Malombi and Lake Chilwa. Lake Malawi is drained by the Shire River which joins the Zambezi River in Mozambique. It forms part of the border between Malawi and Tanzania and between Malawi and Mozambique.

Lancaster House (*Zimbabwe*)
In 1979, Margaret Thatcher, the British prime minister, issued invitations to a constitutional conference on Zimbabwe to be held at Lancaster House, London and chaired by Lord Carrington, then British foreign secretary. The decision followed an agreement made at the meeting of Commonwealth Heads of Government in Lusaka, Zambia in 1979. Invitations were sent to the Patriotic Front leaders and the Muzorewa government (which included the white rebel leader, Ian Smith, and Ndabaningi Sithole, in its delegation).

The conference opened on 10 September 1979 and concluded after 47 plenary sessions. Agreement was reached on a new constitution, arrangements for the transitional period and a ceasefire. Documents were signed on 21 December 1979. On 12 December illegal rule was ended when Zimbabwe-Rhodesia renounced independence and reverted to colonial status as Southern Rhodesia. Elections were held in February 1980 and Zimbabwe achieved its independence.

The Lancaster House conference took place at a time when the liberation struggle had reached a critical stage. The British prime minister was persuaded by Lord Carrington not to recognise Zimbabwe-Rhodesia because of the opposition from the Commonwealth as well as because of a desire to protect British trade and investment in black Africa. The Patriotic Front leaders (the main nationalist force in the country) faced pressure from the 'frontline' states to agree a settlement, because their economies were being wrecked by the guerrilla war. And Ian Smith and Bishop Muzorewa faced sanctions, the hostility of the international community and the growing pressures of the liberation war.
See also **Carrington; 'frontline' states; Muzorewa; Patriotic Front; Sithole, Ndabaningi; Smith; Zimbabwe-Rhodesia**.

Land Apportionment Act (*Zimbabwe*)

The Land Apportionment Act was passed in 1930 and came into effect in April 1931. It introduced the principle of racial segregation into land allocation throughout the country by defining separate categories of land for purchase by Africans and whites. The act was extended over the years and replaced by the more discriminatory Land Tenure Act in 1970.
See also **African National Congress (Zimbabwe); constitution of 1969; Land Tenure Act**.

Land Proclamation Act of 1907 (*Swaziland*)

The 1907 Land Proclamation Act was passed by the British and divided Swaziland into three areas: one third was set aside for occupation by the Swazis in the form of 32 'native areas', later known as Swazi Nation land; the rest was given to individual tenure for individual settlers to use except for a small portion which was held by the British crown for allocation at its discretion.

This partition was implemented from 1914 and had the disastrous effect of reducing the self-reliant, non-capitalist Swazi economy to a level where it could not sustain itself. The situation was made worse by the fact that thousands of cattle had also been killed by disease during this period. Furthermore, the partition of the land changed the nature of the state, opening up the economy to foreign capital, and leading to the development of an orthodox capitalist state in Swaziland.

When King Sobhuza II assumed the throne in 1921 he made the return of Swazi land his major preoccupation, establishing a fund in order to purchase back the land. All Swazis had to contribute an annual sum to this and by the time of independence in 1968 just under one half of the land had been restored to traditional Swazi ownership; the proportion in the early 1980s was about two-thirds. To this day in Swaziland no foreigner can buy land without approval from a land approval board.
See also **Sobhuza; Swazi Nation land**.

Land Tenure Act (*Zimbabwe*)

The Land Tenure Act was passed in 1970 by the white rebel leader, Ian Smith, and his Rhodesian Front party. It replaced the Land Apportionment Act when the constitution of 1969 came into effect. The new act formalised the division of land between white and black in the country, giving half Zimbabwe's land to each group despite the fact that black Zimbabweans made up some 95 per cent of the population.
See also **Land Apportionment Act**.

Lára, Lúcio (*Angola*)

Lúcio Lára was one of the founding members of the MPLA and one of its chief representatives, internationally. His primary role in the party was one of organisation. He was elected to the executive committee in 1962 and took charge of organisation, continuing this role on his election to the Political Bureau and Central Committee in 1974. He was the member of the Central Committee Secretariat responsible for organisation and the Secretary-General of the MPLA until the second national

congress of the MPLA-PT in December 1985. He is now Deputy Defence Minister.
See also **MPLA**.

Latham (*Tanzania*)

Latham and Mafia are the small islands off the coast of Tanzania which were once ruled by the Sultan of Zanzibar. They became part of the United Republic of Tanzania in 1964.

'leadership code' (*Zambia*)

The 'leadership code' was introduced under the 1973 Zambian constitution. It expressly forbids MPs to receive parliamentary salaries as well as incomes from other sources. It was introduced at a time when Kenneth Kaunda was trying to unify his party and come to grips with Zambia's economic crisis.
See also **austerity programme**.

League of Nations Mandate on Namibia (*Namibia*)

At the Treaty of Versailles, in 1919, the South African military occupation of Namibia officially ended. South Africa had persuaded the British cabinet to agree to its annexation of the country, but US opposition led to agreement on British administration of Namibia under a League of Nations class 'C' mandate. Britain then transferred control to South Africa. While the mandate, in theory, required South Africa to prepare the country for self-determination, and to 'promote to the utmost the material and moral well being and the social progress of the inhabitants', the only specific duty under the mandate was an annual report to the League.

In reality South Africa regarded the mandate as 'nothing else but annexation', in the words of South African Prime Minister Jan Smuts. South Africa proceeded to integrate Namibia with its own territory, treating it effectively as a fifth province. The supervision of the mandate was legally transferred to the United Nations, on its formation, although this was contested by South Africa, and became the basis of a long-running jurisdictional dispute between that country and the United Nations. South Africa's continuing defiance of the UN led to the termination of its right to administer the mandate, in October 1966. Since then, while the UN has had *de jure* sovereignty, South Africa has retained *de facto* control.
See also **German colonialism; International Court of Justice; UNCN**.

Leballo, Potlake (*South Africa*)

Potlake Leballo was acting-president and chairperson of the PAC for some 17 years while the PAC's president, Robert Sobukwe, was banned and imprisoned in South Africa. He left the ANC, together with Robert Sobukwe, in 1958 and helped found the PAC a year later. He became acting president in 1962 and president after Robert Sobukwe's death, from 1978 to 1979. He was forced to resign in 1979 because of feuding within the organisation which dogged the PAC throughout his leadership.

He was replaced by a three-person Presidential Council and a subsequent reorganisation of the PAC took place under a new leader, John Pokela. See also **PAC; Pokela; Sobukwe**.

Legislative Council (*Swaziland*)

The Legislative Council was a multi-racial law-making body set up under the 1963 constitution in Swaziland. It consisted of 31 members chosen by different methods including election. Elections took place in 1964 but the council was replaced by Parliament under the 1967 constitution. The Reconstituted European Advisory Council, a conservative white organisation, helped to set up the Legislative Council, hoping it would exclude the emerging nationalist parties from constitutional discussions.
See also **Reconstituted European Advisory Council**.

Lekhanya, Patrick (*Lesotho*)

Major General Patrick Lekhanya took over as the leader of Lesotho on 20 January 1986 after a military coup in which the prime minister, Chief Leabua Jonathan, was deposed. The General rules the country at the head of a six-person Military Council and the Council of Ministers (a 'non-political' body). A decree entitled 'Order 1 of 1986' vested all executive and legislative power in the King, but General Lekhanya governs and the order specifically stipulated 'The King shall act in accordance with the advice of the Military Council'.

Major General Lekhanya was a conservative soldier who was deeply opposed to Chief Jonathan's policy of arming members of his party's Youth Wing and building them into a politicised unit within the armed forces. He also mistrusted the BNP Youth Wing for political reasons; many of its members were close to the ANC of South Africa and provided it with a base inside Lesotho. Major General Lekhanya feared that ANC activity could bring down the military wrath of South Africa on Lesotho.

The most visible change in the new government has been a dramatic decline in anti-South African rhetoric and the deportation of some 100 members of the ANC to Zambia in accordance with South Africa's long-standing demands. The General has also restored relations with South Korea and sent some North Korean technicians home. 'Lesotho's relations with South Africa are characterised by a spirit of neighbourliness', General Lekhanya explained in an interview shortly after the coup. South Africa's imposition of border restrictions in early January 1986 precipitated the coup. These economic sanctions rapidly reduced Lesotho to a state of siege and created the conditions for General Lekhanya's take-over.
See also **Council of Ministers; Jonathan; Military Council; relations with South Africa**.

Lembede, Anton Muziwakhe (*South Africa*)

Anton Lembede was the main inspiration behind the formation of the Congress Youth League, set up within the ANC in 1944. He was a self-educated school teacher who had lived in the Orange Free State and spoke Afrikaans. This gave him insight into Afrikaner nationalism and

helped him to develop the philosophy of 'Africanism', which argued that Africans had to rely on themselves alone for liberation. Both the Youth League and this philosophy, which it adopted, came to have a substantial influence on the development of the ANC.
See also **'Africanism'; Congress Youth League**.

Lenshina, Alice (*Zambia*)
Alice Lenshina was the founder leader of the Lumpa church in Zambia in 1954. She was a controversial and charismatic 'holy woman' and a friend of President Kaunda. The Lumpa Church was involved in an uprising in 1964.
See also **Lumpa Church**.

Leselinyana la Lesotho (*Lesotho*)
Leselinyana is the publication of the Evangelical Church in Lesotho. It supported the opposition parties and reported allegations of police torture during the 1970 state of emergency declared by Chief Leabua Jonathan. The paper was first published in November 1863 and is occasionally translated into English as 'The Little Light of Basutoland'. It was the first newspaper in Lesotho and the first in any African language in southern Africa.

LESOMA (*Malawi*) See **Socialist League of Malawi**.

Lesotho

Official title: Kingdom of Lesotho
Head of State: HM King Moshoeshoe II
Ruler: Maj-Gen. Justin Lekhanya, Chairman of Military Council
Area: 30,355 sq. km
Population: 1,500,000 (1986)
Capital: Maseru
Official languages: Sesotho, English
GDP per Capita: US$273 (1984)
Major exports: diamonds (35–40 per cent of total), wool, mohair, livestock
Currency: Maloti (M); M2.068 per US$ (July 1987)
Political parties: Following the military coup in January 1986 the Military Council banned all political activity. Before the coup the main political parties were the ruling Basotho National Party and the Basutoland Congress Party

Lesotho is the most vulnerable of South Africa's neighbours, as it is completely surrounded by the Republic. South Africa has used this ruthlessly to serve its own ends. For instance, in 1986, an economic blockade was imposed by South Africa forcing Lesotho to expel alleged ANC guerrillas and precipitating a military coup. Lesotho is a mountainous, beautiful and rugged country with a heavy annual rainfall. Economically it is one of the world's least developed countries; one commentator has listed its resources as 'people, water and scenery'. Agriculture, in the shape of smallholdings, employs most of the workforce

although less than 10 per cent of the land area is considered cultivable.

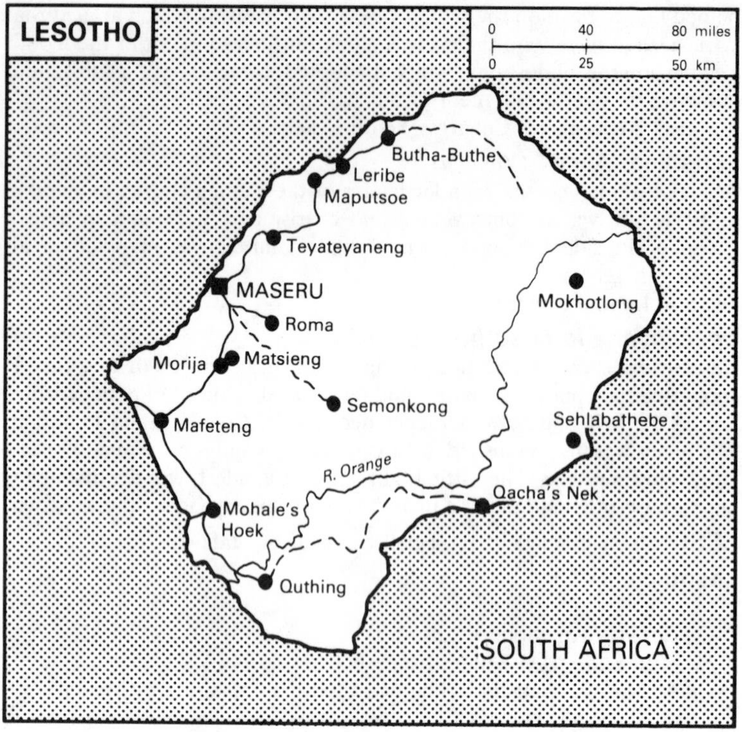

The Kingdom of Lesotho (then Basutoland) became independent on 4 October 1966 as a constitutional monarchy with Chief Leabua Jonathan as prime minister at the head of the Basotho National Party (BNP). However, Chief Jonathan suspended the constitution in 1970 when it seemed as though the opposition Basotho Congress Party might win. A state of emergency was declared and government was conducted by decree until 1973 when Chief Jonathan introduced an Interim National Assembly of 93 nominated members drawn from a number of different parties. In January 1986 Chief Jonathan was deposed in a military coup and Major-General Justin Lekhanya took over at the head of a six-person Military Council. A 'non-political' cabinet was named and, although the military governs the country, the King's influence has increased since the coup.

See also **Basutoland Congress Party; Basotho National Party; Jonathan; Lekhanya; Moshoeshoe II; relations with South Africa**.

Lesotho Liberation Army (*Lesotho*)

The Lesotho Liberation Army (LLA) is the military wing of the Basuto-

land Congress Party. The Lesotho government has repeatedly accused South Africa of harbouring the LLA and of allowing it to launch military operations across the border. South Africa has denied the allegations but it is widely believed that South Africa does support the LLA and even uses it to attack Lesotho when that country acts against South African interests.
See also **Basutoland Congress Party**.

Lesotho United Democratic Party (*Lesotho*)
The Lesotho United Democratic Party was formed in 1966 by Charles Mofeli who had broken away from the Basutoland Congress Party. In 1973 Chief Jonathan established an Interim National Assembly to draw up a new constitution which would be 'compatible with traditional institutions'. Charles Mofeli sat in this Assembly, but in 1974 called on the government to resign after violence had erupted throughout Lesotho. The party soon came to oppose the government absolutely and in 1975 Charles Mofeli asked the United Nations and the rest of the world to refuse all further support for Lesotho until a new government had been elected in a fair election.
See also **Basutoland Congress Party; Jonathan**.

Libandla (*Swaziland*)
The *Libandla* is the council of the *Ngwane* or Swazi nation. It is the National Council and it represents the people and advises the monarch. Its members are the chiefs, princes (*bantfanenkosi*) and other important people in the community. The *Liqoqo*, more recently described as the Supreme Council of State, sets the agenda for debate and often discussion in the *Libandla* is heated. The monarch announces the consensus when it has been reached and approves it.
See also ***Liqoqo***.

liberated zones (*southern Africa*)
Areas of the countries under the control of guerrilla forces during liberation struggles. For instance, the two main areas under FRELIMO control during the early stages of the war in Mozambique were rural tracts in the provinces of Cabo Delgado and Niassa, in the north of the country, and later expanded to include Tete, Manica and Sofala. Once having established military control, the liberation movements set about creating new social, economic and political structures in the liberated zones. These included schools, health clinics, and village committees.

Liberation Front (*Namibia*)
Constituent party of Namibia Democratic Coalition: see **Namibia Democratic Coalition**.

Lilongwe (*Malawi*)
Lilongwe became the new capital of Malawi in 1975. President Banda had announced the plan to move Malawi's capital from Zomba to Lilongwe in 1964. When the British government refused to fund the move, the presi-

dent turned to the South African government for a loan. A sum of some £4.7 million was agreed to finance the first stage of the project. The estimated population of Lilongwe in 1981 was 136,800.

Limpopo River (*southern Africa*)
The Limpopo River rises as the Crocodile (*Krokodil*) River in South Africa and flows for about 1,600 km, eventually passing through Mozambique to the Indian Ocean. The river forms the border between South Africa's Transvaal province and Botswana and the southern border of Zimbabwe separating that country from South Africa.
See also **Zambezi River**.

linkage (*Angola, Namibia, South Africa*)
This is the term applied to US and South African insistence that negotiations for Namibian independence under UNSC Resolution 435 should be linked to and dependent on the withdrawal of Cuban forces and other personnel from Angola. The existence of the policy emerged as a result of a 'leak' in May 1981. Formal acknowledgement did not come until 21 June 1982, however, when P.W. Botha, South African premier, announced: 'We cannot enter into the third phase (implementation) of the agreement (UNSCR 435) with the Western Five unless the Cubans are withdrawn from Angola. I stand by this statement.'

Having demanded Cuban withdrawal, South Africa reinforced Angola's need for Cuban support two months later with a massive invasion of the country. It has continued to occupy parts of Angola since, and has increased its military and logistical support for UNITA. The architect and chief proponent of linkage was US Under-Secretary of State for African Affairs, Chester Crocker.

No other governments have joined the US and South Africa in demanding such linkage, but both the British and the West German governments have expressed opinions that Cuban withdrawal would facilitate negotiations for Namibian independence. The concept has been rejected by both SWAPO and the Angolan Government as interference in the affairs of a sovereign state (Angola) and a ploy to avoid progress towards Namibian independence. Many other Namibian and international organisations, including the Namibia Council of Churches, the 'front line' states, the OAU and the Non-Aligned Movement, have joined them in their rejection of linkage.
See also **Contact Group; Cuban troops; UN Security Council Resolution 435**.

Liqoqo (*Swaziland*)
The *Liqoqo* functions as the executive arm of the Swazi National Council or *Libandla*. It is made up of the king or queen's closest advisers. Membership consists of the monarch's chief uncle, the *Indvuna* (a commoner chosen by the king or queen to serve officially as a counsellor, an intermediary between the people and their ruler) of the chief village, the *Tinvuna* (plural of *indvuna*) of royal villages and others who are appointed for life. The council is supposed to be influential in Swazi

politics as the monarch is expected to heed its advice. Its formal role is to keep the monarch in touch with the people and set the agenda for the *Libandla* which represents the Swazi people.

In recent times the *Liqoqo* has been at the centre of controversy. After the death of King Sobhuza in 1982, the *Liqoqo* was transformed into a public body and its members received state salaries. It also became known as the Supreme Council of State. Some sections of the royal family now feel that the council is turning the monarch into a figurehead, and that Swaziland is increasingly being ruled by *Liqoqo* politicians. Evidence for this was the dispute between the queen regent, Dzeliwe, and the *Liqoqo*. In 1983 the *Liqoqo* presented the queen regent with a document transferring many of her powers to the office of the 'Authorised Person', an already powerful post in the traditionalist structure. Dzeliwe, however, refused to sign, and soon afterwards she was dismissed and removed from office. She was replaced by Ntombi as queen regent, who signed the document. Thus the *Liqoqo* was recognised as more powerful than the regency.

The situation began to change in 1986 when an amendment to the Constitution was announced in the *Government Gazette* of October 10th. The *Liqoqo* would no longer be described as the Supreme Council of State. Furthermore, the *Liqoqo* no longer had the power to appoint the 'Authorised Person'. The 'reign' of the *Liqoqo* in its present form was ended by the new king, Mswati III, when he dissolved it on 27 May 1986, just a month after coming to the throne.
See also **'Authorised Person'; Dzeliwe; Libandla; monarchy; Mswati; *ndvuna*; *nkundhla*; Ntombi**.

LLA (*Lesotho*) See **Lesotho Liberation Army**.

Lobamba (*Swaziland*)
Lobamba is the royal capital of Swaziland. It is situated not far south of Mbabane, the administrative capital. It also houses the modern parliamentary buildings and an official residence of the monarch. The name Lobamba was first used by Ngwane II for his first village. Pilgrimages are still made to a cave where he and other Swazi kings have been buried. The nation's traditional capital is the village of the monarch's mother, the *Ndlovukazi*, and thus changes when monarchs change.

lobola (*southern Africa*)
Women in southern Africa are often still subject to many of the traditional practices which established their inferior status. *Lobola*, or a bride price, is still paid in most parts of the country by men to their prospective wives' families in exchange for their daughters. *Lobola*, especially for male peasants dependent on family labour, represents an investment in the labour of both the wife and future potential offspring.
See also **OMM**.

Luanda (*Angola*)
Luanda is the capital city of Angola and its second most important port after Lobito. Its population in the last Angolan census in 1970 was

480,613, eight times that of the next largest town, Huambo. Current estimates of Luanda's population put it between 1 and 1½ million. Luanda was an important slave trading port and the centre of the Portuguese colonial administration. Prior to independence it became a major stronghold of the MPLA and it was the movement's successful defence of the city against invasions from north and south in 1975 which enabled it to win control of the country.
See also **Kifangondo marshes; MPLA; South African invasions**.

LUDP (*Lesotho*) See **Lesotho United Democratic Party**.

Lumpa Church (*Zambia*)
The Lumpa Church was a separatist, religious sect which was active in the north of the country before independence. Its leader was Alice Lenshina, a charismatic and influential 'holy woman'. The Lumpa Church opposed outside interference in its activities and in the middle of 1964 the Lumpa uprising took place. This was suppressed by Kenneth Kaunda's new government and led to the death of some 700 people in the fighting. The Lumpa uprising was one of the first problems the government had to face and the violence was made more difficult for Kenneth Kaunda because of his non-violent philosophy and his friendship with Alice Lenshina. When she started the church in 1954, Kenneth Kaunda's elder brother was among her first disciples.

Lusaka (*Zambia*)
Lusaka is the capital city of Zambia. It is in the centre of the country along its main north-south road and rail axis. In the 1980 census its population was estimated as 538,830.

Lusaka Agreement, 1974 (*Mozambique*)
This agreement arose out of a meeting between Portuguese Foreign Minister Mario Soares and FRELIMO President Samora Machel on 7 September 1974 in Lusaka, Zambia. It set up a Transitional Government to take power in Mozambique 18 days later, established a ceasefire between the forces of Portugal and FRELIMO, and set 25 June 1975 as the date for full Mozambican independence.
See also **FRELIMO; Machel**.

Lusaka Agreement between South Africa and Angola, 1984 (*Angola, Namibia, South Africa*)
In February 1984 at a meeting between South Africa and Angola, held in Lusaka, Zambia, South Africa agreed to end its occupation of southern Angola. It hoped to persuade Angola, in exchange, to refuse to allow PLAN to obtain logistical supplies via southern Angola. In reports of the agreement, the Angolan and South African governments have been quoted giving different accounts, but no text has been made public to date.

A Joint Monitoring Commission (JMC) of Angolan and South African personnel was set up in February 1984 to monitor the South African withdrawal. By June of that year, however, despite the JMC, Angola

was still complaining of South African troops 40km inside its territory. In April 1985 South Africa claimed that its forces had been completely withdrawn from Angola and the JMC was disbanded the following month. By July South African forces had again attacked FAPLA in strength 40km inside the country.
See also **FAPLA; PLAN; SADF; SWATF**.

Lusaka talks on Namibia, 1984 (*Namibia*)

In May 1984 SWAPO and South Africa (the South African delegation included representatives of the MPC) met in Lusaka for talks on Namibian independence called by Zambian President, Kenneth Kaunda. Despite the fact that the talks were not, unusually, held under UN auspices, they proved a triumph for SWAPO when representatives of the NCDP, the National Independence Party, the Federal Party, the Damara Council and SWANU joined the SWAPO delegation to sit opposite the MPC leadership. The talks collapsed when South Africa insisted on linkage between the issue of Namibian independence and the presence of Cuban personnel in Angola and refused to agree to the immediate implementation of UNSC Resolution 435. Subsequent to the collapse of the talks, an informal alliance emerged among SWAPO, the Damara Council and SWANU.
See also **Damara Council; Federal Party; linkage; MPC; Namibia Independence Party; NCDP; SWANU; SWAPO; UN Security Council Resolution 435**.

Lutuli, Albert John (*South Africa*)

Chief Lutuli was one of the ANC's most respected and charismatic leaders. He had an international reputation and won the Nobel Peace Prize in 1960. He became president-general of the organisation in December 1952 (six years after he joined it) and led it in a new direction. Unlike his predecessors he had been an active local administrator and a working chief of the *Abasemakholweni* Zulu ethnic group, and had the experience and the personal warmth to appeal to ordinary people. His religion – he was a Methodist lay preacher – gave him a considerable inner strength and courage as well as enabling him to give his political convictions the dignity of his faith. He believed passionately in non-violence and his leadership took the organisation into a period of outspoken protest.

Before his election as president-general Chief Lutuli was ordered by the state to resign either from the ANC or from the chieftaincy. He refused and replied in public:

> Who will deny that thirty years of my life have been spent knocking in vain, patiently and modestly at a closed and barred door? What have been the fruits of moderation? The past thirty years have seen the greatest number of laws restricting our rights and progress, until today we have reached a stage where we have almost no rights at all. It is with this background and with a full sense of responsibility that, under the auspices of the ANC, I have joined my people in the new

> spirit that moves them today, the spirit that revolts openly and boldly against injustice and expresses itself in a determined and non-violent manner.

Chief Lutuli suffered at the hands of the South African state for his politics. He was banned several times, arrested in December 1956 with others on a charge of high treason and eventually released one year later. He remained president of the ANC until his death on 21 July 1967, when he was hit by a train. The circumstances of his death have never been explained.

See also **ANC**.

M

MacBride, Sean (*Namibia*)

The first UN Commissioner for Namibia, from 1973 to 1977, was Sean MacBride, formerly Minister of Foreign Affairs in the Republic of Ireland, and Secretary-General of the International Commission of Jurists.

See also **UN Commissioner for Namibia**.

Machel, Samora Moïsés (*Mozambique*)

After winning his reputation as a field commander in the liberation war, and subsequently as Secretary for Defence, Samora Machel was elected the second President of FRELIMO to succeed Eduardo Mondlane after the latter's assassination in 1969. Machel, charismatic, well-liked and a compelling speaker, successfully ended the years of strife in the FRELIMO leadership, and his accession to the presidency marked the start of a period of stability and continuity through to independence.

Machel was sworn in as the first president of independent Mozambique on 25 June 1975, and remained in that position until his death, along with 34 other passengers and crew, in an air crash on 19 October 1986. The passengers were returning from a summit in Zambia which had been called to discuss ways of decreasing dependence on South Africa and containing the rebel MNR in Mozambique. The Russian-built Tupolev aircraft crashed in a remote part of the eastern Transvaal, in South Africa, prompting rumours and accusations that the South Africans had engineered the president's death.

An international investigation team set up by South Africa to ascertain the cause of the accident, but which Mozambique and the USSR boycotted, suggested that the most likely cause was pilot error. The inquiry suggested that the pilot became lost and was searching for the landing lights of Maputo airport, believing that he was in the vicinity of the capital, when he crashed into the hills near the South African-Mozambican border. The inquiry's findings did not end speculation, however, of South African involvement in the tragedy and allegations that it had been engineered by a decoy directional beacon continued to be made.

Samora Machel was hailed, at his death, as a great statesperson who had both led his country to independence and maintained its unity in the face of sustained Rhodesian and South African aggression. Even South

African Foreign Minister 'Pik' Botha expressed his regret at his death, citing the signing of the Nkomati Accord as evidence of Machel's realism and stature. He was widely mourned by his own people, for whom he had been a beloved leader, and by many throughout southern Africa to whom he had exemplified the struggle against colonialism and for self determination.
See also **Chissano; FRELIMO; Nkomati Accord**.

Machungo, Mário (*Mozambique*)
Mário Machungo was Prime Minister of Mozambique under Samora Machel and continued in the post when Machel was succeeded by Joaquim Chissano. A trained economist, he is the only member of the Central Committee who holds no military rank, having worked throughout the liberation war as an underground FRELIMO militant in Maputo.
See also **Chissano; FRELIMO**.

Madaraka (*Tanzania*)
Madaraka or 'responsible government' was the goal TANU, Tanzania's nationalist party, set for itself at its annual conference in 1958. If this were not achieved by the end of 1959 the conference decided to adopt 'positive action' towards that end.
See also **TANU**.

Mafia (*Tanzania*)
One of two small islands off the coast of Tanzania: see **Latham**.

MAFREMO (*Malawi*) See **Malawi Freedom Movement**.

Makhosetive (*Swaziland*) See **Mswati III**.

Makotoko, Seth P. (*Lesotho*)
Dr Seth Makotoko formed the Marema-Tlou Freedom Party in 1962. The new party was a merger of the Marema-Tlou Party and the Basutoland Freedom Party.
See also **Basutoland Freedom Party; Marema-Tlou Freedom Party**.

Malan, Magnus Andre de Merindol (*South Africa*)
General Magnus Malan is a protégé and close political ally of President P. W. Botha. He was appointed Minister of Defence in 1980 to succeed Botha (the first serving soldier to be in the cabinet) and joined the National Party, becoming MP for Modderfontein the following year.

Malan was born in 1930 and was one of the first students to take a university degree in military studies. After receiving officer training he rose through the ranks to become Chief of the Army in 1973 and Chief of the Defence Force in 1976. He is credited with having rationalised the South African Defence Force (SADF) and revamped its management structures. Greatly influenced by P. W. Botha, Malan worked with him to formulate the concept of 'total strategy' and 'total onslaught'. He argued that there could be no purely military solution in South Africa

and that the military could only play a part in bringing about a political solution.

General Malan is one of the most influential members of P. W. Botha's cabinet, and has presided over aggressive and covert military action against neighbouring countries. It is he, as Minister of Defence, who has waved the stick of military intervention at them while his colleague, 'Pik' Botha has offered the carrots of economic co-operation.

See also **Botha, P. W.; Botha, 'Pik'; SADF; 'total strategy'**.

Malawi

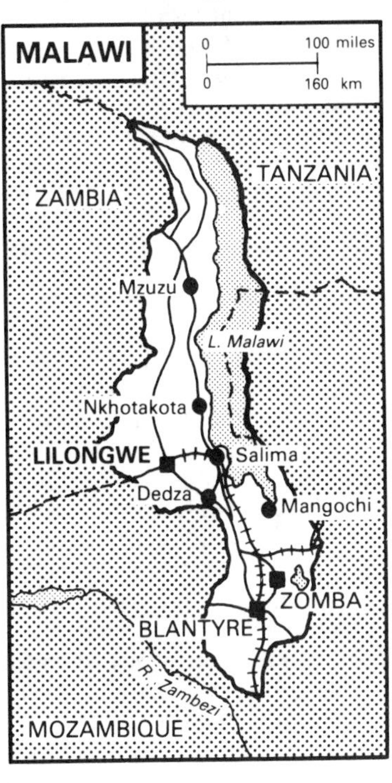

Official title: Republic of Malawi
Head of state: President Ngwazi Dr Hastings Kamuzu Banda
Area: 118,484 sq. km
Population: 7 million (1985)
Capital: Lilongwe
Official languages: English, Chichewa (Tumbuka, Yao, Lomwe spoken)
GDP per Capita: US$151 (1985)
Major exports: tobacco (40–45 per cent); tea
Currency: Kwacha (K) 100 tambala; K2.28 per US$ (July 1987)
Political parties: the only legal party is the ruling Malawi Congress Party – membership is compulsory for all adults; there are several small opposition groups in exile: the Congress for the Second Republic, the Malawi Freedom Movement, the Save Malawi Committee and the Socialist League of Malawi

Malawi is a long, narrow, landlocked country surrounded by Tanzania in the north, Zambia to the east and Mozambique to the south. More than a fifth of its surface area (24,208 sq. km) is covered in water; the north and centre of the country is dominated by Lake Malawi – the third-largest lake in Africa. It is one of the most densely populated countries on the continent and has some of the region's most fertile soils. Agriculture dominates the economy.

The country is a one-party state under the executive presidency of Dr Hastings Banda and his Malawi Congress Party. Banda was appointed

Life President on 6 July 1971, and appoints a cabinet and up to 15 members of the National Assembly. The rest (87) are elected for five-year terms. The president has become increasingly despotic over the years and he brooks no opposition to his rule. Malawi has close relations with South Africa and has been described as its closest friend in the region. It is the only country in the region to maintain formal diplomatic links with South Africa. More recently, however, South Africa's aggressive regional policy has forced Malawi to strengthen its links with the other countries of the Southern African Development Coordination Conference (SADCC), and it has been forced to commit itself to supporting Mozambique in its battle against the South African-supported MNR rebels.
See also **accord between Malawi and Mozambique, 1986; Banda; Congress for the Second Republic; Lake Malawi; Malawi Freedom Movement; MCP; Save Malawi Committee; Socialist League of Malawi**.

Malawi Congress Party (*Malawi*) See **MCP**.

Malawi Freedom Movement (*Malawi*)
The Malawi Freedom Movement (MAFREMO) is a dissident movement based in Dar es Salaam, Tanzania. All opposition parties are banned within Malawi, so all are based outside the country. There are several opposition groups but none have so far (1987) posed a threat to President Banda's regime. MAFREMO claims to have widespread internal support but it derives most of its members from Malawian refugees, exiled intellectuals and professionals. Its political philosophy is socialist.

In December 1981 Orton Chirwa, leader of MAFREMO, was allegedly kidnapped from Zambia together with his wife, Vera. They were both put on trial for treason in Blantyre, Malawi, and sentenced to death. Appeals against the sentences were ignored but after international protests President Banda commuted the sentences to life imprisonment.

Edward Yapwantha took over the leadership from Orton Chirwa in 1983. There are reports that MAFREMO has been revived, partly through the efforts of its regional officials in Bulawayo, Zimbabwe. The group made an armed attack on a police station in the Malawi border town of Kaporo on 10 January 1987, and it is said to have support from the authorities in Zimbabwe, Tanzania and Mozambique.
See also **Chirwa; Congress for the Second Republic; Save Malawi Committee; Socialist League of Malawi**.

Mandela, Nelson Rolihlahla (*South Africa*)
Nelson Mandela is the Life President of the ANC of South Africa. He is in prison for life, charged with plotting to overthrow the government with violence. In 1982 he was moved from Robben Island, together with Walter Sisulu and others, and transferred to Pollsmoor prison in Cape Town. He was arrested in 1962 and sentenced to five years' imprisonment with hard labour (three years for incitement to strike and two years for leaving the country without valid travel documents). In 1963 many fellow leaders of the ANC and its military wing, *Umkhonto we Sizwe*, were

arrested at a house in Rivonia near Johannesburg. Nelson Mandela was brought to stand trial with them for plotting to overthrow the government by violence. His statement from the dock received considerable international publicity. On 12 June 1964 eight of the accused, including Nelson Mandela, were sentenced to life imprisonment.

Nelson Mandela began his career as a lawyer and set up the first African legal partnership in South Africa in 1952 with Oliver Tambo, the ANC's current president-general. He joined the ANC in 1944 and became national president of the Youth League (which he had helped to found) in 1950 and the deputy national president in 1952. He, Walter Sisulu and Oliver Tambo rose to senior positions in the ANC together and were to dominate its leadership for many years. Nelson Mandela controlled the volunteer organisation of the 1952 Defiance Campaign (he was National 'Volunteer-in-Chief') and this led to a suspended prison sentence and several banning orders. In 1956 he was charged with treason and acquitted in 1961.

After the banning of the ANC in 1960 Nelson Mandela argued strongly for the setting up of the ANC's military wing, *Umkhonto we Sizwe*. He explained his views in his statement at the Rivonia Trial: 'Unless responsible leadership was given to canalize and control the feelings of our people, there would be outbreaks of terrorism which would produce an intensity of bitterness and hostility between the various races of this country'. In June 1961 the ANC executive considered his proposal on the use of violent tactics and agreed that those members who wished to involve themselves in Nelson Mandela's campaign would not be stopped from doing so by the ANC. This led to the formation of *Umkhonto we Sizwe*. Throughout the following year, during the sabotage campaign, Nelson Mandela became the most wanted man in South Africa, but he proved so elusive that the English-speaking newspapers referred to him as 'the Black Pimpernel'.

During his years in prison Nelson Mandela's reputation has grown steadily. He is now widely accepted as the most significant black leader in South Africa and he has become a potent symbol of resistance as the anti-apartheid movement gathers strength. The possibility of his release has become the subject of international speculation after the launching of a 'Release Mandela' campaign in 1982. He has consistently refused to compromise his political position to obtain his freedom.

See also **ANC; Defiance Campaign; Mandela, Winnie; Rivonia Trial; Sisulu, Walter; Tambo; Treason Trial; *Umkhonto we Sizwe*.**

Mandela, Nomzamo Winnie (*South Africa*)

Winnie Mandela is a member of the national executive of the African National Congress Women's League and a member of the national executive of the Federation of South African Women. She is a leading spokesperson against apartheid and has become a symbol of black resistance over the years. She has been actively involved in the ANC since its early days and married Nelson Mandela, the ANC leader now in prison for life. She helped found the Black Women's Federation and the Black

Parents' Association, both banned in 1977. She has been banned and arrested many times by the South African state for her anti-apartheid activities.
See also **ANC; Federation of South African Women; Mandela, Nelson**.

MANU (*Mozambique*)
The Mozambique African National Union (MANU) was established in Kenya in 1961 by Mozambican expatriates and was modelled on TANU and the Kenya African National Union. MANU was one of the constituent parties of FRELIMO when it was formed the following year.
See also **FRELIMO; TANU**.

Maputo (*Mozambique*)
The principal port and city of Mozambique and formerly known as Lorenço Marques, Maputo was renamed shortly after independence. It had an estimated population of 785,512 in 1981, and is the administrative and legislative capital of the country. While South African destabilisation of the Mozambican economy has meant the diversion of trade away from Maputo, it remains one of the primary ports of the region and central to SADCC plans to reduce dependence on South African controlled trade routes.
See also **SADCC**.

Marais, Jacob Albertus 'Jaap' (*South Africa*)
Jaap Marais was elected deputy leader of the *Herstigte Nasionale Party* (HNP) on its formation in 1969, shortly after his expulsion along with other rebel right-wingers from the National Party. In 1977, following the retirement of Albert Hertzog from politics, Jaap Marais succeeded him as leader of the HNP.

Like the policies advocated by his party, Marais is on the extreme right of the Afrikaner political spectrum in South Africa, advocates a return to the apartheid of Verwoerd, and attacks the National Party government for its 'liberalism'. Marais became distanced from the leadership of the National Party under Vorster as a result of his opposition to what he saw as the liberal policies of the Vorster regime, and, in particular, the decision to establish diplomatic relations with Malawi. His expulsion from the National Party in 1969 followed his refusal to support Premier Vorster, in a parliamentary vote of confidence, unless the motion was linked to a condemnation of the Afrikaans press for its mischievousness. Marais held the parliamentary seat of Innesdal for the National Party from 1958 to 1969, and for the HNP from 1969 to 1970.
See also **HNP; National Party; Verwoerd; Vorster**.

Marema-Tlou Freedom Party (*Lesotho*)
The Marema-Tlou Freedom Party was formed in 1962 by a merger between the Marema-Tlou Party and the Basutoland Freedom Party. The new party soon began to disintegrate and the two leaders, Chief S. S. Matete and B. M. Khaketla, withdrew. The Marema-Tlou Freedom Party continued under Dr Seth Makotoko and won just four seats in the 1965 elections. There have been no further effective elections in Lesotho since that date.

See also **Basutoland Freedom Party; Marema-Tlou Party**.

Marema-Tlou Party (*Lesotho*)

The Marema-Tlou Party was formed by Chief S. S. Matete in 1957. Its name comes from the Sesotho proverb *Marema-Tlou a ntsoe leng*, 'in order to bring down an elephant you have to push together and at the same time'. It was a conservative party loyal to the future King Moshoeshoe II. It came second in the 1960 elections, becoming the main opposition party. In 1962 it merged with the Basutoland Freedom Party to form the Marema-Tlou Freedom Party. Chief Matete revived the old Marema-Tlou Party but it won no seats in the elections of 1965.
See also **Basutoland Freedom Party; Marema-Tlou Freedom Party**.

Maseru (*Lesotho*)

Maseru is the capital city of Lesotho. It lies on the edge of the disputed boundary with the Orange Free State, a province of South Africa. Its name means 'the place of red sandstone', and its population was estimated as 63,000 in 1981.

Mashonaland (*Zimbabwe*)

Mashonaland is the term used for the Shona territories east of Matabeleland, although many Shona-speakers live in Matabeleland and vice versa.
See also **Matabeleland; Ndebele; Shona**.

Masire, Quett K. J. (*Botswana*)

Dr Quett Masire is the President of Botswana. He had been Sir Seretse Khama's vice-president and minister of finance and development planning since independence in 1966 and was an old friend and ally. Together with Seretse Khama he founded the ruling Botswana Democratic Party and became its first secretary-general. After Seretse Khama's death Dr Masire was elected president by secret ballot in the National Assembly on 18 July 1980, five days after Seretse Khama's death.

Quett Masire is regarded as an intellectual and the architect of Botswana's economic policies. He is an effective speaker and an effective politician. In 1981 he became President of the Southern African Development Coordination Conference (SADCC) for three years. Politically he is less conservative than his predecessor and is more sympathetic to guerrilla movements such as SWAPO, the Namibian nationalist group fighting for independence from South Africa. Relations have improved with Zimbabwe under President Masire, despite the continuing problem of Zimbabwean refugees in Botswana.

The pressures from South Africa for Botswana to sign a non-aggression pact similar to those signed by Mozambique and Lesotho have increased in recent years. On 14 June 1985 South African commandos launched an attack on the capital city, Gaberone. They raided ten houses and killed twelve people, including a Dutch national and a child. The South Africans claimed that they were attacking houses occupied by activists of the African National Congress of South Africa; the ANC denied this. In May 1986 the South African defence force launched further land and air attacks on Botswana, Zambia and Zimbabwe, causing one death in

Botswana. South Africa once again claimed its action was aimed at ANC bases despite the fact that Botswana had just agreed it would not be used as a base for attacks on South Africa and in March the Botswana government had expelled ANC representatives. President Masire has so far refused to sign a treaty with the South Africans.
See also **Botswana Democratic Party; Khama; refugees**.

Massinga, José (*Mozambique*)
Massinga was a leader of Mozambican students studying in the USA who strongly opposed the decision by the movement that students should return to Mozambique to join the armed struggle and work in the liberated areas. When Eduardo Mondlane went to the USA in the mid to late 1960s to persuade students to return, some of them responded with public attacks on him and FRELIMO.

Massinga later took advantage of FRELIMO's policy of reconciliation to return to Mozambique, however, and after independence joined the Ministry of Foreign Affairs and became the head of staff planning. In reality, far from having become reconciled to FRELIMO, he had been working for the CIA, a fact he admitted in 1981.
See also **FRELIMO; Mondlane**.

Matabeleland (*Zimbabwe*)
Matabeleland is the name given to the region around Bulawayo in which Ndebele-speakers traditionally live, although many live elsewhere in the country as well. Gradually the term has been used to indicate a wider area including many non-Ndebele-speaking areas, until Matabeleland makes up about half the country.
See also **Mashonaland; Ndebele; Shona**.

Matante, Philip (*Botswana*)
Philip Matante was a founder and the first Vice-President of the Bechuanaland People's Party which was to become the Botswana People's Party. The BPP split, its factions influenced by the South African Pan-Africanist Congress and the African National Congress; Philip Matante supported the PAC. He later became the party's president and Leader of the Opposition after Kgaleman Motsete was expelled.
See also **Botswana People's Party**.

Matenje, Dick (*Malawi*)
Dick Matenje was secretary of Malawi's only political party, the MCP (Malawi Congress Party), and a prominent minister (Minister Without Portfolio) in President Banda's government in 1983. It is thought that a power struggle for the succession developed between him and John Tembo, the then governor of the Reserve Bank of Malawi. President Banda, it is rumoured, preferred John Tembo. Dick Matenje and three others died in May that year under mysterious circumstances. They were reported to have been killed in a car accident, but exiled opposition politicians claimed they had been shot while trying to leave Malawi. These allegations were denied by the government.
See also **Banda; MCP; Tembo**.

Matete, Samuel Seephephe Mphosi (*Lesotho*)

Chief Matete founded and led the Marema-Tlou Party in 1959. Soon after the emergence of a new party, the Marema-Tlou Freedom Party, Chief Matete withdrew from the leadership together with B. M. Khaketla. He failed to revive his old party's fortunes and died soon afterwards.
See also **Marema-Tlou Freedom Party; Marema-Tlou Party**.

Matthews, Zachariah Keodirelang (*South Africa*)

Professor Matthews was a prominent academic and political figure in southern Africa who became a leader of the ANC. He was South Africa's first black law graduate and continued to study widely in other fields. In 1936 he was appointed professor and head of the Department of African Studies at the University College of Fort Hare in South Africa. Throughout this period he was active in black politics and in 1942 he joined the ANC. He was immediately elected to the national executive and in 1949 became president in the Cape province. In 1956, while the Acting Principal of Fort Hare University College, he was arrested on a charge of high treason and stood trial until 1958 when the charge was withdrawn. In March 1960 he was detained under the State of Emergency and imprisoned for almost six months without charge or trial. Professor Matthews originated the idea of the 1955 Congress of the People, and proposed that it drew up a 'Freedom Charter for the democratic South Africa of the future'.
See also **ANC; Congress of the People; Freedom Charter**.

Mbabane (*Swaziland*)

Mbabane is the administrative and judicial capital city of Swaziland; the royal capital is Lobamba. Mbabane is situated to the north-west of the country some 15 km from the South African border. In August 1982 its population was estimated as 38,646.

Mbanderu Council (*Namibia*)

This small community-based group, led by Chief Munjuku, was part of the Namibia National Front. Since May 1984 it has supported SWAPO.
See also **National Convention of Namibia**.

Mbeki, Govan and Thabo (*South Africa*)

Govan Mbeki is an ANC leader and *Umkhonto we Sizwe* activist who was arrested on 12 July 1963, together with others including Walter Sisulu, at a house in Rivonia near Johannesburg. They were all sentenced to life imprisonment for plotting to overthrow the government by violence and Govan Mbeki was sent to Robben Island. He was a leading theorist and organiser for the ANC, and was released, ill, in late 1987.

Thabo Mbeki is Goven Mbeki's son. He is the ANC's director of information and is considered one of the most influential members of the younger generation.
See also **ANC; Rivonia Trial; *Umkhonto we Sizwe***.

MCP (*Malawi*)

The MCP (Malawi Congress Party) is Malawi's sole legal political party.

It was formed in 1959 after the banning by the colonial authorities of the Nyasaland African Congress (NAC), and led by acting president Orton Chirwa. Hastings Banda, the President of the NAC, remained in prison and was kept informed of MCP activities. The main aim of the party was to break up the Central African Federation which joined together Northern and Southern Rhodesia (later Zambia and Zimbabwe) and Nyasaland (later Malawi). The Federation, it was felt, would benefit the white-ruled south and damage the interests of Northern Rhodesia and Nyasaland.

By 1960 Britain, the colonial power, decided that the only alternative to increasing violence was independence. A state of emergency was in operation and some 52 people had been killed since its declaration. Ian Macleod, then the British Colonial Secretary, released Hastings Banda from prison and summoned him to London for talks. Dr Banda returned home, took over the leadership of the MCP and fought the election campaign for the country's first elections based on universal suffrage in 1961. The MCP won an overwhelming victory, and in 1963 the country attained full self-government. On 6 July 1964 the British protectorate of Nyasaland became the independent state of Malawi, with the MCP as its ruling party.

The MCP was essentially a conservative political party although it had emerged as the ruling party after an extremely militant nationalist campaign. The reason for this was that the MCP was not concerned to effect a revolution, but chiefly to break up the Central African Federation. Its leader, Hastings Banda, was a conservative politician with no desire to introduce great changes. In addition Malawi is a small, poor country, enormously dependent on white South Africa for its economic survival. Thousands of the country's migrant workers had jobs in Southern Rhodesia and South Africa and these two white-ruled countries were Malawi's main African trading partners.

In 1964 Hastings Banda and the MCP met their first real opposition. Six senior ministers resigned or were dismissed following their failure to persuade Dr Banda to pursue more radical anti-colonial policies. As a result an MCP convention in 1965 adopted a constitution which made the country a one-party state. The new constitution, which came into force in 1966, established Malawi as a republic with a president who was the head of the army as well as the government. The president's power is considerable and his control over the party absolute. According to the 1974 party constitution, the MCP is the government and party officials actually rank above members of government (although these are often the same people).

See also **Banda; Central African Federation; Chirwa; Malawi; Northern Rhodesia; Nyasaland; Nyasaland African Congress**.

MDIA (*Angola*)

The *Movimento de Defesa dos Interesses de Angola* (MDIA) was a small ethnically based political group in the north of the country which was temporarily allied to the MPLA in the FDLA.

See also **FDLA; MPLA**.

mestiço (*Angola, Mozambique*)
Mestiço is a Portuguese-derived term for people of mixed European and African ancestry. They formed a high proportion of the *assimilado* community in the Portuguese colonies.

MFA (*Angola, Mozambique*)
The MFA, or Armed Forces Movement (*Movimento das Forças Armadas*) was a Portuguese movement which was instrumental in creating the conditions for the decolonisation of Angola and Mozambique in that it removed from power those committed unequivocally to the retention of Portugal's colonies. On 25 April 1974 General Antonio Spinola took over the Portuguese Government in a coup on behalf of the MFA and aimed at 'democratisation at home and decolonisation in Africa'. Spinola's opposition to independence for the colonies led to his overthrow, and, by July 1975, both Guinea Bissau and Mozambique had been recognised as independent. Angola took a little longer. The last ceasefire with its nationalist groups was signed in November 1974, two months after the overthrow of Spinola, and independence was agreed for November 1975.
See also **Alvor Agreement; Angola; Mozambique**.

MIA (*Angola*)
The *Movimento para a Independência de Angola* or MIA was a Marxist party which joined with the PLUA to form the MPLA.
See also **MPLA; PLUA**.

migrant workers (*Botswana*)
Migrant labour to South Africa – either in the mines or as farm workers – is still the dominant feature of the Botswana employment pattern. It involves some 30–40,000 people (some 21,000 of these work in the mines). This is a substantial proportion of the estimated 60,000 paid workers in the country. With a serious shortage of jobs in Botswana (some one in ten school-leavers can expect paid employment according to one estimate) migrant labour becomes the only choice for many people. The workers' remittances sent home contribute substantially to the total national income of Botswana, and this pattern of work creates all the familiar problems of dependence on South Africa and misery and separation for many families.

migrant workers (*Lesotho*)
Lesotho's economic dependence on South Africa is well illustrated by the earnings of the Basotho migrant workers. About half of the adult male labour force (some 145,000 men in 1984) are employed in South Africa's gold and coal mines or as domestic servants. They are driven to adopt this difficult way of life because of low wages and the lack of employment opportunities at home. In 1983 the money earned by migrant workers made up almost 52 per cent of Lesotho's gross national product. The government uses these earnings to finance development spending.

Since February 1975, 60 per cent of miners' cash earnings have to be remitted to the Lesotho Bank. Half can be drawn by workers' families, but the rest has to wait until the miners return at the end of their contracts. This led to serious rioting at some mines but the arrangement still stands. During the period in the early 1980s when South Africa was imposing border controls on Lesotho to force that country not to shelter alleged guerrillas of the ANC of South Africa, it used the threat of repatriating all migrant workers to Lesotho.
See also **relations with South Africa**.

migrant workers (*Malawi*)

Malawians have consistently sought work beyond the borders of their own country, in South Africa, Zimbabwe and Zambia. In 1974, 74 Malawian mineworkers were killed in a plane crash on their way home from South Africa, and Malawi's president, Hastings Banda, reacted by forbidding any further recruitment of labour by South Africa. By 1977 the number of Malawian workers in South African mines had fallen from some 123,000 in 1973 to a few hundred. The ban was lifted later in the year and soon thousands of Malawians were working in South Africa again, although the numbers have never been as high as in earlier years.

The effect of migrant labour on the home country is considerable. Migrant workers are mainly male and most work in the mines of southern Africa. Their earnings boost their home country's foreign exchange. However, they leave their families for long periods of time returning only for short holidays and the pattern of work causes great hardship.
See also **Banda**.

migrant workers (*Namibia*) See **contract labour**.

migrant workers (*South Africa*)

Migrant labour is an integral part of the successful operation of apartheid. At the start of the development of bantustan or 'Homeland' ideology in June 1959, its architect, Dr Verwoerd, told Parliament: 'The Bantu homelands . . . may be areas which to a large extent (although the people live within their own areas and are governed there) are dependent on basic incomes earned in the adjoining white territory. In that case the whole of the threat to the white areas falls away'.

Basic survival for Africans in the often impoverished rural 'Homelands' depends on access to the 'White' areas for work. Unable to live in 'White' areas, they have either to commute to work from often distant 'Homelands' or live in solitude in barrack-like compounds in African townships, travelling home at long intervals to see their families.

Migrant labour dominates the life of the 'Homelands', both from an economic and a social point of view. At any one time nearly 35 per cent of employable males theoretically living in the 'Homelands' are absent while many others are desperately waiting for jobs in the 'White' areas. Over 70 per cent of the economically active population in the 'Homelands' is dependent on the migrant labour system. Furthermore, studies suggest that the most able and the strongest are those who leave to seek work

elsewhere. In Venda, for example, the population between the ages of 30 and 39 is 84 per cent women who cannot easily find work elsewhere. These women are left to cultivate what food they can, care for the home and bring up children alone, while their men work many miles away for the white economy.
See also **apartheid; 'Homelands'**.

migrant workers (*Swaziland*)
Many people in Swaziland are migrant workers. However the pattern of migrant labour is different from, say Malawi, where the vast majority of workers leave the country and seek work in other southern African countries, particularly South Africa. In Swaziland many migrant workers live in the rural areas known as Swazi Nation land, and work in the towns.

The Swazi Nation land is vested in the monarch, 'in trust for the nation'. In the 1980s Swazi Nation land accounted for 57 per cent of the total land area. Within this area there are widely differing patterns of landholding, ranging from subsistence farming to commercial farming. A study has shown that some 80 per cent of all rural homesteads had one or more members of the family earning outside the farm – about a quarter of the population of these areas in the 1976 census, with three times as many men as women. About 77.5 per cent of these migrant workers remain in the towns of Swaziland while some 17 per cent work in South Africa (about half of these being employed on the mines).
See also **Swazi Nation land**.

Military Council (*Lesotho*)
Lesotho is now ruled by a six-person Military Council headed by Major General Justin Lekhanya. As well as Maj-Gen. Lekhanya its members are (June 1986): Col. Elias Ramaema, Col. Khethang Moseunyane, Col. Michael Tsotetsa, Lt-Col. Thaabe Letsie and Lt-Col. Joshua Letsie. The Military Council took over after the military coup which deposed Chief Leabua Jonathan in 1986. General Lekhanya has named a cabinet which he terms 'non-political' (Council of Ministers), and King Moshoeshoe II remains the Head of State.

Two brothers within the Military Council, Thaabe and Joshua Letsie, are reputed to be very powerful. They are said to be cousins of the King, pro-South African and to have been trained by British officers at Sandhurst. There are reports of divisions emerging within the council, with some members on the right-wing being held responsible for the death squads which are operating in Lesotho at present. The King too is increasing his powers under the council's rule and he is said to be unlikely readily to accept too much South African control over Lesotho.
See also **Lekhanya; Moshoeshoe**.

MINA (*Angola*)
The MINA or *Movimento de Independência Nacional de Angola* was a small party which joined the MPLA in 1958.
See also **MPLA**.

Mineworkers' Union of Zambia (*Zambia*)

The Mineworkers' Union of Zambia is the most powerful trade union in the country and represents the mineworkers of the Copperbelt whose industry forms the base of Zambia's economy. In its relationship with the nationalist movement the union has always kept its organisational structure separate and guarded its independence.

The African Mineworkers' Union (AMU) first started in 1949 at the same time as the African National Congress. It was led by Lawrence Katilungu who encouraged other industries to unionise and join together in 1950 in the Trades Union Congress (TUC). For some ten years Lawrence Katilungu's presidency of the AMU and the TUC was unchallenged and his strong personality dominated union affairs. However, he became involved in politics and this led to a split and a new Reformed Trade Union Congress (later the United Trades Union Congress) was set up in 1960. This was to become the Zambia Congress of Trade Unions in 1965. The African Mineworkers' Union withdrew from the United Trades Union Congress and renamed itself the Zambia Mineworkers' Union in 1965. The Zambia Mineworkers' Union was suspicious of the Zambia Congress of Trade Unions and the government's intentions in setting this up, and it did not join until 1966. The ruling party, UNIP, tried to exert control over the mineworkers' leaders but failed. After this the number of strikes increased and the Brown Commission was set up in 1966 to report on all aspects of the mining industry. In 1967 the Zambia Mineworkers' Union merged with two other mining unions to form the Mineworkers' Union of Zambia (MUZ). Tension between the government and the miners' union continues, and in 1981 some miners' leaders were suspended from UNIP.

See also **Brown Commission; Copperbelt; Katilungu; Zambia Congress of Trade Unions**.

Mingas, Saydi (*Angola*)

Saydi Mingas joined the MPLA in the mid–1960s and held various posts within the organisation. He was elected to the Central Committee in 1974 and in 1975 became the first post-independence Minister of Finance. He was assassinated in May 1977 by supporters of the Nitista coup attempt.

See also **Alves; coup attempt, 1977; MPLA; Nitistas**.

Mishra, Brajesh (*Namibia*)

Brajesh Mishra, a former Indian diplomat at the UN, was the UN Commissioner for Namibia from 1982 to 1 July 1987, when he was replaced by Bernt Carlsson.

See also **Carlsson; UN Commissioner for Namibia**.

Mlambo, Johnson (*South Africa*)

Johnson Mlambo became the new leader of the PAC on 15 July 1985 following the sudden death of John Pokela. He is reported to have been elected unanimously and has set about trying to prevent the ANC from establishing itself as the sole nationalist organisation outside South Africa.

Johnson Mlambo was in prison in South Africa for 20 years, from 1963 to 1983, convicted for his political activities.
See also **Lembede; PAC; Pokela**.

MNA (*Angola*)
The *Movimento Nacional Angolano* (MNA) was a small, ethnically based party in the north of Angola which joined the FDLA.
See also **FDLA**.

MNR (*Mozambique*)
Established as an anti-FRELIMO organisation to destabilise Mozambique, the organisation was originally named the *Movimento Nacional de Resistência de Moçambique* (MNR), but this was later changed to *Resistência Nacional Moçambicana* (Renamo, RNM or Mozambique National Resistance). The MNR, as it is still best known, was a creature of the last years of the Smith regime in pre-independence Rhodesia. The head of the Rhodesian Central Intelligence Organisation recruited members of anti-FRELIMO groups who fled to Zimbabwe in 1974, and with the help of the founder of the notorious Selous Scouts moulded them into a force which could be used to destabilise Mozambique and harass and gather information on the Mozambique-based forces engaged in the struggle for Zimbabwean liberation.

In 1977 an ex-FRELIMO fighter, Andre Matzangaissa, who had broken with the movement after being punished for corruption, was installed as head of the MNR by the Rhodesians. He was instrumental in recruiting to the organisation other ex-guerrillas who were disillusioned at their failure to be rewarded on independence. By 1979 the MNR was operating to some effect inside Mozambique, supplied by air by the Rhodesian military. It suffered a serious setback in October 1979 when the Mozambican army overran its main base and Matzangaissa was killed. Independence in Zimbabwe further undermined the MNR and its leaders and members based in that country were forced to move to South Africa. MNR members inside Mozambique were ordered to gather at a base on the Sitatonga mountain near the border with Zimbabwe, and the organisation suffered further when a Mozambican military operation successfully captured the base in a battle in which the MNR suffered 272 casualties with 300 of its members captured. Leaderless and battered, the MNR disintegrated into bands of directionless marauders.

It was left to South Africa to take up where the Rhodesians had left off and rebuild the MNR. It took over control in 1980 and backed as the new leader Alfonso Dhlakama, another ex-FRELIMO fighter who had fled Mozambique after being disciplined for corruption. Dhlakama had taken control of the organisation after a bloody shoot-out with his former comrades in mid–1980. The MNR radio resumed broadcasts from the Transvaal where a safe rear base was established for the training of MNR personnel. Supplies were stepped up, delivered through air-drops and, later, by fishing boats and even submarines. For specialist operations South Africa occasionally risked its own military personnel, and one member of the SADF, a Briton called Alan Gingles, was killed in October

1981 while attempting to sabotage the Beira to Mutare (in Zimbabwe) railway line. Evidence of South African involvement became so overwhelming that in 1983 even the US State Department acknowledged, publicly, that the MNR received the bulk of its support from South Africa.

The effectiveness of the MNR increased dramatically once the South Africans took over, reflecting the increased resources available to it and the wider objectives of the South Africans who sought to destabilise Mozambique as a whole. In August 1982 South Africa established MNR bases in Malawi, for the first time, and by early 1983 the MNR was actively attacking targets in nine out of ten of Mozambique's provinces. These included railway lines, roads, ports, and the Beira to Mutare oil pipeline. In addition, in 1982 and 1983 the MNR destroyed 840 schools, 212 health establishments, over 200 villages and 900 shops. In March 1982, as part of its campaign against the MNR, Mozambique appointed military commanders to each of its ten provinces.

Apart from their catastrophic effect on Mozambique, MNR activities have resulted in severe costs to Zimbabwe and Malawi owing to increased transport costs. It was partly in recognition of this, and of the fact that far from being a genuine internal opposition movement, the MNR was one arm of a South African strategy to undermine Mozambique and SADCC, that Zimbabwe sent its own troops into Mozambique. The move was made at the request of the Mozambican government and the troops were used to guard the Beira-Mutare railway and oil pipeline. In 1986 and 1987, as the prospect of Zimbabwean sanctions against South Africa loomed and it became clear that the country would become almost totally dependent on routes through Mozambique, more Zimbabwean troops were sent into Mozambique to help in the war against the MNR. A 1986 estimate suggested that there were 10,000 Zimbabwean troops in Mozambique, and Mozambican government sources confirmed that Tanzanian troops were being brought in to join the counter-offensive. In addition, discussions were held on the establishment of an international military force from either African or non-aligned countries to combat MNR attacks on the main trade routes through Mozambique.

The depradations of the MNR were a major factor in the collapse of Mozambique's foreign earnings from $260 million in 1980 to $90 million five years later. In 1987 the Government was forced to seek the approval of the newly elected People's Assembly for severe economic measures agreed with the International Monetary Fund in exchange for the extension of financial credit to the country. MNR disruption was also one of the main causes of the famine which gripped Mozambique in 1987.

See also **FRELIMO; SADCC; Selous Scouts; South African destabilisation of Mozambique**.

Mofeli, Charles (*Lesotho*)

Founder and leader of the Lesotho United Democratic Party after the 1966 break from the Basutoland Congress Party: see **Lesotho United Democratic Party**.

Mokhehle, Ntsu (*Lesotho*)
Ntsu Mokhehle founded the Basutoland Congress Party (BCP) and became its president in 1952. The formation of this party marked the beginning of modern party politics in Lesotho and it won the 1960 elections. Ntsu Mokhehle was influenced by the ANC of South Africa (despite the fact that he supported the rival PAC in later years) and Kwame Nkrumah's Pan-Africanism. His attitude towards the chieftanship changed from one of nationalist support in the 1950s to hostility as his party developed along more modern lines. In contrast to Chief Leabua Jonathan, who was later to become Prime Minister of Lesotho, he showed a strong hostility to South Africa while conceding the need for a pragmatic relationship. He described himself as a socialist who wanted above all to integrate his country with the rest of black Africa.

The crucial pre-independence elections of 1965 were, however, not won by the BCP, but by the Basotho National Party led by Chief Jonathan and backed by the South Africans. After this Ntsu Mokhehle and his party began to lose influence, and in 1970, after the suspension of the constitution, he was imprisoned. He was released at the end of 1971 and placed under house arrest. In 1974 after renewed violence Ntsu Mokhehle fled Lesotho and went into exile in Zambia.
See also **Basotho National Party; Basutoland Congress Party; Jonathan**.

monarchy (*Swaziland*)
The Kingdom of Swaziland is a monarchy and its present ruler is King Mswati III. However, in more recent years three distinct power centres have come to operate within the country. The first, the monarchy, is dominant. It consists of the monarch, the 'Authorised Person' (a powerful office which, together with the Elders of the Nation, oversees Swazi politics) and the other Elders of the Nation. The other centres of power are the *Liqoqo* or Supreme Ruling Council (now formally abolished by the new king) and the cabinet and Parliament.
See also **'Authorised Person'; Elders of the Nation; Liqoqo; Mswati**.

Mondlane, Eduardo (*Mozambique*)
Eduardo Mondlane was elected as the first President of FRELIMO at its initial congress in September 1962.

Despite a relatively privileged start as the son of a traditional chief, he had very soon reached the limits of the education available to Africans in Mozambique and was forced to travel to South Africa where he completed his secondary education on a scholarship. Following his expulsion from that country for political activities, he won a further scholarship, to the USA, where he obtained a PhD.

His international experience stood him in good stead, and he was offered work with the Trusteeship Department of the United Nations in 1957. He was able to use his United Nations credentials to maintain contact with Mozambican nationalists both inside and outside the country. In 1961 he resigned his post with the UN to devote his time to nationalist politics, and was instrumental in persuading UDENAMO, MANU and

UNAMI to form a joint organisation, FRELIMO, in 1962. Three months later he became its first president.

Mondlane remained President of FRELIMO until he was assassinated by a parcel bomb in 1969. He was succeeded as president by Samora Machel.
See also **FRELIMO; Machel; UDENAMO; UNAMI**.

Moshoeshoe II (*Lesotho*)
King Moshoeshoe II has been King of the Basotho since the independence of Lesotho in 1966. He is a direct descendant of Moshoeshoe I, the founder of the Basotho nation, and was christened Constantine Bereng Seeiso when he was born on 2 May 1938. He was educated in Britain (the Benedictine College of Ampleforth and Corpus Christi College, Oxford University, where he read Politics, Philosophy and Economics) and was installed as paramount chief in 1960; he was recognised as king shortly before independence on 4 October 1966.

Relations between the prime minister, Jonathan, and the king up until the military coup of 1986 were not easy. In 1970 the king was forced into exile in Holland after the constitution had been suspended. He returned in December that year having accepted a proclamation which prohibited him from taking part in politics, while acknowledging him as Head of State.

Since the military coup in January 1986 the position of King Moshoeshoe II has altered. Under the Lesotho Order (1986) all executive and legislative powers were vested in the king although in effect the Military Council governs the country. Nevertheless, the king has increased his political presence and has been building up his power over the past year.
See also **Jonathan; Lekhanya**.

Motlana, Nthato Harrison (*South Africa*)
Dr Nthato Motlana is the chair of the Soweto Committee of Ten (he was elected in 1977) and the leading figure in the Soweto Civic Association which was set up by his Committee. Both these organisations were founded after the Soweto uprising of 1976. Dr Motlana was also Vice-Chairman of the Black Parents' Association and a founder member of the Black Community Programmes. More recently he has been involved in setting up the street committees of the townships. He is a director of a privately-owned clinic in Soweto and a respected local figure.
See also **Committee of Ten; Soweto uprising, 1976; street committees**.

Motsete, Kgaleman T. (*Botswana*)
Founder and first president of the Bechuanaland People's Party: see **Botswana People's Party**.

Movimento das Forças Armadas (*Angola*) See **MFA**.

Movimento de Defesa dos Interesses de Angola (*Angola*) See **MDIA**.

Movimento de Independência Nacional de Angola (*Angola*) See **MINA**.

Movimento Nacional Angolano (*Angola*) See **MNA**.

Movimento Nacional de Resistência de Moçambique (*Mozambique*)
Later renamed *Resistência Nacional Moçambicana*, an anti-FRELIMO force backed by South Africa: see **MNR**.

Movimento para a Independência de Angola (*Angola*) See **MIA**.

Movimento Popular de Libertação de Angola (*Angola*) See **MPLA**.

Movimento Popular de Libertação de Angola-Partido de Trabalho (*Angola*) See **MPLA-PT**.

Moyo, Jason J.Z. (*Zimbabwe*)
Jason Moyo was a close ally of Joshua Nkomo and an important ZAPU leader. He was originally a union leader among railway workers and helped to form the African National Congress in 1957. He accompanied Joshua Nkomo to the Geneva Conference and was described as head of the ZAPU guerrillas based in Botswana and Zambia. He was killed in Lusaka, Zambia, by a letter bomb in January 1977.
See also **African National Congress (Zimbabwe); Nkomo; ZAPU; ZIPRA**.

Mozambique

Official title: People's Republic of Mozambique
Head of state and government: President Joaquim Chissano
Area: 783,030 sq. km
Population: 13,426,604 (official estimate, 1984)
Capital: Maputo
Official language: Portuguese (local languages, including Ronga, Shangaan and Muchope also widely spoken)
GDP per capita: US$152 (1985)
Major exports: shrimps and prawns (29 per cent of total value), cashew nuts (16 per cent), tea (11 per cent)
Currency: Metical (MT) = 100 centavos; MT40 per US$ (Oct. 1986). By August 1987, following pressure from the IMF, the metical had been devalued to about MT400 per US$.
Political parties: *Frente de Libertação de Moçambique* (FRELIMO), the ruling party; opposed by *Resistência Nacional Moçambicana* (MNR).

Mozambique lies on the south-eastern edge of Africa, bordered on its east by the Indian Ocean. A long, narrow country, it shares land borders with six other countries. To the north is Tanzania, to the north-west Malawi and Zambia, to the west Zimbabwe, and to the south, South Africa and Swaziland. Its shape and position makes it of crucial strategic importance to the region. It controls outlets to the sea for all six countries, is pivotal to SADCC, and is regarded by South Africa as an essential target in its attempts to dominate the region. It has three important ports,

and railway lines running largely east to west, linking it to the countries to its west. Much of its road and rail network has been damaged or become dangerous to travel, however, as a result of the depradations of the South African-backed MNR. Mozambique is one of the 'frontline' states.

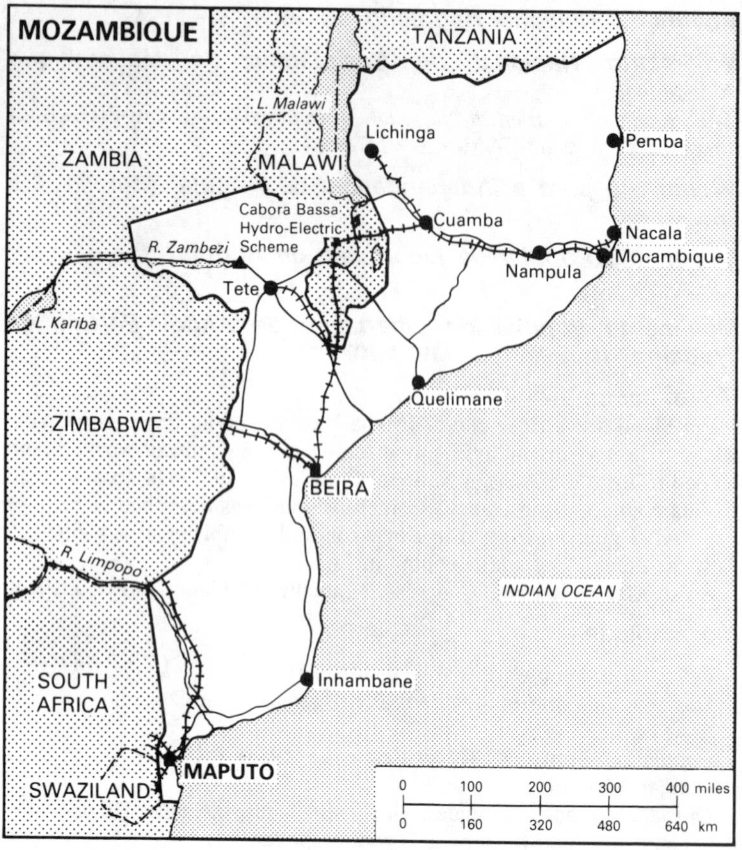

Mozambique is well-watered with 25 major rivers flowing to the Indian Ocean. Always a desperately poor country, Mozambique has few mineral resources and was reliant on the export of migrant workers to the South African gold mines. Its economy was dealt a crushing blow by the departure of the Portuguese settlers at independence. With them went most of the country's skilled people, and they left behind them smashed equipment and ruined estates. Of the working population, over 80 per cent are engaged in agriculture, most of it subsistence. Mozambique exports hydro-electricity to South Africa from the massive Cabora Bassa complex, but most of the payment goes to Portugal which built the scheme. The main export-earning sectors are fishing and agriculture, with prawns, cashew nuts and tea being most important. This dependence on agricul-

ture has meant that recent floods and droughts, combined with the activities of the MNR, have further disrupted an already fragile economy.

After years of a liberation war fought by FRELIMO against the Portuguese colonial regime, the People's Republic of Mozambique, or *República Popular de Moçambique* (RPM) became independent on 25 June 1975. Its constitution provides for power to be exercised by the workers and peasants led by FRELIMO, the last being the leading force of the state and society.

The RPM is a secular, socialist, one-party state, the supreme organ of which is the People's Assembly. The executive head of the RPM is the president, currently Joaquim Chissano, who is also head of the ruling party and the armed forces. The Council of Ministers of the People's Assembly constitutes the government of the RPM.
See also **FRELIMO; MNR; People's Assembly**.

Mozambique African National Union (*Mozambique*)
One of the constituent parties of FRELIMO at its foundation: see **MANU**.

Mozambique Democrats (*Mozambique*)
Following the MFA coup in Portugal in 1974, restrictions on political activity in Mozambique were eased and a number of political parties were formed. One was the Mozambique Democrats, a progressive organisation which included a number of clandestine FRELIMO activists in its membership. The party worked to try to persuade white people in Mozambique that FRELIMO would not be hostile to them and that they had nothing to fear from a FRELIMO government, but that an end to hostilities would be in everyone's interests.
See also **FRELIMO; MFA**.

Mozambique National Resistence (*Mozambique*)
Resistência Nacional Moçambicana, an anti-FRELIMO force backed by South Africa: see **MNR**.

Mozambique Revolutionary Committee (*Mozambique*)
A former political rival to FRELIMO in Mozambique: see **COREMO**.

Mpakati, Attati (*Malawi*)
Dr Attati Mpakati was the president of the Socialist League of Malawi. He was murdered in Harare, Zimbabwe in March 1983.
See also **Socialist League of Malawi**.

MPC (*Namibia*)
The Multi-Party Conference (MPC) is the latest in a series of attempts by South Africa since the 1970s to promote client politicians and produce sympathetic 'internal' administrations. Following the collapse of the State Council a conference of 17 small internal parties was called in September 1983 to formulate alternative independence proposals to those contained in UN Security Council Resolution 435. The Multi-Party Conference (MPC), as it became known, was eventually launched on 12 November 1983 with only seven participating parties, having been denounced as a South African Government ploy by parties on both the left and the right.

The constituent parties of the MPC were the DTA, SWANU, the Damara Council, SWAPO-D, the NCDP, the Labour Party and the Rehoboth Liberation Front, none of which commanded mass support. The MPC was later joined by SWANP, but left by the NCDP which denounced it for working with the security police and revealed that it was funded by the Hans Seidl Foundation, an organisation closely associated with the right-wing Christian Socialist Union of West Germany.

The MPC's decision to draw up a new constitution for Namibia counter to UN Security Council Resolution 435 led to growing internal dissention and in March 1984 the Damara Council withdrew, declaring support for Security Council Resolution 435 and denouncing the MPC as an anti-SWAPO front. Opposition grew within SWANU leading to a split and the ousting of the existing leadership followed by the withdrawal of all but a rump of SWANU from the MPC. The MPC was embarrassed at the Lusaka talks with SWAPO in May 1984 when a majority of parties, including some constituent parties of the MPC, crossed the floor to join the SWAPO delegation.

Despite set-backs for the MPC, South Africa pressed on with the formation of what was termed the 'Transitional Government of National Unity' (TGNU) based on the MPC's constitutional proposals.
See also **AG8; AG9; Damara Council; DTA; interim government; Lusaka talks on Namibia, 1984; NCDP; State Council; SWANU; SWAPO; SWAPO-D; Transitional Government of National Unity; UN Security Council Resolution 435**.

Mpho, Motsamai (*Botswana*)

Motsamai Mpho was one of the founders of the Botswana People's Party (then the Bechuanaland People's Party). He became its first secretary-general in 1960. When he was expelled from the BPP in 1962 he founded the Botswana Independence Party.
See also **Botswana Independence Party; Botswana People's Party**.

MPLA (*Angola*)

The MPLA (*Movimento Popular de Libertação de Angola*), or People's Movement for the Liberation of Angola, the country's first nationalist organisation, was founded in December 1956. Two organisations, the *Partido da Luta Unida dos Africanos de Angola* (PLUA) and the *Movimento para a Indepêndencia de Angola* (MIA) joined forces to form it, and were later joined by the *Movimento de Independência Nacional de Angola* (MINA).

The MPLA began organising an armed struggle against Portuguese colonialism in April 1961, following mass arrests of its leaders by the Portuguese security police, PIDE. In 1962 the MPLA held its first national conference, electing Agostinho Neto to replace Mário de Andrade as president. By 1963, as the guerrilla war grew, the MPLA had been forced to close its external headquarters in Zaire (where President Mobutu favoured the rival UPA) and move to Congo-Brazzaville. From the beginning the MPLA gave priority to the mass mobilisation of

communities, building up grass-roots support and politicising the population in areas under its control.

The MPLA found itself under attack from two rival nationalist groups almost from the start, as it struggled to establish itself in Angola. The FNLA (and its predecessor the UPA) and UNITA adopted different tactics to those of the MPLA and attacked MPLA fighters at frequent intervals.

The MPLA organised subsidiary organisations to represent and mobilise different groups in the Angolan population. The *Organização das Mulheres de Angola* (OMA) organised among women, the *União Nacional dos Trabalhadores de Angola* (UNTA) represented workers, *Juventude da MPLA* (JMPLA) was the youth organisation, the *União dos Estudantes Angolanos* (UEA) organised students and the *Serviço de Assistência Médica do MPLA* (SAM) was the medical wing of the movement.

Like many other liberation movements, the MPLA suffered periodic splits as a result of factionalism and other stresses in its ranks. The 1963 da Cruz split resulted from Neto's election as president and allowed the rival GRAE to win recognition from the OAU. Ironically, the MPLA was at its most divided in 1973, shortly before the Portuguese coup, suffering from Daniel Chipenda's so-called 'Eastern Revolt' and severe military pressure from Portuguese forces. A conference in 1974 consolidated the party, however, reinforcing Neto's position as president and electing a Central Committee and Political Bureau.

By November 1974 the MPLA had signed a ceasefire with the Portuguese, and in January 1975 participated with the FNLA and UNITA in the Alvor independence conference. The political situation in Angola subsequently deteriorated into civil war. Massive arming of the FNLA and UNITA by the CIA and Western countries was followed by the intervention of mercenaries in the north of the country, the South African invasion of the south of the country and the MPLA's request to Cuba to send its forces to assist it in its struggle for survival.

In December 1977 the First Party Congress transformed the MPLA into the MPLA-PT (*Partido de Trabalho*) or MPLA-Workers' Party with formalised membership selection, training of militants and division of the party leadership structures into different departments.

See also **Chipenda; CONCP; de Andrade; Eastern Revolt; FAPLA; FNLA; GRAE; MFA; Neto; South African invasions; Transitional Government; UNITA; UPA**.

MPLA-PT (*Angola*)

On 10 December 1977 the MPLA was transformed into the MPLA-PT or MPLA-Workers' Party, officially the *Movimento Popular de Libertação de Angola-Partido de Trabalho*, part of the transformation of the liberation movement into a Marxist-Leninist party.

See also **MPLA**.

Mswati III (*Swaziland*)
King Mswati III, formerly Crown Prince Makhosetive, was crowned king of Swaziland on 25 April 1986 at the royal residence, *Lusaseni*, near the capital, Mbabane. He was born in April 1968, the second youngest son of King Sobhuza, and one of some 70 sons. He was completing his secondary education at Sherborne College in Dorset, England, when recalled for his coronation. In 1983 he was ceremonially introduced to the Swazi people as their future king, helping to confirm Queen Regent Ntombi – his mother – in power.

King Mswati came to the throne early (it was planned to crown him when he was 21) in order to try to put an end to the power struggles which followed King Sobhuza's death in 1982. Soon after ascending the throne the new king moved to consolidate his power by dismissing Prime Minister Prince Bhekimpi Dlamini and appointing Sotja Dlamini in his place. He also dissolved the *Liqoqo* (Supreme Ruling Council).
See also **Dlamini, Bhekimpi; Dlamini, Sotja; *Liqoqo*; Ntombi**.

mtfanenkosi (*Swaziland*)
An *mtfanenkosi* is a prince or a male child of a king (the plural form is *bantfanenkosi*).

Mueda massacre (*Mozambique*)
In the late 1950s peasant farmers of the Mueda plateau near the northern Mozambican border with Tanzania formed the African Voluntary Cotton Society of Mozambique (SAAVM), a cotton-growing co-operative which allowed them to avoid forced labour while keeping the profits of their cotton growing for themselves. The formation of the co-operative had been stimulated through contact with similar developments in Tanzania, and through such contacts, as well as through their own experience in the co-op, the members became politicised and began to demand changes from the administrator.

On 16 June 1960, following the arrest of some co-op members, a large crowd gathered to protest to the provincial governor who had arrived on a visit to the area. He responded by ordering his troops to fire on the crowd, killing almost 600 people. SAAVM subsequently collapsed, but the experience radicalised many of those involved and several of the leaders became central figures in the subsequent nationalist movement.
See also **forced labour; Nkavandame**.

Mugabe, Robert Gabriel (*Zimbabwe*)
Robert Mugabe became the first prime minister of independent Zimbabwe in 1980, after his party, ZANU-PF, won the first free elections in the country. It also won the 1985 elections overwhelmingly, gaining 63 of the 79 contested seats and some 77 per cent of the total vote.

Robert Mugabe was a founder and the secretary-general of the original ZANU party in 1963. He was detained by the rebel Ian Smith's regime for ten years from 1964 in the Salisbury (as the capital, Harare, was then known) Prison. Before this he was Publicity Secretary of the National Democratic Party and then of the other main nationalist party, ZAPU.

In 1974 he was released from prison and became a founder member and the joint leader of the Patriotic Front with Joshua Nkomo. In October 1976 he was elected president of ZANU and commander-in-chief of its army, ZANLA.

Since independence Robert Mugabe has consolidated his power and he is widely perceived as highly intelligent, articulate and charismatic. He is a strong leader although he has an academic style and is softly spoken. He has led Zimbabwe in the direction of a mixed economy with a proclaimed socialist ideology. In 1987, after years of mutual hostility, ZANU and ZAPU united, creating an effective one-party state along the lines envisaged by Mugabe.
See also **National Democratic Party; Nkomo; Patriotic Front; ZANLA; ZANU; ZAPU**.

Muldergate (*South Africa*)
A common term for the so-called 'Information Scandal' which broke in the late 1970s and ruined the career of Vorster's expected successor, Connie Mulder, then Minister of Information: see **Information Scandal**.

musseques (*Angola*)
The popular name for the poor suburbs of colonial Luanda. The term derives from Kimbundu and means 'sandy place', a reference to the sandy nature of the hills surrounding Luanda on which the houses were built. The *musseques* were home to most black inhabitants of Luanda, while Portuguese settlers lived in the more prosperous areas of the city.

Mutuswa, Solomon (*Zimbabwe*)
The real name of Rex Nhongo, a prominent commander of the nationalist forces: see **Nhongo**.

MUZ (*Zambia*) See **Mineworkers' Union of Zambia**.

Muzorewa, Abel Tendekayi (*Zimbabwe*)
Bishop Abel Muzorewa is one of Zimbabwe's best-known religious politicians. He was prime minister of Zimbabwe-Rhodesia for a year in 1978 – the key member of the rebel leader Ian Smith's internal settlement for Rhodesia. His political career has gone hand in hand with his religious calling. He was consecrated bishop in 1968 and became resident bishop for the United Methodist Church (Rhodesian area). In the same year he was elected leader of the African National Council (later reformed into the UANC). In the 1980 elections he was elected member of parliament for Mashonaland East. His party, the UANC, won only three seats in these elections, and it won no seats in the 1985 elections.

Bishop Muzorewa is seen by most Zimbabweans as a collaborator with the white minority Smith regime, as a result of his involvement in the internal settlement. In 1984 he was detained by Robert Mugabe's government for some months after he had visited Israel and South Africa. Many of his former armed supporters (auxilliaries) are alleged to have gone to South Africa and he was suspected of conspiring against the ZANU(PF) government.

See also **African National Council; internal settlement; Smith; UANC; Zimbabwe-Rhodesia**.

Mwafrika (*Tanzania*)
Mwafrika is the name given to TANU's Swahili newspaper. It was launched in 1977.
See also **TANU**.

Mwalimu (*Tanzania*)
Mwalimu is a title used to signify respect in Tanzania. It means 'the teacher' in Swahili and is often used to refer to the former president, Julius Nyerere, who was once a schoolteacher.
See also **African Association; Nyerere**.

Mwinyi, Ali Hassan (*Tanzania*)
Ali Hassan Mwinyi is the new President of Tanzania, who took over from President Nyerere in 1985. In elections held on 27 October Ali Hassan Mwinyi, the sole candidate, received 92.2 per cent of the votes. He had already been adopted by the ruling party, *Chama cha Mapinduzi* (CCM), on 15 August 1985 as the presidential candidate, and had the support of Julius Nyerere.

The new president had been Vice-President of the Republic and President of Zanzibar since early 1984. He is a quiet man, once a schoolteacher, and his politics tend to seek consensus rather than confrontation. He served in the Zanzibar Ministry of Education from 1963 until the revolution and the union with Tanganyika the following year, becoming Permanent Secretary to the Minister. He became Minister of State in the President's Office in the Tanzanian Cabinet in 1970, and has held various senior government posts ever since. As President of Zanzibar he was effective in handling the potentially explosive situation surrounding the resignation of the former President of Zanzibar, Aboud Jumbe. He sought reconciliation with island dissidents while liberalising political and economic measures thereby defusing the tensions. In August 1984 he was elected Vice-Chairman of the CCM.

Mwinyi will have more limited power than Nyerere had while president, since the 1984 amended constitution makes the president more accountable to parliament and prevents him from holding the post for more than two consecutive terms. Furthermore, Nyerere was concurrently chairperson of the ruling party, the CCM, and he has said he plans to retain that post for some time. Ali Hassan Mwinyi is likely to continue along the same political lines as Julius Nyerere, not abandoning socialism but making some changes to seek greater efficiency.
See also **Nyerere**.

N

NAC (*Malawi*) See **Nyasaland African Congress**.

NACTU (*South Africa*)
NACTU (the National Confederation of Trade Unions – known as CUSA-AZACTU until April 1987) was formed in 1986 by merging CUSA, a union grouping with a policy of black leadership, and AZACTU, a strongly black consciousness group of unions which admitted only black members. NACTU was formed in response to the emergence of the non-racial COSATU, which is far bigger and more influential. NACTU, in contrast, holds an 'anti-racist' position. CUSA was one of the main groups involved in the negotiations to form COSATU, but felt it could not compromise on the issue of black leadership. CUSA-AZACTU claimed to have 23 affiliates in 1986. NACTU's leaders are Phirowshaw Camay and Pandelani Nefolovhodwe.
See also **AZACTU; COSATU; CUSA**.

NAFCOC (*South Africa*)
The National African Federated Chambers of Commerce (NAFCOC) represents African business people in South Africa and works to mobilise African capital and to build an African middle and capitalist class. Founded as the National African Chambers of Commerce (NACOC) in 1964, it became NAFCOC in 1969 when it became a federation of regional groups based on the provinces and the bantustans.

Since 1969 NAFCOC has had an equivocal attitude to apartheid, using some provisions such as group areas to seek exclusive trading rights for African capitalists, and opposing other provisions which discriminate against them. Its essentially pragmatic approach has also been reflected in its close relationship with the state under P. W. Botha and with white capitalists, while attempting to mobilise African capital and use a nationalist ideology to attract support and build an African class of entrepreneurs.
See also ***Afrikaanse Handelsinstituut; ASSOCOM; bantustans***.

Naicker, G.M. 'Monty' (*South Africa*)
Dr Naicker was President of the Natal Indian Congress for many years from 1945. He signed the Xuma-Dadoo-Naicker Pact in 1947 which set up a working alliance between the ANC and the South African Indian Congress.

See also **Dadoo; Natal Indian Congress; South African Indian Congress; Xuma-Dadoo-Naicker Pact**.

n

Nakuru Agreement *(Angola)*
Reached after five days of talks in Nakuru, Kenya, among Holden Roberto, Agostinho Neto and Jonas Savimbi, and with Kenyan President Jomo Kenyatta in the chair, the agreement was an attempt to save the Transitional Government in Angola. Faced by the danger of the transitional arrangements breaking down, the leaders of the three nationalist movements agreed to renounce the use of force against each other, and attempted to find an agreed basis for peaceful progress to independence. In the event, and despite the agreement, armed hostilities among the groups subsequently escalated into open warfare.
See also **Alvor Agreement; FNLA; MPLA; Neto; Roberto; Savimbi; Transitional Government; UNITA**.

Nama Chiefs' Council *(Namibia)*
Constituent group of NCN: see **National Convention of Namibia**.

Namib Desert *(Namibia)*
The Namib Desert runs the length of the Namibian coast, covering about 20 per cent of its total land area, and is one of the driest deserts in the world. The word Namib derives from the Nama language and means 'enclosure'. The desert was so named because for centuries it formed a natural barrier against intrusion from the sea, and was one of the reasons for the relatively late colonisation of Namibia.

In a statement published in 1968, SWAPO gave this reason for its decision to use the word as the basis for a new name, Namibia, for the country formerly known as South West Africa. The statement quoted a poem about the Namib by Noel Mostert:

> It was this Coastal Namib Desert
> Running like a narrow white moat
> Along almost the entire length of the
> South West shore, that helped seal off
> The interior and those living there
> From any prying new-comers.

See also **German colonialism; Namibia; SWAPO**.

Namibia

Official title: Namibia (South African preferred title: South West Africa/Namibia)
Head of state and government: Legal head – Bernt Carlsson, UN Commissioner for Namibia; Popular head – Sam Nujoma, President of SWAPO (recognised by UN as authentic representative of Namibian people); Head of South African occupation forces – Louis Pienaar, Administrator General for the Territory of South West Africa
Area: 824, 269 sq. km

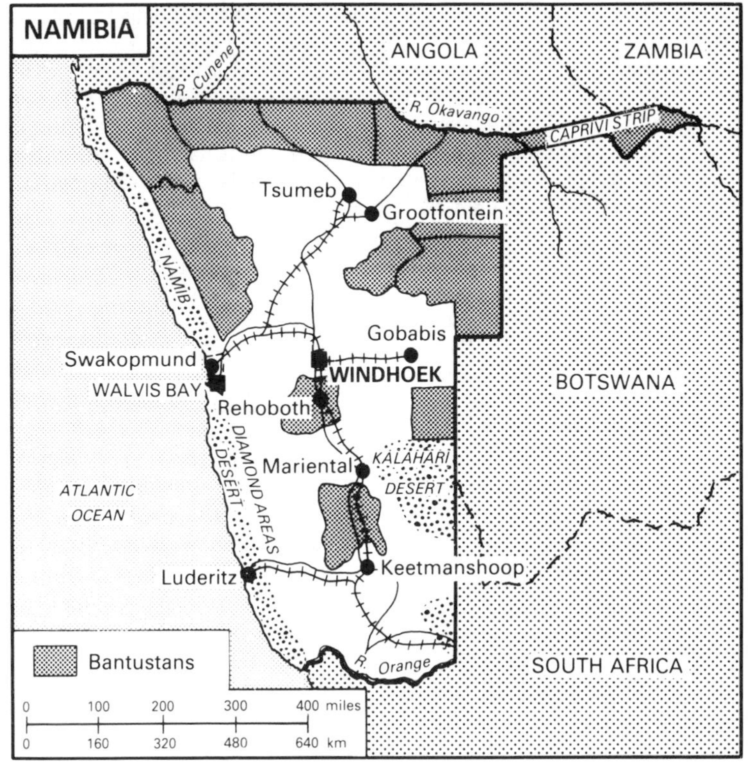

Population: 1,033,196 (official census, 1981); estimated at 1.5m, 1985
Capital: Windhoek
Official languages: English (SWAPO); English and Afrikaans (South Africa); German and many local languages also spoken
GDP per capita: US$1,162 (1984)
Major exports: uranium (30 per cent of total value), diamonds (22 per cent), other minerals (19 per cent), fish, meat and cattle (15 per cent)
Currency: South African rand (R) = 100 cents; R2.07 per US$ (July 1987)
Political parties: Main parties are SWAPO (South West Africa People's Organisation of Namibia), South West Africa National Union, Democratic Turnhalle Alliance, National Party of South West Africa.

Adopted by the UN in 1968 and accepted everywhere except in South Africa, the name 'Namibia' was formulated by SWAPO. It is derived from the Namib Desert which, for centuries, saved the country from colonisation. The country has also been known as South West Africa and

German South West Africa. It is termed South West Africa/Namibia by South Africa.

Namibia lies on the south-western edge of Africa, bordered on the west by the Atlantic Ocean and on the east by Botswana. It shares a common border with South Africa in the south, and with Angola in the north. Its north-east corner extends eastwards in the narrow Caprivi Strip which shares borders with four countries, Angola, Botswana, Zambia and Zimbabwe. The Namib Desert runs along the entire western edge of the country and is uncrossed by a single perennial river. Much of Namibia is desert or semi-desert, the wettest parts being the well-populated north.

Namibia's economy is based on extractive industries. Mining contributes about a quarter of GDP and is followed by agriculture and fishing at about one tenth. The economy is highly dependent on foreign trade, exporting about 90 per cent of goods produced locally, and importing about 85 per cent of those consumed. It is a net food importer. Within Namibia, material inequalities are extreme, with most black people reliant on small incomes from migrant labour or subsistence agriculture, in contrast to widespread affluence among whites. Massive spending on 'Homeland' bureaucracies, and on developing the infrastructure in the north of the country to facilitate military activities, has resulted in a massive debt burden for the country. SWAPO has indicated that an independent Namibia will not accept liability for the debt.

Originally colonised by Germany, and subsequently administered by South Africa, Namibians have been resisting non-indigenous rulers for over a century. Namibia's contemporary history has been dominated by SWAPO's liberation struggle against the illegal South African occupation. The war, and the intense oppression of the population by South Africa, has forced almost 10 per cent of the population into exile, and many have died or been maimed. South African militarisation of the north of the country has been on such a scale that it is claimed to have the highest proportion in the world of military and paramilitary personnel relative to the civilian population.

The current internal administration is termed the Transitional Government of National Unity (dominated by the Multi-Party Conference), an administration formed from those politicians prepared to collaborate with South Africa. The legal head of state, however, is the UN Commissioner for Namibia, acting on behalf of the UN Council for Namibia, the legal administering body. Most Namibians would choose a government led by SWAPO and its president, Sam Nujoma, but are denied both independence and a free choice of leaders and government.
See also **Contact Group; MPC; SWAPO; UN Security Council Resolution 435**.

Namibia Christian Democratic Party (*Namibia*) See **NCDP**.

Namibia Democratic Coalition (*Namibia*)
A coalition of three smaller groups, the (Rehoboth) Liberation Front, the National United Democratic Organisation Progressive Party and the

Liberation Party, the Namibia Democratic Coalition is led by Hans Diergaardt.

Namibia Democratic Turnhalle Party (*Namibia*)
Constituent party of the Democratic Turnhalle Alliance: see **DTA**.

Namibia National Convention (*Namibia*)
Association of parties formed in 1975 to replace discredited NCN: see **National Convention of Namibia**.

Namibia National Front (*Namibia*)
Formed in 1977 as an anti-SWAPO, anti-Turnhalle, umbrella group: see **National Convention of Namibia**.

Namibian National Students' Organisation (*Namibia*)
The Namibian National Students' Organisation (NANSO) held its first public meeting, attended by 500 people, in mid–1986. Speaking at the meeting, NANSO president Paul Katenga said, 'We have had enough of our schools being used as recruitment centres for a colonial army . . . of indoctrination, false propaganda, brainwashing and injustices.'
See also **SWAPO Youth League**.

Namibia Peace Plan 435 (*Namibia*)
The Namibia Peace Plan 435 (NPP435) was formed in November 1986 by a group of white people in Namibia as a study and contact group for the implementation of UN Security Council Resolution 435. Headed by former Federal Party leader, Bryan O'Linn, its aim was to encourage a settlement on the basis of Resolution 435. O'Linn stressed, at the time of its formation, that it was not a political party and would seek to supplement the activities of existing pro-Resolution 435 organisations. While it was a non-racial group, he added, its executive was all-white because the group's main objective was to influence the white population in Namibia.
See also **Federal Party; UN Security Council Resolution 435**.

Namibia People's Liberation Front (*Namibia*)
One of the parties in the Democratic Turnhalle Alliance: see **DTA**.

Namwi Foundation (*Namibia*)
A propaganda organisation linked to the military in Namibia: see **SWATF**.

NANSO (*Namibia*) See **Namibian National Students' Organisation**.

NAPDO (*Namibia*)
National African People's Democratic Organisation, one of the six parties which formed the NCN in 1971, and subsequently joined SWAPO: see **National Convention of Namibia**.

Nasionale Party (*South Africa*) See **National Party**.

Natal Indian Congress (*South Africa*)
The Natal Indian Congress (NIC) was founded by Mahatma Gandhi in

1894 and developed the tactic of passive resistance which was used later in the struggle for Indian independence. The NIC, together with the Transvaal Indian Congress formed the South African Indian Congress in 1920, and this later became part of the Congress Alliance. The NIC was revived in 1971 by Mewa Ramgobin because state harassment had paralysed the activities of the South African Indian Congress. In recent years the NIC has been active in campaigning against the government-organised South African Indian Council (SAIC) and in setting up the UDF to oppose apartheid. The current president of the NIC is George Sewpersadh.
See also **SAIC; Sewpersadh; South African Indian Congress**.

Natal Option (*South Africa*) See ***Indaba***.

National African Federated Chambers of Commerce (*South Africa*) See **NAFCOC**.

National African People's Democratic Organisation (*Namibia*) See **NAPDO**.

National Coalition Party (*Mozambique*)
The successor to Gumo as the main opposition to FRELIMO in the post-coup period in Mozambique, the National Coalition Party brought together the main anti-FRELIMO groups. Its leaders included Lázaro Nkavandame and Rev. Uria Simango.
See also **FRELIMO; Gumo; Nkavandame; Simango**.

National Confederation of Trade Unions (*South Africa*) See **NACTU**.

National Conservative Party (*South Africa*)
The National Conservative Party (NCP) was formed by Connie Mulder and his followers in 1979, after his expulsion from the National Party. Mulder had been the favourite to succeed John Vorster as leader of the National Party, and was P. W. Botha's main opponent in the leadership battle which followed Vorster's decision to resign his premiership. The so-called 'Information Scandal', or 'Muldergate' as it later became known, was used to discredit both Mulder and Vorster and probably cost Mulder victory in the leadership contest.

The NCP failed to win any seats in the 1981 general election, and in 1982 merged with the newly-formed Conservative Party under the leadership of Andries Treurnicht, another former cabinet minister and leader of the National Party in the Transvaal.
See also **Botha, P. W.; Conservative Party; Information Scandal; National Party; Treurnicht; Vorster**.

National Convention of Namibia (*Namibia*)
Formed on the initiative of SWAPO in November 1971, the NCN comprised six organisations united by a common opposition to apartheid and the South African occupation of Namibia. Its component groups were SWAPO, NUDO, SWANU, Voice of the People, the *Rehoboth*

Volksparty and NAPDO (the National African People's Democratic Organisation). It also had the support of the largest churches in Namibia and of the Herero, Damara and Nama Chiefs' Councils.

The NCN had ceased to be active by early 1974, and tensions were generated by the decision of the Herero leader, Clemens Kapuuo to participate in the Turnhalle Constitutional Conference. In July 1974 Kapuuo tried to promote the NCN internationally as an alternative to SWAPO. Six months later he denounced SWAPO as an Ovambo organisation, and advocated a political system based on apartheid divisions. Shortly thereafter, on 17 January 1975, SWAPO announced its withdrawal from the NCN and immediately set about forming a replacement organisation based on common opposition to South Africa's Turnhalle initiative.

The creation of the Namibia National Convention (NNC), was announced in March 1975. Organisations involved included SWAPO, SWANU, NAPDO and the *Rehoboth Volksparty*. In August 1976 the *Rehoboth Volksparty* disbanded and joined SWAPO. In October the four largest groups in the south of the country, led by Pastor Hendrik Witbooi, did likewise. They were followed into SWAPO by NAPDO in November and by the Herero Royal House in April 1977. SWANU remained aloof. With most of the constituent parties of the NNC now in its own ranks, SWAPO announced its withdrawal in December 1976.

The rump of the NNC joined together in April 1977 to form the Namibia National Front (NNF), dominated by SWANU and united by a common opposition to SWAPO and the Turnhalle initiative. Apart from SWANU, the NNF included the Damara Council, the National Independence Party (NIP) and several smaller groups. The white Federal Party joined the NNF in March 1978, but left again in June the following year. The history of the NNF was marked by a series of splits and disagreements. In May 1978 it entered into a working arrangement with SWAPO-D. Attempts to merge the two at a congress in April 1980 collapsed in acrimony and two other groups, the NIP and the Rehoboth Group left the NNF. Only SWANU, the 'Coloured' Progressive Party, the Damara Council and the Mbanderu Group remained. In October 1980 the NNF dissolved, to all intents and purposes. At the Lusaka talks on Namibia in 1984, SWANU and the Damara Council joined the SWAPO side and subsequently entered into an informal alliance with SWAPO.

See also **Damara Council; DTA; Federal Party; Kapuuo; Lusaka talks on Namibia, 1984; NUDO; SWANU; SWAPO**.

National Democratic Party (*Zimbabwe*)

In 1961 Joshua Nkomo was called back from campaigning overseas to take over as president of the National Democratic Party (NDP). The party, the first mass nationalist movement in Zimbabwe, was formed in January 1960 by a group including Michael Mawema as a caretaker party for the African National Congress which had been banned. Its programme was similar: majority rule and an end to the white-dominated Central African Federation.

n

Sir Edgar Whitehead, then prime minister of Southern Rhodesia, announced that Joshua Nkomo would not be arrested on his return and he was greeted by thousands of supporters in Harare. The NDP rejected the 1961 constitutional conference proposals made by Duncan Sandys, Secretary of State for Commonwealth Relations in the British government. However, Joshua Nkomo had at first gone along with the new constitution, which, in effect, guaranteed white rule for the forseeable future, and, although he and his party soon denounced it, the constitution became law in November 1962. A campaign of non-cooperation developed and there was some urban violence. In December 1961 the NDP was banned and ZAPU, the Zimbabwe African People's Union, was formed in its place.
See also **African National Congress (Zimbabwe); Central African Federation; Nkomo; ZAPU**.

National Federation of Trade Unions (*South Africa*)
The National Federation of Trade Unions was formed on 1 April 1987 from the original eight affiliates of TUCSA. It claims to have a membership of 150,000.
See also **TUCSA**.

National Forum (*South Africa*)
The National Forum was launched in June 1983 in Hammanskraal north of Pretoria by some 500 delegates representing over 100 organisations. It is a black consciousness organisation set up after the formation of the UDF in order to provide a similar umbrella group but with a distinct political philosophy based on black consciousness. AZAPO was the driving force behind its formation and the National Forum set out its aims in a 'Manifesto of the Azanian People'. This reaffirmed that the struggle was against 'racial capitalism' and committed its adherents to 'worker control of the means of production, distribution and exchange'. It committed the National Forum to 'anti-sexism' as well as opposition to imperialism and racism, and resolved to promote 'anti-collaboration' rather than the original 'non-collaboration' with the ruling class and its allies.
See also **AZAPO; Black Consciousness; UDF**.

National Front for the Liberation of Angola (*Angola*) See **FNLA**.

National Independence Party (*Namibia*)
A small anti-SWAPO party which joined the NNF in 1977: see **National Convention of Namibia**.

National Industrial Development Corporation of Swaziland (*Swaziland*)
The National Industrial Development Corporation of Swaziland (NIDCS) is the government body officially concerned with developing industrialisation. It was set up in 1971 as a 95 per cent government-owned parastatal organisation. However NIDCS began to make losses – accumulated losses

had reached more than SL (lilangeni) 20 million by 1985. A report on the corporation was sponsored by the European Investment Bank and this concluded that the parastatal was 'a lame duck within the financial system of Swaziland', having lost its 'capital, leadership and purpose' and 'being without new cash resources'. As a result a new company, the Swaziland Industrial Development Company (SIDC) was launched on 3 September 1986 to take over much of the work of NIDCS, particularly in investments. SIDC is to be a private company with the government holding only 37.5 per cent of its shares.

See also **SIDC**.

National Intelligence Service (*South Africa*) See **NIS**.

National Party (*South Africa*)

The roots of the National Party lie in the defeat of the two Afrikaner republics by Britain in the South African War of 1899 to 1902. It had been a bitter war, and despite relatively generous peace terms, Afrikaners resented the British domination which followed in South Africa. It was in response to this that Afrikaner nationalism was fostered by intellectuals to mobilise the *volk*.

In 1934, the National Party under Albert Hertzog, joined with Jan Smuts' South African Party to form the United South African National Party (better known as the United Party). A rump broke away to form the *Gesuiwerde Nasionale Party* (Purified National Party). This party later joined with other elements of Afrikaner nationalism to form the *Herenigde Nasionale Party of Volksparty* in 1940, (HNP/V or Reunited National Party or People's Party) and it, together with the milder Afrikaner Party (formed by a breakaway from the HNP/V in 1941), won the May 1948 general election in South Africa (in a poll confined to 'White' and a few so-called 'Coloured' voters). In 1951 the two parties merged to form the National Party which has ruled South Africa ever since.

As well as being the party of Afrikaner nationalism, the National Party originated the apartheid system which has made South Africa notorious throughout the world. The party progressively removed what limited political rights black South Africans had shared with their white compatriates, and conferred on them, instead, rights in separate apartheid structures. It entrenched Afrikaner power in South Africa, and promoted the growth of Afrikaner capitalists, at the expense of the black population who were forged into a dispossessed proletariat to provide cheap labour for the country's mines, factories and farms.

Beginning in the late 1960s, the National Party began to soften its exclusivist image in an attempt to attract conservative English speaking white voters. This it did with increasing success, winning 57 per cent of the vote in the ('White') general election of 1981, a new high, and two thirds of the votes for its new constitutional proposals in the constitutional referendum (again 'White' only) of 1983.

Since 1948 the leaders of the National Party (and hence prime ministers of the country) have been D. F. Malan, J. G. Strydom, H. F. Verwoerd, B. J. Vorster and P. W. Botha. The 1983 constitution established an executive presidency to replace the position of prime minister. After over 25 years of political unity, internal stresses in the party led to an open power struggle in the mid-1970s over who should succeed the then prime minister, John Vorster. The party divided into a so-called *verligte* (enlightened) camp of reformists and a *verkrampte* (loosely translated, narrow-minded) or reactionary camp. The favoured candidate of the reformists, P. W. Botha was victorious in the party leadership elections of 1978 and became prime minister in succession to Vorster.

Botha used his position as party leader to continue the campaign against those opposed to his reformist policies, and in 1982 eighteen former National Party MPs resigned and formed the opposition Conservative Party. Together with the *Herstigte Nasionale Party* (HNP), formed by four former National Party MPs after a smaller schism in 1969, the Conservative Party presented the first credible electoral threat from the right of the National Party since its formation. In the 1987 elections the National Party share of the vote dropped to 52.3 per cent as the Conservative Party drew away some of its supporters and became the main opposition party.

See also **apartheid; Botha, P. W.; Conservative Party; constitution of 1983; HNP; total strategy; United Party; *verkrampte*; *verligte*; Verwoerd; *volk*; Vorster**.

National People's Party (*South Africa*)
The National People's Party led by Amichand Rajbasani has emerged as the dominant party in the House of Delegates – the chamber of parliament reserved for 'Indians' under South Africa's new constitution. In the election held on 28 August 1984 for the House of Delegates the party won 29,930 votes and 18 seats. Many voters boycotted this election and the UDF was set up in order to campaign against participation. The turnout was very low, 20 per cent overall.
See also **constitution of 1983; House of Delegates**.

National Service of Popular Security (*Mozambique*)
Mozambique's internal security organisation: see **SNASP**.

National Union for the Total Independence of Angola (*Angola*) See **UNITA**.

National Union of Angolan Workers (*Angola*) See **UNITA**.

National Union of Mineworkers (*South Africa*) See **NUM**.

National Union of Namibian Workers (*Namibia*) See **NUNW**.

National Union of South African Students (*South Africa*) See **NUSAS**.

National Union of Tanganyika Workers (*Tanzania*)
In 1964 the National Union of Tanganyika Workers' Act allowed the National Union of Tanganyika Workers (NUTA) to exist as the only legal trade union. This took over from a divided Tanganyika Federation

of Labour, some members wanting closer links with TANU, Tanganyika's political party, and others rejecting this political control of workers' organisations. It was replaced in 1978 by the Union of Tanzania Workers (JUWATA).
See also **Tanganyika Federation of Labour; Union of Tanzania Workers**.

National Union of Tanganyika Workers Act (*Tanzania*)
Legislation controlling trade unions in Tanganyika: see **Tanganyika Federation of Labour**.

National United Democratic Organisation Progressive Party (*Namibia*)
One of the groups in the Namibia Democratic Coalition: see **Namibia Democratic Coalition**.

National Unity Democratic Organisation (*Namibia*) See **NUDO**.

Natives Land Act of 1913 (*South Africa*)
The 1913 Natives Land Act laid the foundation for apartheid by retaining most of South Africa's land for the 'White' community. It prohibited Africans, except those in the Cape, from acquiring land outside particular 'native reserve' areas, without special permission from the Governor-General. These reserves constituted 7.3 per cent of the land area of the country (about nine million hectares). The act also restricted the number of African families allowed to remain on 'White'-owned farms.
See also **apartheid, Natives Trust and Land Act of 1936**.

Native Trust and Land Act of 1936 (*South Africa*)
The 1936 Native Trust and Land Act incorporated the 1913 Natives Land Act and extended it. It is a cornerstone of the government's apartheid policy and divides South Africa's land into racially exclusive areas, giving minority 'Whites' 87 per cent of the land area. This basic land division was retained in legislation passed under the National Party government after 1948. Amendments in 1954 and 1964 tightened up the clauses of the act, and the government's current policy of bantustans and 'independent Homelands' is directly based on the 'native reserves' created in 1913.

The 'reserves' had been officially designated as the only areas in which Africans could acquire land. The 1936 act provided for an additional area of land to be added to them – some 6.2 million hectares. This expanded them to 13 per cent of the total land area. The act also extended the provisions of the 1913 act to the Cape province and laid down stricter controls on the numbers of Africans allowed on 'White' farms.
See also **apartheid; 'Homelands'; Natives Land Act of 1913**.

Nats (*South Africa*)
The term 'Nats' is a term used particularly by the English-language press and English-speaking whites to refer to the National Party or its members and supporters. It has slightly disparaging connotations, and is associated with an anti-Afrikaner racism prevalent among many English-speaking whites.
See also **National Party**.

n

Naudé, Christiaan Frederick Beyers (*South Africa*)
Beyers Naudé is a dynamic and free-thinking Afrikaner cleric. After completing a theology degree at Stellenbosch in 1939 he became an assistant minister in the NGK and joined the *Afrikaner Broederbond*. For twenty years his ministry remained comfortably within the mainstream of NGK policies until, after Sharpeville in 1960, he began to question the role of his church in South Africa. Ironically, the start of this questioning coincided with his appointment as moderator of the church. At the same time, he became a founder member of the Christian Institute, established to bring together Christians of all denominations and backgrounds and later to take a crusading role in the struggle against apartheid.

In 1963, following criticism of his involvement in the Christian Institute, Naudé resigned as moderator and became director of the Institute. In the next few months he was defrocked as a minister in the NGK and resigned from the *Afrikaner Broederbond*. Over subsequent years both Naudé and the Christian Institute were subjected to continuous harassment and persecution by the authorities. He was deprived of his passport and his organisation was banned from receiving funds from abroad for its work. In 1975 he was banned for five years, and in 1982 the order was renewed for a further three years, only being lifted in 1984 after widespread internal and international protests. Naudé's final break with the white NGK came in 1982 when he and three other prominent NGK theologians joined the black and progressive *NGK in Africa*.
See also ***Afrikaner Broederbond*; banning; churches**.

NCDP (*Namibia*)
The Namibia Christian Democratic Party, or NCDP, is a small party led by Hans Rohr, and is based mainly in the so-called Kavango region. It contested the internal elections of 1978, winning one seat. In 1982 Rohr denounced the campaign of killings and abduction waged in the north of the country by *Koevoet*. In 1984 the NCDP was one of the parties which sat with SWAPO in the Lusaka talks.
See also ***Koevoet*; Lusaka talks on Namibia, 1984**.

NCN (*Namibia*) See **National Convention of Namibia**.

NCP (*South Africa*) See **National Conservative Party**.

Ndebele (*Zimbabwe*)
Zimbabwe's indigenous population is divided into two broad linguistic groups, the Ndebele or Matabele and the Shona or Mashona. There are about four times as many Shona as Ndebele-speakers in the country. Shona-speakers live in the north and centre of the country around the capital, Harare, and they have tended to support the ruling ZANU party. Ndebele-speakers live in Matabeleland, an area in the west of Zimbabwe, and they have tended to support Joshua Nkomo's ZAPU party. ZAPU tended to be more broadly based during the 1960s, but in subsequent years it has received its main support from Ndebele-speakers. This

political pattern emerged despite strong opposition to tribalism in both nationalist parties.
See also **Mashonaland; Matabeleland; ZANU; ZAPU**.

Ndlovukazi *(Swaziland)*
The *Ndlovukazi* ('she-elephant') is the Queen Mother. The present *Ndlovukazi* is Queen Regent Ntombi.
See also **Dzeliwe; Ntombi**.

ndvuna *(Swaziland)*
An *ndvuna* (the plural is *tinvuna*) is a commoner chosen by the king to serve officially as a representative, an intermediary between the monarch and the people in each village. The title may also be given to individuals serving under chiefs.

'necklacing' *(South Africa)*
'Necklacing' is the practice of killing someone by tying their hands and feet and placing a petrol-filled tyre around their neck before setting them alight. The practice has become widespread in recent years in the South African townships where it is used by militant black activists on fellow blacks whom they suspect of collaborating with the authorities. Those 'necklaced' have included police ('blackjacks'), local councillors, and police informers. According to the South African Bureau for Information in an October 1986 press report, 269 people had been burned to death in this way since 1984.

The 'necklace' has become controversial since Winnie Mandela was reported to have told mourners at a funeral in 1986 that such violence could help to liberate South Africa: 'Together, hand in hand with our boxes of matches and our necklaces, we shall liberate this country'. Oliver Tambo, the president of the ANC, is reported to have said, 'We would rather there was no necklacing . . . We are not happy with the necklace but we will not condemn people who have been driven to such extremes . . . The necklace is a recent appearance and its cause is apartheid.'
See also **'blackjacks'; Mandela, Winnie; Tambo; townships**.

neighbourhood committees *(Angola)* See ***bairro***.

Neto, Agostinho *(Angola)*
A founder member of the MPLA and its first president, Dr Agostinho Neto also became the first President of the People's Republic of Angola, when it was established on 11 November 1975. He remained head of party and government until his death in a Moscow hospital on 10 September 1979. The medical doctor, poet (his first book of poetry led to his arrest and rustication from a Portuguese university in 1955), intellectual and revolutionary leader who had come to symbolise the struggle for independence was widely mourned in Angola. Having survived several challenges to his leadership of the MPLA, Neto was succeeded on his death by one of his close associates. José Eduardo dos

Santos was elected president on 20 September 1979. Dos Santos had joined the MPLA in 1961 and was a veteran of the liberation war.
See also **Dos Santos; MPLA; People's Republic of Angola**.

New Republic Party (*South Africa*)
In 1977 the United Party held a dissolution congress in a futile attempt to form a united opposition to the National Party. The party split three ways, its remaining liberal representatives forming the Committee for a United Opposition (and eventually merging with the Progressive Reform Party), its right forming the South African Party (and eventually joining the National Party), and its centre merging with the tiny Democratic Party to form the New Republic Party (NRP). The NRP was the largest of the three fragments of the defunct United Party, and had 23 MPs led by Radclyffe Cadman.

When Cadman lost his parliamentary seat to the Progressive Federal Party in the subsequent general election, he was replaced by the present NRP leader, W. Vause Raw. The NRP held only 10 of its 23 seats in the 1977 general election, 8 in the 1981 election and 1 in the 1987 election. It became largely a Natal party, where it continued to dominate provincial politics despite its poor performance elsewhere. This reflected the basic conservatism of white English-speakers in Natal, who, while refusing to espouse the mild liberalism of the Progressive Federal Party, found the National Party unpalatable owing to deep-rooted prejudices against Afrikaners. The continued conservatism of the NRP has shown it to be successor in both spirit and fact to the United Party.
See also **National Party; Progressive Federal Party; South African Party; United Party**.

Ngoyi, Lilian (*South Africa*)
Lilian Ngoyi was appointed the president of the South African Federation of Women in 1956. She led the mass anti-pass demonstrations organised by FSAW in 1955 and 1956. She joined the ANC in 1952 and became the national and Transvaal provincial president of its Women's League until the ANC was banned in 1960. She was also elected to the ANC national executive committee in 1956. She was arrested on a charge of treason in 1956 after attending the Congress of the People, and acquitted with others when the charges collapsed. She was banned and imprisoned by the South African authorities, spending 71 days in solitary confinement during one prison term. She died in 1980 while serving her 1975 banning order at her home in Orlando, near Johannesburg. Lilian Ngoyi was a brilliant speaker and her charismatic personality inspired many South African women in the politics of resistance.
See also **Federation of South African Women**.

Ngwane (*Swaziland*)
Ngwane III was one of the most important of Swaziland's early rulers during the mid-eighteenth century. The term *KaNgwane*, used to mean Swaziland, refers to Ngwane III.
See also ***KaNgwane***.

n

Ngwane National Liberatory Congress (*Swaziland*)

The Ngwane National Liberatory Congress (NNLC) was formed from splits in the Swaziland Progressive Party (SPP). It was created in April 1963 by its leader Dr Ambrose Zwane. Before its actual formation it was often referred to as the SPP (Zwane). The new party advocated a democratic, socialist, multi-racial Swaziland based on universal suffrage, and emerged as the leading opposition party. It sought to build up a mass working-class base and gained much support from workers engaged in the series of strikes which occurred during the period 1962–3; many trade unionists were NNLC members.

However, its very success led to its downfall. In the 'constitutional crisis' of 1973, King Sobhuza banned all parties because the NNLC had won three seats in parliament, which had hitherto been regarded as the sole domain of the 'King's party', the Imbokovdo National Movement. The NNLC ceased to exist, and the king ruled by decree until 1978 when he introduced a new constitution based not on political parties but on the traditional structures.

See also **'constitutional crisis' of 1973; Imbokovdo National Movement; Swaziland Progressive Party; Zwane**.

Ngwiri, John (*Malawi*)

John Ngwiri was Cabinet Secretary in Malawi for many years. He had often played the leading administrative role in running the country, particularly in the 1980s when negotiations with aid agencies had been so important. In December 1985 President Banda suspended him from office, giving his age as the official reason.

See also **Banda**.

Ngwizako (*Angola*)

Ngwizako Ngwizani a Kongo was a small northern, ethnically based political party which was part of the FDLA.

See also **FDLA**.

Nhongo, Rex (*Zimbabwe*)

Rex Nhongo was active in ZAPU, one of Zimbabwe's main nationalist organisations, from an early age. His real name was Solomon Mutuswa and he took the name Rex Nhongo as his *chimurenga* (or fighting) name when he became involved in the armed struggle. He left Zimbabwe in 1967 to join the guerrillas and the next year went to Moscow for a course in radio communication and administration. In 1970 he went to Tanzania because of troubles within ZAPU in Lusaka, Zambia. There he decided to join ZANU, ZAPU's rival nationalist party, mainly because of anxieties over what he saw as inadequacies in ZAPU's fighting programme. He became a senior member of ZANLA, ZANU's army, in 1971, in time for the decisive campaign against the white Rhodesian regime. Later he became commander of ZIPA, a joint army consisting of cadres from ZANLA and ZIPRA (ZAPU's fighting force).

See also ***Chimurenga*; ZANLA; ZANU; ZAPU; ZIPA; ZIPRA**.

n

NIBMAR (*Zimbabwe*)
In 1966 Britain became committed to the principle of 'no independence before a majority rule' at the Commonwealth conference. This became widely known as NIBMAR. The expression was first coined by Harold Wilson, then the British prime minister, in December 1966.

There were five NIBMAR principles: unimpeded progress towards majority rule; guarantees against retrogressive amendments to the constitution that would retard African advancement; an increase in African political representation; the progressive end of racial discrimination; British satisfaction that proposals for independence were acceptable to the people of Zimbabwe as a whole.

All the talks between the white rebel leader Ian Smith and various British prime ministers about independence foundered on NIBMAR. In 1964 Ian Smith went to London to demand independence from the then Prime Minister Sir Alec Douglas-Home. He was prepared to threaten a Unilateral Declaration of Independence if he failed in his objective. Britain refused to accept the outcome of a referendum in which only whites would participate. Nor would it accept Ian Smith's assertion that the views of selected African chiefs represented majority black opinion in the country. Negotiations continued with Harold Wilson who became British prime minister in 1964. In December 1966 Ian Smith met him on board the British ship HMS *Tiger*. There a sixth principle was added to NIBMAR to please the Rhodesian leader: that there would be no oppression of the minority by the majority or vice versa. The terms discussed on board HMS *Tiger* were rejected by the white Rhodesian cabinet and more talks took place in 1968 on board HMS *Fearless*. The rebel Rhodesians again rejected the proposals.

After this the British government began to back down and NIBMAR came to feature less in negotiations. The 1971 Anglo-Rhodesian agreement between Sir Alec-Douglas-Home, then British foreign secretary, and Ian Smith was not based on the six principles. However, one of the principles did play a part. The Pearce Commission was formed after the 1971 Anglo-Rhodesian agreement, and it found that majority black opinion was against the settlement terms (the fifth principle). As a result Britain refused to implement the 1971 agreement.

NIBMAR continued to represent the basis for international opposition to the illegal Smith regime until Zimbabwean independence in 1980.
See also **Anglo-Rhodesian agreement; *Fearless*, HMS; Pearce Commission; Smith; *Tiger*, HMS; Unilateral Declaration of Independence**.

NIDCS (*Swaziland*) See **National Industrial Development Corporation of Swaziland**.

NIP (*Namibia*)
National Independence Party: see **National Convention of Namibia** and **Lusaka talks on Namibia, 1984**.

NIS (*South Africa*)
The precursor of the National Intelligence Service (NIS) was the Bureau

of State Security (BOSS). BOSS, a name still popularly applied to the NIS, was established in 1969, replacing the smaller Republican Intelligence (RI) which had functioned as part of the Security Police (SP). BOSS was set up as an autonomous organisation under the former head of the Security Police, General H. J. van den Bergh, who was intensely loyal to the then prime minister B. J. Vorster. Under Vorster and Van den Bergh, BOSS dominated the South African intelligence field and was used both against those perceived as 'subversives', and against Vorster's internal opponents in the National Party.

In 1978, however, with the victory of P. W. Botha in the leadership elections, BOSS was toppled from its position of dominance in favour of Military Intelligence, Botha's favoured agency. The excuse was provided by the disastrous results of the invasion of Angola in 1975 (an adventure which was pushed by BOSS in the face of military opposition), and BOSS's total failure to predict, or offer solutions to, the 1976 Soweto uprising.

In September 1978 Van den Bergh resigned and Boss was renamed the Department of National Security (DONS). It was purged and restructured and in 1980 was again renamed, becoming the National Intelligence Service (NIS). A 31–year-old former academic, Lukas Barnard, was appointed as its new head. It was subsequently announced that the NIS would concentrate its efforts abroad, taking over control of external agents of the Security Police, and that the latter would take responsibility for internal security. Military Intelligence would continue to operate in both jurisdictions. The NIS appears to have retained an overall coordination role for all the intelligence services, however, and is the dominant presence on the State Security Council.

Both the military and the NIS were involved in the abortive Seychelles coup attempt in 1981, and in 1982 and 1983 South African agents were brought to trial in London for conspiracy to burgle SWAPO, ANC and PAC offices. At the same time, the head of the Washington NIS station was expelled for activities which 'went beyond normal intelligence gathering as it is understood between governments'.

See also **Security Police; State Security Council**.

Nitistas (*Angola*)

The so-called Nitistas were supporters of the movement led by Nito Alves which culminated in the coup attempt of 27 May 1977.

See also **Alves; coup attempt, 1977.**

Nkala, Enos Mzombi (*Zimbabwe*)

Enos Nkala is minister of home affairs and secretary for finance for the ZANU-PF Politburo and Central Committee. He spent most of the twenty years before independence in detention for his opposition to minority rule. He was elected caretaker general-secretary of the National Democratic Party and later deputy secretary-general. He was a founder member of ZANU which was launched at his house and he was elected treasurer-general in 1964.

See also **Zanu**.

n

Nkavandame, Lázaro (*Mozambique*)
Having come to prominence as a leading figure in the SAAVM Co-op, Lázaro Nkavandame became head of FRELIMO's Provincial Co-operatives Committee during the liberation war. He was a controversial figure who abused his post for personal benefit, even using FRELIMO vehicles to transport goods for his trading stores in Tanzania.

Nkavandame led a group of conservative FRELIMO leaders who opposed the decision to integrate the military and political wings of the movement. The department of defence in FRELIMO had been radicalised by the direct involvement of its members in the armed struggle, and Nkavandame and others attempted to prevent its politics from dominating the movement by insisting that the fighters end their political involvement and concentrate on the military struggle. Divisions grew so strong that his group eventually began urging peasants to refuse to feed the guerrillas.

Nkavandame boycotted the second congress of FRELIMO over the involvement of military represenatives, and when the congress confirmed the unity of the military and political struggles, began campaigning against FRELIMO in Tanzania. He won an initial success when Tanzania closed the border with Cabo Delgado province in Mozambique to FRELIMO. A meeting was organised to reconcile the two sides. When the meeting failed, however, TANU supported the FRELIMO leadership and the border closure was ended.

Nkavandame responded by recruiting his own army and killing a FRELIMO representative in Cabo Delgado, while campaigning for the province to secede. When this failed he abandoned FRELIMO and went over to the Portuguese in April 1979.
See also **FRELIMO; Mueda massacre**.

Nkomati Accord (*Mozambique, South Africa*)
The Nkomati Accord was a controversial agreement reached between Mozambique and South Africa on 16 March 1984. It took its name from the Inkomati River, on the banks of which it was signed under the canvas of huge marquees erected to accommodate the host of dignatories and the world's press invited to witness the unexpected event.

The Nkomati Accord was in effect a mutual non-aggression pact between South Africa and Mozambique, according to which South Africa would end its support for the MNR while Mozambique would expel all ANC personnel associated with the armed struggle, leaving only a small diplomatic delegation. In addition South Africa won agreement to increased economic opportunities for its corporations in Mozambique, and a public non-aggression treaty, something which was part of its plan for a 'constellation of states' in southern Africa. The agreement was to be monitored by a joint security commission.

Within weeks of the signing ceremony Mozambique had met its side of the bargain, welcoming South African business representatives who came with proposals for different projects, and expelling all but a small number of ANC members and thereby severely straining its relations with the liberation movement. South Africa gave the initial appearance of abiding by its side of the bargain, and very quickly closed down the MNR radio station.

But despite Mozambique hailing the pact as marking the end of the war of destabilisation, this proved to be far from the reality of the situation. After an initial lull, the MNR operations resumed with increased ferocity until the war was as severe as it had been at its previous height. Initial reactions from South Africa suggested that this reflected either the fact that the MNR had become a truly indigenous organisation which South Africa could no longer control, or that the military in South Africa was operating in support of the MNR in defiance of the political authorities.

In fact South Africa's destabilisation had simply become too effective for it to give it up, and having achieved the expulsion of the ANC it pressed on towards its real ambition, the replacement of the FRELIMO government in Mozambique with a compliant MNR administration which would turn the country into the client state for which South Africa had always hoped. Total domination of Mozambique would make the SADCC unworkable, and end attempts by the countries of southern Africa to end their infrastructural dependence on South Africa.

See also **FRELIMO; MNR; SADCC**.

Nkomo, Joshua (*Zimbabwe*)

Joshua Nkomo is known as the 'Old Man' of Zimbabwean politics. However, the 1980s have seen his decline from power. His party, ZAPU, one of Zimbabwe's leading nationalist organisations in the 1960s and 1970s, has come under increasing pressure from the prime minister, Robert Mugabe, and his ZANU party, until it is now more or less confined to a traditional base among Ndebele speakers in Matabeleland.

Joshua Nkomo was president of the African National Congress from 1952 to 1957 and secretary-general of the Railway African Workers' Union before that. In 1960 he became president of the National Democratic Party and in 1961 president of ZAPU. He was repeatedly restricted by the rebel regime of Ian Smith and imprisoned for some years. In 1975 he was co-founder and president of the Patriotic Front. He also became supreme commander of ZIPRA, ZAPU's army. In the 1980 elections he was elected member of parliament for Midlands and refused the position of president of Zimbabwe. Instead he was minister of home affairs for a while and then minister without portfolio with special duties.

In 1981, together with many of his colleagues, Joshua Nkomo was dismissed from the cabinet following the discovery of illegal arms caches on properties belonging to ZAPU in Matabeleland. Violence and political unrest in Matabeleland developed, with allegations against the government of political repression and brutality. In the 1985 elections Joshua Nkomo's party held all its Matabeleland constituencies, winning 15 seats. This caused the prime minister some political anxiety since it indicated a developing split along ethnic lines. However, since then Robert Mugabe has led Zimbabwe towards a one-party state and in December 1987

ZANU and ZAPU agreed to merge, and in 1988 Nkomo rejoined the government.
See also **African National Congress (Zimbabwe); Mugabe; Ndebele; National Democratic Party; Patriotic Front; ZANU; ZAPU; ZIPRA.**

n

Nkumbula, Harry (*Zambia*)
Harry Nkumbula was president of the African National Congress (ANC) in Northern Rhodesia (later to become Zambia) in 1951. He was the first campaigning nationalist leader and took up the fight against the Central African Federation. However, when federation took place in 1953 the ANC lost much of its momentum and its leader's popularity declined. In 1958 Harry Nkumbula's position was challenged by a group of young radicals who left the ANC to form another party. They were led by Kenneth Kaunda who was to become independent Zambia's first president. The feeling among the younger members of the ANC was that Harry Nkumbula did not have the style of leadership suitable to a liberation struggle. This, they felt, should be undertaken by men of asceticism, commitment and self-sacrifice. Harry Nkumbula, in contrast, led his party on a very personal basis, was not particularly interested in party discipline and very much enjoyed the pleasures of life. Once Kenneth Kaunda's party, UNIP, took over as the main nationalist movement, Harry Nkumbula became a less significant figure in the politics of the country.
See also **African National Congress (Zambia); Central African Federation**.

nkundla (*Swaziland*)
Tinkundla (singular *nkundla* or *inkundla*) are rural district councils. Twenty nine were established throughout the country when they were started in 1956. Then they consisted of local chiefs and an appointed chairperson. Under the new constitution of 1978 the *tinkundla* were reorganised and given a new and vital role. The *Libandla* or parliament is now 'elected' through these structures in such a way that the monarch nominates it entirely. The influence of the councils is reduced where the chiefs feel threatened by their existence, but usually they serve to bring local dissent and political activity within the control of the traditional structure.
See also ***Libandla*; monarchy**.

NNC (*Namibia*)
Namibia National Convention: see **National Convention of Namibia**.

NNF (*Namibia*)
Namibia National Front: see **National Convention of Namibia**.

NNLC (*Swaziland*) See **Ngwane National Liberatory Congress**.

'no independence before majority rule' (*Zimbabwe*) See **NIBMAR**.

Norman, Denis (*Zimbabwe*)
Denis Norman was the only white member of Robert Mugabe's cabinet after the 1980 elections. A well-known Zimbabwean farmer and president

of the Commercial Farmers' Union, he was elected senator and appointed minister of agriculture in 1980. This appointment was regarded as politically astute as it did much to reassure and reconcile white minority opinion in Zimbabwe to the new government. It also helped the government to keep agricultural production at a high level just after independence.
See also **Mugabe**.

Northern Rhodesia *(Zambia)*
Cecil John Rhodes, a rich nineteenth-century entrepreneur and explorer set off from the Cape and founded the colony of Rhodesia in 1893, administering it through his British South Africa Company. This area north of the Limpopo River was to become Southern Rhodesia and later Zimbabwe. Rhodes then became interested in reports of the rich mining deposits which existed north of the Zambezi river in what is now Zambia, so he sent explorers north to investigate. In 1889 the British government had granted a charter to the British South Africa Company to make treaties and administer the territory north of the Limpopo River. Soon Cecil Rhodes had taken over most of what became Northern Rhodesia, placing it under British control. In 1953 the territory of Northern Rhodesia was joined with Southern Rhodesia and Nyasaland (later to become Malawi) in the Central African Federation. The Federation subsequently collapsed, and on 24 October 1964 the colony of Northern Rhodesia became the Republic of Zambia.
See also **Rhodes; British South Africa Company**.

Northern Rhodesia African Congress *(Zambia)*
In 1948 the Federation of African Societies renamed itself the Northern Rhodesia African Congress. This in turn was renamed the African National Congress in 1951 and became the first popular nationalist party campaigning for independence in what was then Northern Rhodesia and is now Zambia.
See also **African National Congress (Zambia); Federation of African Societies**.

NP *(South Africa)* See **National Party**.

Nquku, John June *(Swaziland)*
J. J. Nquku was the head of the first Swazi-controlled political party, the Swaziland Progressive Party, from 1960 to 1972, when he retired.
See also **Swaziland Progressive Party; Zwane**.

Nqumayo, Albert Muwalo *(Malawi)*
Albert Muwalo Nqumayo was a nationalist active in the central committee of Malawi's only political party, the MCP. After independence he entered government and in 1966 he was appointed Minister of State in the President's Office where he remained for over a decade. However, in January 1977 a secret trial was conducted under rules and procedures determined by the president, and Albert Nqumayo, together with the head of Malawi's special branch, was found guilty of plotting to have the president

assassinated. He was sentenced to death. Other powerful men in President Banda's government – like Gwanda Chakuamba and Bakili Maluzi – have also been arbitrarily removed from office.
See also **Banda; Chakuamba**.

NRAC (*Zambia*) See **Northern Rhodesia African Congress**.

NRP (*South Africa*) See **New Republic Party**.

Ntombi Latfwala (*Swaziland*)
Queen Mother Ntombi was the Queen Regent of Swaziland until her son was crowned King Mswati III on 25 April 1986. She took over from Queen Regent Dzeliwe on 9 August 1983 after a power struggle in which the *Liqoqo*, or Supreme Ruling Council, ousted Dzeliwe because she would not relinquish some of her powers. Queen Regent Ntombi, the mother of the heir to the throne, Prince Makhosetive, agreed to the *Liqoqo's* conditions and was installed in her place. Soon, however, the new Queen Regent began to downgrade the *Liqoqo*. In 1985 she dismissed two of its most powerful members, Prince Mfanasibili Dlamini and George Msibi, and later took away some of its powers by legislation.
See also **Dlamini, Mfanasibili; Dzeliwe; Liqoqo; Mswati**.

Nucleus of Mozambican Secondary Students (*Mozambique*)
This was a multi-racial organisation formed in the early 1960s which brought together a number of those who later became important figures in FRELIMO. These included Mozambique's first two presidents, Samora Machel and Joaquim Chissano, Central Committee members Armando Guebuza and Mariano Matsinhe and health minister Pascoal Mocumbi.
See also **Chissano; FRELIMO; Machel**.

NUDO (*Namibia*)
The National Unity Democratic Organisation (NUDO) was formed in 1964 by the Herero Chiefs' Council when it left SWANU, which it regarded as too radical. Despite this history, NUDO, along with SWANU, was one of six political parties which joined the SWAPO-led National Convention of Namibia (NCN). Subsequently, however, its leader, Chief Clemens Kapuuo, attempted to portray the NCN as an alternative to SWAPO. When this failed, NUDO joined the Turnhalle Constitutional Conference, an anti-SWAPO alliance.

By 1977 SWAPO was accusing NUDO of violent attacks on its meetings and supporters. Because NUDO members were almost exclusively Herero in origin, the South African administration attempted to represent the conflict as 'inter-tribal fighting'.

Following the assassination of Kapuuo in 1978, Kuaima Riruaka became the NUDO leader, and subsequently, President of the DTA.
See also **DTA; Kapuuo; National Convention of Namibia; SWANU**.

Nujoma, Sam (*Namibia*)
President of SWAPO since its formation, Sam Nujoma, with Andimba Toivo ja Toivo, was instrumental in its founding. He went into exile in 1960, lobbied the UN, and in 1961 established SWAPO's provisional

headquarters in Dar es Salaam, Tanzania. He subsequently returned to Namibia, but was forced out in 1966, since when he has led the movement from outside Namibia. In 1973 SWAPO was recognised by the UN as the authentic representative of the Namibian people.

Nujoma is a widely respected leader both within Namibia and in international circles. A relaxed, pragmatic politician, he has guided SWAPO through the minefield of international diplomacy until it now enjoys almost unprecedented recognition and support from countries not usually sympathetic to liberation movements. In delicate negotiations involving the UN and South Africa, Nujoma has repeatedly outmanoeuvred the South Africans so that they stand exposed as the illegal occupiers of Namibia, and as brutal and intransigent aggressors in the war there.

In contrast to other liberation struggles in Africa, SWAPO's has been marked by the continuity of leadership. Sam Nujoma has led SWAPO virtually unchallenged for 27 years, and, with other leading figures in the organisation, has established a base of support which includes white and black in Namibia, northerners and southerners, Ovambo migrant workers and Windhoek intellectuals. South Africa's refusal to countenance free and fair elections in Namibia reflects at least one thing it has in common with the liberation movement – the assessment that in any such elections, SWAPO would command the overwhelming support of the people.

See also **SWAPO; Toivo ja Toivo**.

NUM (*South Africa*)

The NUM (National Union of Mineworkers) is the largest and most powerful single trade union in South Africa. In 1986 it was reported to have 360,000 members. Its general-secretary, Cyril Ramaphosa, is a leading figure in COSATU, the main federation of trade unions in the country. The NUM was founded in 1982 by CUSA, the Council of Unions of South Africa (which the NUM later left in order to affiliate to COSATU). Before its foundation African mine workers had not been organised since the collapse of the African Mine Workers' Union after the 1946 African mineworkers' strike.

Mineworkers had proved particularly hard to organise in South Africa since some 60–70 per cent of them are migrant workers. Nevertheless the NUM made rapid progress and within three years it was able to claim just under half the black labour force on the mines as signed-up members – some 220,000. In mid–1987 NUM members were engaged in a widespread and long-running strike for better conditions and wages.

See also **COSATU; CUSA; migrant workers; Ramaphosa**.

NUNW (*Namibia*)

Trade unions in Namibia have suffered from the hostility of both the South African occupation administration and local management. Although Namibia has a long history of militant action by workers, including a successful general strike in 1971–2 against the contract labour system, it does not have a similar history of militant trade unionism.

Most trade unions are narrowly based occupational unions in semi-skilled areas of work, many of them dominated by conservative white leadership. The exception is the National Union of Namibian Workers (NUNW), a country-wide general union which was conceived by the Labour Department of SWAPO, formally constituted on 24 April 1970, and began open organisation in 1977. It concentrated on the organisation of contract workers in the major mining enterprises and had achieved wide support by the end of 1978. Its affiliation to SWAPO, however, prompted legislation in July 1978 banning political links for officially recognised unions.

By 1979 the NUNW had been forced underground, and SWAPO, rooted among contract workers and their families, was the only effective representative organisation of working people in Namibia. Perhaps as a result, SWAPO Secretary for Labour in Namibia, Jason Angula, was banned in 1980. In April 1986 the NUNW was relaunched above-ground in Namibia as an umbrella body for Namibian industrial unions. The job of National Organiser was taken up by Ben Uulenga following his release from prison on Robben Island. Two industrial unions have been launched, the Namibian Food and Allied Union (NAFAU) in September 1986, led by Alfons 'John' Pandeni (also following his release from Robben Island), and the Mineworkers' Union of Namibia (MUN), in November 1986 and led by Ben Uulenga.

The 'Aims and Objectives' of the NUNW as spelt out in its draft constitution in 1977, include: 'to organise and protect from exploitation . . . all workers in various job categories in Namibia'; 'to pave the way and prepare for participation by the workers of Namibia in the government of an independent Namibia'; 'to take part and contribute to a complete change in the present social, economic and political order'; 'to oppose all tribalism and ethnic grouping . . . among Nambian workers'; 'to fight for just wages, good working conditions and to protect the interests of all workers'; and 'to regulate, negotiate and settle disputes between workers and employers'.

See also **SWAPO; trade unions (Namibia)**.

NUSAS (*South Africa*)

The National Union of South African Students (NUSAS) was formed in 1924 to promote unity between Afrikaans and English students in the white universities in South Africa. In 1933 most Afrikaans universities left the organisation following a campaign for the admission of the black Fort Hare University (Fort Hare's application was rejected). The last Afrikaans campus, Stellenbosch, left in 1936. In 1945 Fort Hare was allowed to affiliate, and affiliations by other black campuses followed.

While NUSAS campaigns during the 1950s had been limited mainly to opposition to the segregation of tertiary education and to other educational issues, by the 1960s it began to raise wider questions and to agitate generally against apartheid. Despite this, the radical critique by black consciousness organisations of its liberal politics and white domination led to the disaffiliation of black campuses and the formation of an exclusivist black student organisation, the South African Students' Organisation (SASO) in 1969.

A period of crisis followed, which led to the radicalisation of NUSAS and its involvement in the formation of non-racial trade unions. While NUSAS became heavily infiltrated by agents of the Security Police and BOSS during this period, it also rediscovered a role for itself as a radical white voice allied to other organisations engaged in the struggle for democracy in South Africa. While this injected a new dynamism into the organisation, it also resulted in the disaffiliation of a number of more conservative English-speaking campuses including that at Rhodes University and the University of Natal, Pietermaritzburg. In 1983 the then NUSAS president, Kate Phillip, proposed 'an ideological and supportive role for the organisation designed to complement the practical role facing black democratic students' working for change in South Africa. In recent years NUSAS has participated in joint campaigns with the Azanian Students' Organisation (AZASO) and the Congress of South African Students (COSAS).

Throughout the 1960s and 1970s NUSAS found itself the subject of repression and dirty tricks by the state including the arrest, banning and trial for subversion of its leaders and members, the subjection of the organisation to a major commission of inquiry which resulted in the banning of overseas funding for its projects and operations, and the distribution of forged documents designed to undermine the organisation. See also **ASB; AZASO; BOSS; COSAS; POLSTU; SASO; Security Police**.

NUTA (*Tanzania*) See **National Union of Tanganyika Workers**.

Nxumalo, Simon Sishayi (*Swaziland*)
Sishayi Nxumalo was the managing director of the *Tibiyo Taka Ngwane* fund, a powerful institution controlled by the monarchy which plays a central role in Swaziland's economy. He was the driving force behind setting up the fund and became its first managing director. He resigned this post only to become finance minister in 1983, and continued to be the secretary of the fund.

While King Sobhuza was alive, Sishayi Nxumalo was reputed to be the second most powerful man in the country. He founded the Swaziland Democratic Party in 1962 and was its president and then its secretary general. After his defeat in the 1964 elections he joined the ruling monarchist party, the *Imbokodvo* National Movement, and became very influential within it. He had been a government minister since the 1967 elections. In 1984 he was involved in a power struggle based around corruption charges, launching a campaign against Prince Mfanasibili Dlamini of the *Liqoqo*. This may also have been a personal struggle to hold power within the influential *Tibiyo Taka Ngwane* group in government. Sishayi Nxumalo lost and was dismissed from the *Liqoqo* on 8 June 1984, together with three others. They were named the 'gang of four' by Prince Mfanasibili and accused of threatening a 'bloody revolution'. On 31 December 1985 it was reported that Dr Nxumalo and the others had been released in an amnesty granted by the Queen Regent. See also **Dlamini, Mfanasibili; Swaziland Democratic Party; Tibiyo Taka Ngwane**.

n

Nyasaland (*Malawi*)

The present Republic of Malawi was known as the British protectorate of Nyasaland until its independence in 1964. The protectorate was declared by the British in 1891.
See also **Malawi**.

Nyasaland African Congress (*Malawi*)

Before the formation of the Nyasaland African Congress (NAC) many people belonged to various political 'associations'. In 1944 the associations joined together to form the NAC. This was to become the main nationalist movement campaigning for independence and it was renamed the Malawi Congress Party (MCP) in 1959.

At first, however, the NAC was neither campaigning nor radical, but a rather weak, reformist body. The imposition of the Central African Federation in 1953, against the wishes of the majority of the population, undoubtedly increased support for the party and helped to radicalise it. Most Africans from the three countries of the Federation, Nyasaland (later to become Malawi), Northern Rhodesia (later Zambia) and Southern Rhodesia (later Zimbabwe), opposed its formation as it was felt it would benefit the white rulers of Southern Rhodesia above all else. The NAC received the support of the Nyasaland Chief's Union over the issue of the Federation and this was a crucial factor in its growing popularity.

By 1955 the party came under the influence of a group of young radicals headed by Henry Chipembere and Kanyama Chiume and began to demand universal suffrage and self-government. In 1957 the NAC invited Hastings Banda to return to Malawi and work for independence. He arrived in 1958 and became the NAC's president. Soon his campaigning style and denunciations of the Federation provoked clashes between NAC members and the colonial officials. A state of emergency was declared in March 1959, the NAC banned, and many of its leaders arrested. Not long after this, however, the NAC's mantle was taken over by the MCP.
See also **Associations; Banda; Central African Federation; Chipembere; Chiume; Malawi; MCP**.

Nyasaland Chief's Union (*Malawi*)

An association of chiefs which supported the Nyasaland African Congress in its opposition to the Central African Federation: see **Nyasaland African Congress**.

Nyazonia massacre (*Mozambique*)

On 9 August 1976, the forces of Smith's Rhodesia launched an attack on the refugee camp of Nyazonia in Manica province, near what is now called Mutare in Zimbabwe. 875 people were massacred.
See also **MNR; South African destabilisation**.

Nyerere, Julius Kambarage (*Tanzania*)

President Dr Julius Nyerere is a leading African statesperson of the post-colonial era. He ruled Tanzania from independence in 1962 until 1985 when his appointed successor took over. He is one of Africa's great nationalist leaders who is also known as a socialist and intellectual.

Inside Tanzania he is called *Mwalimu*, 'the teacher', and his personality has dominated the political scene for over twenty years. A charismatic and gentle man, believing deeply in his socialist ideals, he has led by example, living simply and eschewing the pomp of office. He was a schoolteacher when he founded TANU, the Tanganyika African National Union, in 1954. In 1960 he became prime minister and in 1962 was elected president. Soon he implemented his *ujamaa* policy which became the cornerstone of Tanzanian socialism. It is derived from the Swahili word 'familyhood' and it is an idea of socialism based on specifically African concepts. Later *ujamaa* was extended to include the concept of an *ujamaa* village which involved communal farming. This policy provoked much criticism, largely owing to its bureaucratic implementation, and has done much to discredit President Nyerere's leadership.

Julius Nyerere has also been a major influence in wider African affairs, including the Organisation of African Unity (he was Chairman from November 1984 to July 1985) and SADCC, the Southern African Development Coordination Conference. He is an outspoken critic of the white South African regime and in recent years has advocated international sanctions against the apartheid government. In 1985 President Nyerere stepped down but remained party leader. Ali Hassan Mwinyi, President of Zanzibar, took over on November 5th as Tanzania's new president.

See also **Mwinyi; SADCC; TANU; *Ujamaa***.

Odendaal Commission (*Namibia*)
In 1962 South Africa appointed a Commission of Inquiry into South West Africa, chaired by F. H. Odendaal, the administrator of the province of the Transvaal. Popularly known as the Odendaal Commission, the inquiry was to investigate ways of further promoting 'the welfare' of Namibians in terms of apartheid.

The commission's report, published in 1964, laid the foundations for the implementation of full apartheid in the country. It divided Namibians into twelve apartheid categories according to notions of 'race': Rehoboth Basters, Namas, Damaras, Hereros, Kaokovelders, Ovambos, Kavangos, Caprivians, Tswanas, 'Bushmen' (San), 'Coloureds' and 'Whites'. Although the black population was divided into eleven groups on the basis of language, colour, culture, history and other criteria, the white population, with distinct German, Afrikaans, and English communities, was judged to be a single group. Each of the black groups, with the exception of the 'Coloured' and 'Bushmen' categories, was allocated a 'Homeland' or bantustan. Together the bantustans constituted 40 per cent of the land area of Namibia. The rest was reserved for white occupation.

The report recommended that most of the important branches of administration then under the SWA Administration should be transferred to South Africa. Remaining areas were to be transferred to the 'legislative assemblies' of the bantustans. Even the decisions taken by the bantustan assemblies, however, were to be subject to the approval of the State President of South Africa. Following the implementation of the commission's recommendations, the areas set aside for the jurisdiction of the bantustan administrations included education, health and welfare, aspects of business, roads, administration of justice, agriculture, labour bureaux and taxation. Areas such as the police, military and foreign affairs were retained by South Africa. Once each bantustan became 'self-governing', provision was made for a 'cabinet' headed by a 'chief minister'. This process was consolidated by proclamation AG8 of 1980 which brought the Odendaal proposals in line with the client 'internal administration' established by South Africa under the DTA.
See also **AG8; apartheid; DTA; 'Homelands'**.

Odendaal Plan (*Namibia*) See **Odendaal Commission**.

ODP (*Angola*)
The *Organização de Defesa Popular* or People's Defence Organisation is Angola's popular militia which functions as an adjunct of FAPLA. Its role is to assist with the maintenance of internal peace and defence against external invasion. The ODP is organised on a cell basis and its members are given both military and political training.
See also **FAPLA**.

Ohanga, J.B.A. (*Tanzania*)
Leader of the Tanganyika Railway African Union and later the President of the Tanganyika Federation of Labour: see **Tanganyika Federation of Labour**.

OJM (*Mozambique*)
The Organisation of Mozambican Youth was established in 1977 to involve the young people of the country in FRELIMO's development plans as well as to encourage and organise recreational activities for them.
See also **FRELIMO**.

Okello, John (*Tanzania*)
In January 1964 there was an armed revolution in Zanzibar. This was led by Ugandan-born John Okello, known as 'Field Marshal' Okello. He overthrew the government and set up a 'Revolutionary Council' to run the islands. But John Okello did not have an effective political base and in a few weeks the main party leaders Abeid Karume and Abdul Rahman Babu deposed him.
See also **Babu; Karume; Zanzibar**.

O'Linn, Brian (*Namibia*)
Leader of the white opposition Federal Party: see **Federal Party**.

OMA (*Angola*)
The *Organização das Mulheres de Angola* (OMA), or Organisation of Angolan Women, is the women's section of the MPLA.

omakakunya (Namibia)
A derogatory term meaning 'scavenger' and applied to members of South African security forces in Namibia: see ***aamati***.

OMM (*Mozambique*)
The *Organização da Mulher Moçambicana* (Organisation of Mozambican Women) was set up by FRELIMO in 1973 as the organisation for the mass of women not organised into the Women's Detachment of the guerrilla army. It was re-organised by the third FRELIMO congress in 1976. Its role is to function as one of the links between FRELIMO and the people, and it is subject to the decisions and discipline of the party. It may not, as of right, act independently of FRELIMO, and while functioning as the organisation of women, has been encouraged to focus its activities on party priorities and issues affecting Mozambicans as a whole.
See also **FRELIMO**.

Operational Area (Namibia)
This term is used by the South Africans to refer to the 'war zone' in Namibia, the area in the north of the country in which most of the fighting in the liberation struggle has been concentrated.
See also **Caprivi Strip; Namibia; Police Zone**.

Operation Maduka (*Tanzania*)
The government launched Operation Maduka in 1976. It was designed to replace private retailers in or near state farms, industries and *ujamaa* villages by co-operatives. However, the speedy implementation of Operation Maduka caused distribution problems and shortages and its implementation has since been slowed down.
See also ***Ujamaa*; villagisation**.

Operation Production (*Mozambique*)
Operation Production was a controversial policy, following the Fourth Congress of FRELIMO in 1983, which entailed the expulsion from the towns and cities of those who were unemployed. Its objective was to return people to rural areas to work on the land and thereby to boost production while reducing the increasing corruption and growth of unofficial markets in the cities which were perceived to flow from the presence of the unemployed.

It was applied very unevenly, and in some areas was used as an excuse to ship out of urban areas elderly people, the disabled and single women who were suspected of being prostitutes. Thousands of people without the correct cards spent days in 'verification centres' while their right to be in the cities was established. Operation Production is generally considered to have been an expensive failure.
See also **FRELIMO**.

Oppenheimer, Harry Frederick (*South Africa*)
Harry Oppenheimer was, until his retirement, Chairman of the Anglo American Corporation and Chairman of De Beers Consolidated Mines. The two posts gave him complete control of the giant multinational Anglo American Group. His retirement has reduced his day-to-day involvement in the companies, but not his enormous influence in South African business circles. Oppenheimer's astute business sense turned Anglo from one of the largest gold mining companies into a multi-sectoral conglomerate which dominates the South African economy.

Oppenheimer was born into an already burgeoning family business created by his father, Ernest Oppenheimer. He served as an opposition United Party MP in the South African Parliament from 1948–58, resigning to take over Anglo American on his father's death. Having been a major financer of the United Party, Oppenheimer backed the liberal breakaway from that party in 1959 and became the chief source of finance for the Progressive Party from its formation in that year. Having begun as one of the very few leaders of the business community with faith in the party, he had the satisfaction of seeing it grow into the official parliamentary opposition after the final collapse of the United Party in 1977. It was displaced by the Conservative Party after the 1987 white general election.

Harry Oppenheimer was instrumental in the founding of the Urban Foundation in South Africa and is its chair. An essentially conservative man, he has become, in the society in which he lives, a symbol of enlightened liberal capitalism. While always deprecating his political influence, he has seen six prime ministers come, and five go, since he began his political career in 1948, and throughout that time has remained a figure of influence who all, despite his political stance, have chosen to humour.

See also **Anglo American Group; Progressive Federal Party; Relly**.

OPV (*Mozambique*)

The Voluntary Police Organisation or OPV was an adjunct to PIDE made up exclusively of black Mozambicans. It worked with PIDE to suppress FRELIMO, using intimidation, killing and torture in its campaign. Whereas most PIDE members fled at independence, most OPVs remained in Mozambique. Many of those who were identified by their former victims, together with members of the other notorious anti-FRELIMO organisations, became known as 'the compromised'. None were executed, but the worst offenders were sent to re-education camps. For most their punishment was the loss of civil and political rights. Many changed their attitudes in time.

See also ***flechas*; PIDE**.

Orange Free State (*Lesotho, South Africa*)

The Orange Free State is a province of South Africa. Many Sesotho speakers live there and over the years the Lesotho government has laid claim to large areas of land in this part of South Africa.

Organisação da Mulher Moçambicana (*Mozambique*) See **OMM**.

Organisação de Defesa Popular (*Angola*) See **ODP**.

Organisation of Angolan Women (*Angola*) See **OMA**.

Organisation of Mozambican Women (*Mozambique*) See **OMM**.

Organização das Mulheres de Angola (*Angola*) See **OMA**.

Oshakati (*Namibia*)

The site of a notorious military detention centre: see **detention centres**.

Ovamboland People's Organisation (*Namibia*)

The precursor of the Namibian liberation movement SWAPO: see **SWAPO**.

P

palmatoria (*Mozambique*)
This is an unpleasant instrument for corporal punishment used to punish Africans who attempted to avoid, or absconded from, forced labour (or *chibalo*). It is a paddle-shaped piece of wood pierced by holes. When used to beat people, it raises particularly painful weals on the skin.

The *palmatoria* was used with great frequency during the colonial period as one of the main weapons to maintain the forced labour system. See also **forced labour**.

Pan-Africanist Congress (*South Africa*)
The Pan-Africanist Congress (PAC) was formed in 1959 when Robert Sobukwe led a breakaway group from the ANC. The new organisation was based on a philosophy of 'Africanism', the view that Africans should take their destiny into their own hands and not turn to other racial groups for a resolution of South Africa's political problems. Africanism had been an important current of thought inside the ANC ever since the Congress Youth League was formed. The Africanists became more and more critical as the ANC developed its political struggle through the Congress Alliance, a common front of different racial groups opposed to the state. They called the ANC 'Charterist' in a reference to the Freedom Charter which they opposed as multi-racial liberalism. They even published their own newspaper, *The Africanist*. By 1959 differences proved too great and they decided to leave the ANC and form their own organisation.

The PAC's inaugural conference was held in Orlando near Johannesburg and it elected Robert Sobukwe as president and Potalke Leballo as national secretary. In his presidential address Robert Sobukwe outlined the new organisation's basic principles. Above all the PAC stood for government by Africans for Africans. This could not at present include the whites because they were favoured by the existing situation and thus could not properly identify with the African cause. The manifesto aimed towards 'an Africanistic Socialistic democratic social order' and described its philosophy as the pan-Africanism of the All-African Peoples' Organisation's conference at Accra, Ghana, held in December 1958.

In August 1959 Robert Sobukwe announced a 'status campaign' which was designed to demand more courteous treatment by whites of Africans in shops. It was also intended to try and remove traces of 'slave mentality' and encourage assertiveness by Africans. This idea had begun in *The*

Africanist where it was argued that a battle for status should replace the ANC's persistent economic campaigns. The PAC also formed its own trade union organisation, the Federation of Free African Trade Unions of South Africa (FOFATUSA) in June 1959 with the help of the international body, the ICFTU, to which it affiliated. FOFATUSA attacked SACTU, the trade union movement in the Congress Alliance, for associating politics with trade unionism in South Africa. FOFATUSA did not last long as an organisation in South Africa.

On 16 March 1960, following the ANC's plan to step up its anti-pass campaign, the PAC announced a campaign against the pass laws to begin on 21 March. At Sharpeville in the Transvaal some 5,000 Africans gathered outside the police station and 67 were officially reported killed when the police opened fire on the crowd. Soon afterwards on 30 March 1960 the PAC and the ANC were banned and a state of emergency declared. This led to exile and the creation of a brief united front with the ANC, the South African Indian Congress and the South West African National Union (SWANU) to campaign internationally against the white South African state.

Following its banning the PAC set up its military wing, *Poqo*, 'ourselves'. This began a campaign of attacks inside South Africa. In 1963 its underground organisation was destroyed and the PAC has been riven with internal disputes since. In 1981 John Pokela was appointed as the new chairperson, and his efforts were directed to welding the organisation together into an effective resistance body. He died suddenly in Harare, Zimbabwe, in 1985 and Johnson Mlambo was elected as leader in his place. Mlambo is working to try to re-establish the PAC as a rival nationalist organisation to the ANC.
See also **ANC; Congress Youth League; Leballo; Mlambo; Pokela; *Poqo*; SACTU; Sharpeville; Sobukwe**.

PARD (*Zimbabwe*) See **People Against Racial Discrimination**.

Partido Democrático Angolano (*Angola*) See **PDA**.

Partido da Luta Unida dos Africanos de Angola (*Angola*) See **PLUA**.

'pass laws' (*South Africa*)
In South Africa various laws bar people from living and working in certain areas, and they must acquire permits to visit areas reserved for other races. Africans can live and work in 'White' South Africa only if they have the correct documents. The acts enforcing this were known as the 'pass laws'. A pass was designed to prove that a person is legally in a specific area. For instance, the Native Laws Amendment Act, Number 54 of 1972, only permits a rural-born African to visit an urban area for 72 hours without obtaining a special permit. The first passes were carried by slaves in the Cape in 1760.

In 1946 the United Party appointed a commission under Mr Justice Fagan to inquire into the laws relating to 'natives' in urban areas, the pass laws and the system of migratory labour. The commission reported in 1948 recommending various changes to the system. For the first time a 'pass' was mentioned. The report read:

> We think it would be correct to say that in the mind of the Natives a document is a pass, to which they object, if it is a document
>
> (a) which is not carried by all races, but only by people of a particular race; and which either
>
> (b) is connected with restriction of the freedom of movement of the person concerned; or
>
> (c) must at all times be carried by the person concerned on his body, since the law lays the obligation on him of producing it on demand to the police and certain other officials and the mere failure to produce it is by itself a punishable offence.

Under the National Party government all male Africans had to carry passes from 1 February 1958 and African women from 1 February 1963 until passes were abolished from 1 July 1986. Now Africans who are not citizens of an 'independent Homeland' have to have instead a new identity document which they do not have to carry on them. Soon after the repeal of the 'pass laws' there was an increase in the use of the Trespass Act and the Prohibition of Illegal Squatting Act to prosecute Africans moving to the urban areas. Pass laws have been the cause of much resentment over the years among the African population and sparked off the 1960 Sharpeville massacre. Hundreds of people were arrested every year under the laws – in 1984, 163,862. The usual sentence was a fine of R30 or thirty days in prison and R90 or ninety days in prison.
See also **apartheid; Black Sash; Federation of South African Women; 'reforms'; Sharpeville; urban 'Blacks'**.

Patriotic Front (*Zimbabwe*)

The Patriotic Front (PF) was formed on 9 October 1976 as a tactical political alliance between Zimbabwe's two main nationalist organisations, ZANU and ZAPU. Joshua Nkomo, ZAPU's leader and Robert Mugabe, ZANU's leader, intended to go together to the Geneva conference at the end of the month with an agreed list of demands under the banner of the Patriotic Front. The Organisation of African Unity as well as the 'frontline' states had pressed hard for the alliance and gave it their full support in the struggle for independence. The PF denounced the internal settlement in Zimbabwe involving Bishop Muzorewa as an attempt to install a 'puppet' government, and threatened to intensify the guerrilla war. ZANU and ZAPU retained their separate identities and armies throughout the period and fought the 1980 elections separately, Joshua Nkomo adopting the name Patriotic Front as ZAPU's official name. Now ZANU continues to be known as ZANU-PF to distinguish it from Ndabaningi Sithole's very small breakaway ZANU party. However, the PF itself was formed to bring about independence and since then it has ceased to have any relevance.
See also **'frontline' states; Geneva conference; internal settlement; Mugabe; Muzorewa; Nkomo; Sithole, Ndabaningi; ZANU; ZAPU**.

PCC *(Zimbabwe)* See **People's Caretaker Council**.

PDA *(Angola)*
Formed from ALIAZO, the *Partido Democrático Angolano* (PDA), or Democratic Party of Angola, was one of the constituent parties of the GRAE and the FNLA. It continued to function separately within the FNLA until Holden Roberto, FNLA leader, consolidated his power in 1972–3, re-organising the FNLA, purging some PDA leaders and merging the PDA and UPA within the FNLA.
See also **ALIAZO; FNLA; Roberto; UPA**.

Pearce, Edward Holroyd *(Zimbabwe)*
Head of the Pearce Commission which was sent out to Southern Rhodesia to test majority opinion over the Anglo-Rhodesian agreement of 1971. See **Pearce Commission**.

Pearce Commission *(Zimbabwe)*
The Pearce Commission was set up after the Anglo-Rhodesian agreement of 1971. The agreement called for the British government to be satisfied that the basis for settlement was acceptable to 'the people of Rhodesia as a whole' before Britain would grant recognition to the rebel Smith regime. This notion was conceded by Ian Smith in much earlier negotiations during 1964 when the then British Prime Minister, Sir Alec Douglas-Home, outlined the British terms for independence – the so-called five principles or NIBMAR. These included making certain that any settlement was acceptable to the country's black majority.

Hence Lord Pearce arrived in what was still in international law the colony of Southern Rhodesia in January 1972 with the aim of sounding out black opinion. He sent seven teams accompanied by translators out to the rural areas in cars flying the British flag. Riots and noisy crowds greeted the commissioners with lists of grievances against the rebel regime. During these disturbances fourteen people were shot dead by police and many nationalists were arrested. Despite the violence, the commissioners went about their meetings – only eight of the planned 56 had to be postponed.

The message was clear: dissatisfaction with the Smith government and anxiety over the possibility of the proposals being accepted. As far as the whites were concerned, thousands wrote to Lord Pearce favouring the settlement proposals as the last chance for peace and racial harmony. Lord Pearce's verdict was handed to Sir Alec Douglas-Home on 4 May 1972. 'In our opinion', he concluded, 'the people of Rhodesia as a whole do not regard the proposals as acceptable as a basis for independence'. As a result, the Anglo-Rhodesian agreement was never put into operation.
See also **Anglo-Rhodesian agreement; NIBMAR; Smith**.

PEBCO *(South Africa)*
PEBCO, the Port Elizabeth Black Civic Organisation, developed very quickly into a significant mass community organisation. Branches were set up all over the townships near Port Elizabeth, and in 1980 the ratepayers' associations of Malabar and Gelvandale also joined, bringing with them

'Indian' and 'Coloured' supporters. Public meetings were held and these were soon attended by thousands of people. The aim of PEBCO – founded in October 1979 as an amalgamation of civic associations in the area – was to attack apartheid and fight for equal civic rights for the area's black population. It opposed the government-sponsored community councils and was seen by the population as a relevant and significant force which could affect their daily lives.

PEBCO became widely known outside the area of Port Elizabeth after its involvement in two strikes at the Ford Motor Company factory from November 1979 to January 1980. The first followed an attempt to dismiss Thozamile Botha, a work study trainee at the factory, and one of PEBCO's most prominent leaders. He was reinstated but another strike followed at the Cortina plant over workers' grievances. This led to the formation of a new trade union, the Motor Assembly and Component Workers' Union of South Africa (MACWUSA). In 1980 the state moved in earnest against PEBCO, detaining and banning Thozamile Botha and other members of the organisation's executive.

Pemba (*Tanzania*)
Pemba is an island off the Tanzanian coast which joined with the larger island of Zanzibar and Tanganyika to form the United Republic of Tanzania in 1964. Zanzibar and Pemba together with several islets had become a British protectorate in 1890 under the Sultan of Zanzibar, Sayyid Jamshid ibn Abdullah. They became independent as Zanzibar in 1963.
See also **Tanzania; Zanzibar**.

People Against Racial Discrimination (*Zimbabwe*)
People Against Racial Discrimination (PARD) was launched by whites inside Southern Rhodesia (later Zimbabwe) in support of the 1971 Anglo-Rhodesian settlement. It attracted the support of prominent church leaders and businessmen. However, although it was intended to be multi-racial, very few Africans joined. PARD agreed that the settlement was the best thing that could be achieved at the time. The organisation was discredited after the Pearce Commission concluded that majority black opinion was against the settlement proposals, and it was disbanded.
See also **Anglo-Rhodesian agreement; Pearce Commission**.

People's Armed Forces For the Liberation of Angola (*Angola*) See **FAPLA**.

People's Assembly (*Mozambique*)
The People's Assembly, or *Assembleia Popular*, is the parliament or supreme organ of the Mozambican state and meets in ordinary session at least twice per year. Between meetings of the Assembly its functions are performed by a 15-member Permanent Commission (comprising six Politburo members and nine government officials) elected from among its number. It has 210 members who include the Central Committee of FRELIMO, ministers and vice-ministers (appointed by the President of Mozambique), provincial governors (also appointed by the President),

representatives of the armed forces, representatives from each province and ten other Mozambican citizens.

The Council of Ministers, headed by the President, is the government of the RPM and answers to the People's Assembly. Its actions must be in accordance with the resolutions of the Assembly and the decisions of the President.

During the early 1980s the role of the People's Assembly became unclear and it failed to have even the required two meetings per year. It became largely a rubber-stamp for the decisions of the Permanent Commission.

There are also local and provincial people's assemblies or councils, the *Assembleias do Provo* which perform similar functions at local and provincial levels.

See also **Council of Ministers; FRELIMO; Permanent Commission; RPM**.

People's Caretaker Council (*Zimbabwe*)
Loyal ZAPU supporters inside Zimbabwe set up the People's Caretaker Council (PCC) in August 1963 after ZAPU had been banned. ZAPU had moved its headquarters outside Zimbabwe (then Southern Rhodesia). Joshua Nkomo was still regarded as the head of the PCC and the organisation continued to be known as ZAPU, despite the new name. On 26 August 1964 the PCC was banned together with ZANU and many nationalist leaders detained.

See also **Nkomo; ZANU; ZAPU**.

People's Defence Organisation (*Angola*) See **ODP**.

People's Liberation Army of Namibia (*Namibia*) See **PLAN**.

People's Movement for the Liberation of Angola (*Angola*) See **MPLA**.

People's Power (*Angola, Mozambique*) See ***Poder Popular***.

People's Republic of Angola (*Angola*) See **Angola**.

People's Republic of Mozambique (*Mozambique*) See **Mozambique**.

People's United Democratic Movement (*Swaziland*)
The People's United Democratic Movement (PUDEMO) became known in early 1985 when it claimed responsibility for a campaign of anti-government painted slogans and underground pamphlets. It has not been heard of since.

Permanent Commission (*Mozambique*)
Elected from among the members of the People's Assembly on the proposal of the Central Committee of FRELIMO, the 15 members of the Permanent Commission fulfil the functions of the People's Assembly in between its sessions. The meetings of the Permanent Commission are called by the President who also presides over them.

See also **FRELIMO; People's Assembly**.

PFP (*South Africa*) See **Progressive Federal Party**.

picador (*Mozambique*)
This was the term given to black soldiers in the Portuguese colonial army in Mozambique who, in the latter stages of the liberation war, preceded troop convoys on foot, probing the dirt roads with long poles to locate mines before they exploded under the vehicles.

PIDE (*Angola, Mozambique*)
PIDE or the *Polícia Internacional de Defesa do Estado* was the colonial security police force, notorious for its brutality and the torture of detainees. It was renamed the *Direcção de Seguranca* (DGS) in 1969, but continued to fulfil the same function with the same notoriety. PIDE was used by the Portuguese to try to stifle dissent and opposition to their regimes, and it was PIDE's crack-down on nationalist leaders in Angola in the late 1950s which provoked the launch of the armed struggle by the MPLA.
See also **MPLA; UNITA**.

Pinto de Andrade, Joaquim and Mario (*Angola*)
Joaquim and Mario were brothers active in the nationalist struggle and early supporters of the MPLA. Both later fell foul of the movement and became part of the Active Revolt breakaway. Joaquim was a prominent priest who, following his arrest and imprisonment by the Portuguese, was made honourary president of the MPLA in 1962. He was later imprisoned by the Angolan Government in 1976, but was subsequently released. Mario was a founding member of the MPLA and its president from 1960 to 1962, when he was replaced by Agostinho Neto. A poet and writer, Mario was one of the leading intellectuals of the nationalist movement in Angola.
See also **Active Revolt; MPLA; Neto**.

Plaatje, Solomon Tshekiso 'Sol' (*South Africa*)
Sol Plaatje was one of the founders and the first secretary-general of the ANC of South Africa (or the South African National Native Congress as it was then known). He is one of the organisation's best known historical figures and was a journalist and writer as well as a political activist. He led the organisation through its first major campaign against the Land Act of 1913 in which the government curbed the rights of Africans to own land in most of South Africa. This took him and other ANC leaders to London to lobby the Colonial Office. The protest was unsuccessful but Sol Plaatje stayed in Britain for some years lecturing about the plight of black South Africans and publishing books.
See also **ANC; Dube**.

PLAN (*Namibia*)
Through its military wing, the People's Liberation Army, SWAPO has been waging a war against the illegal South African occupation forces for over 20 years. By contrast, the liberation struggle in Mozambique lasted ten years. The armed struggle was launched on 26 August 1966 with an

attack on a military base near Ombulumbashe in the north of Namibia. Since then SWAPO has maintained a constant military pressure on South Africa, while pursuing all available diplomatic avenues to maximise external pressure.

SWAPO's military wing was re-organised in 1973 and given the name of PLAN. Despite periodic South African claims that it has been defeated, and repeated raids into neighbouring Angola, ostensibly in pursuit of PLAN soldiers, PLAN's pressure has been unrelenting. It has forced South Africa to maintain a huge force permanently stationed in Namibia, creating enormous costs in financial and political terms.

Current estimates of PLAN strength suggest that it has about 8,000 active members, including both men and women. They rely on the support and sympathy of local people, mainly peasants and the families of migrant workers, and have been able to tie up an army over ten times their number. Despite large rewards for the capture or killing of PLAN members, and for information about PLAN, few people collaborate with the South Africans.

South Africa's arming and supplying of UNITA, the organisation waging a war against the Angolan government, has had the dual purpose of destabilising that country and disrupting SWAPO's lines of communication and supply. Although SWAPO's relationship with UNITA was a workable one during the Angolan liberation struggle, the latter now actively opposes SWAPO and attacks PLAN detachments.
See also **SWAPO; SWATF**.

p

PLUA (*Angola*)

The PLUA or *Partido da Luta Unida dos Africanos de Angola* was a Marxist party which was one of the forerunners of the MPLA.
See also **MPLA**.

Poder Popular (*Angola, Mozambique*)

Meaning 'People's Power', this is the term for popular participation in the system of government in Angola and Mozambique. In Angola this was organised through street, neighbourhood, village and district committees, while in Mozambique, 'dynamising groups' were set up in workplaces, neighbourhoods and villages. *Poder Popular* was used by the MPLA and FRELIMO, initially, to provide for democratic participation and control in the areas under their aegis. After independence it was organised into national structures with the people's assemblies at their heads.
See also ***bairro*; dynamising groups; FRELIMO; MPLA**.

Pokela, John Nyati (*South Africa*)

John Pokela became the leader of the PAC in January 1981. However, he died suddenly in Harare on 13 July 1985 and was succeeded by Johnson Mlambo. John Pokela took over from a three-man presidential council set up after Potlake Leballo had been forced to resign as president in 1979. The organisation was riven by internal feuding and Vus Make, one of the members of the presidential council, admitted in 1979, at the

Organisation of African Unity summit conference, that 'gangsterism' was draining the movement of its vitality.

John Pokela was imprisoned in 1967 for thirteen years under the Sabotage Act and the Suppression of Communism Act. He served his sentence on Robben Island and was released in 1980 and confined to a town in the Transkei. He subsequently fled South Africa, arriving in Dar es Salaam, Tanzania, in 1981. He began to try to weld the PAC together again and set about reorganising the movement but died before he had been able to achieve his objectives.

See also **Leballo; Mlambo; PAC; Sobukwe**.

p

Police Zone (*Namibia*)

The Police Zone is an area of central and southern Namibia bounded by a line drawn by the German colonial administration to denote areas under its direct control. It comprised, largely, the pre-colonial territories of the Nama and Herero groups. The Germans never established complete control over the north of the country.

See also **German colonialism; Namibia; Operational Area**.

Polícia International de Defesa do Estado (*Angola*) See **PIDE**.

Political Students' Organisation (*South Africa*) See **POLSTU**.

POLSTU (*South Africa*)

The Political Students' Organisation (POLSTU) was formed in July 1980 by a group of liberal Afrikaans students who broke away from the conservative *Afrikaanse Studentebond* (ASB). Its membership is open to all those who profess Christianity and 'loyalty to South Africa', and it campaigns for an end to statutory discrimination, for full citizenship and for equal political participation irrespective of colour.

By the end of its second year POLSTU had 600 members drawn from most white campuses, but had failed to achieve affiliations from any of the student bodies. It characterised itself as the moderate alternative to the radical politics of NUSAS and the reactionary politics of the ASB.

See also **ASB; NUSAS**.

Popular Forces for the Liberation of Mozambique (*Mozambique*)

The armed forces of Mozambique: see **FPLM**.

Poqo (*South Africa*)

Poqo is a Xhosa expression meaning 'alone' or 'pure'. It was used sometimes in 1960 by leaders of the Pan-Africanist Congresss to describe the character of the PAC in contrast to the multi-racial ANC. Gradually it developed as the term used for the fighting wing of the PAC, set up after the PAC and the ANC were banned in 1960. *Poqo's* message was direct. For instance in 1962 when addressing farmworkers it advocated taking the land away from whites and giving it to Africans. The message read as follows: 'the black men of *Poqo* must attack whites and take their possessions and advantages for themselves'. The philosophy of the organisation was different to that motivating the ANC's armed wing, *Umkhonto we Sizwe*, in that it advocated a leading role for the masses in its campaigns; people should be given choices about the attacks and the weapons used. This stood in contrast to *Umkhonto's* strategy of careful planning of military ventures by the leadership.

In 1962 and 1963 *Poqo* violence escalated. Informers and policemen were killed; some whites were killed as a warning; assassination attempts were made on chiefs in the Transkei who were seen to be going along with apartheid policies; and in Paarl in the Cape a plan for a general uprising by some 250 crudely armed men began but was stopped by the police. Mass arrests broke the power of *Poqo* in the early 1960s and after that time it ceased to be a strong force in the country, but operated and organised from exile.

See also **Pan-Africanist Congress; *Umkhonto we Sizwe*.**

p

Port Elizabeth Black Civic Organisation (*South Africa*) See **PEBCO**.

Povo (*Mozambique, Zimbabwe*)

Povo is the Portuguese word for 'people' or 'the masses'. It was widely used in Zimbabwe during the guerrilla war, having been spread by Zimbabwean fighters who trained in Mozambique. It is increasingly used to refer to rural people or peasants.

President's Council (*South Africa*)

The President's Council was appointed in February 1981 to examine possible constitutional changes. It had 56 members (41 'White', 7 'Coloured', 7 'Indian', 1 'Chinese') and was chaired by the Vice-President of the Republic. It set up five permanent committees, dealing with planning, finance, community relations, constitutional and scientific matters. The 1981 President's Council was disbanded on 30 June 1984 having fulfilled one of its major briefs in that the Republic of South Africa Constitution Act of 1983 had been enacted.

A new 60–person President's Council came into being in September 1984. It represented the eight major parties in Parliament as follows: 41 'Whites', 13 'Coloureds' and 6 'Indians'. The Progressive Federal Party agreed to participate in the new council, having refused to be part of the original President's Council. Dr Koornhof, the former Minister of Cooperation and Development, was elected Chairman. 'The offering of advice to Parliament in regard to current affairs should still remain the council's primary function', he said in his acceptance speech.

See also **constitution of 1983**.

press (*Namibia*)

The majority of newspapers in Namibia serve the white minority and are controlled by whites. Only two papers, *Omaweto* (The Friend), the paper of the Lutheran Church in Namibia, and SWAPO's clandestine *Ombuzeya Namibia*, are published in indigenous languages and controlled by black Namibians.

The *Windhoek Advertiser*, the main daily paper, was taken over by a publishing firm in West Germany in April 1978 and was an outspoken supporter of the DTA and its 'transitional government'. A weekly paper, the *Windhoek Observer*, was more independent editorially, but highly idiosyncratic and consistently hostile to SWAPO. It has since toned down its criticism of the South African presence. A new weekly paper first published in 1985, *The Namibian*, has taken a more robustly independent and objective editorial approach than others, and is the only reliable commercial paper. All non-clandestine newspapers in Namibia are subjected to official censorship. They may not publish articles on the military situation, prisons or the police without official approval.

All radio and television broadcasts from within Namibia are the prerogative of the SWA Broadcasting Corporation (SWABC), under the direct control of the Administrator General. Like the SABC, it is largely propaganda and is used to convey official views and accounts of events. The SWABC has attempted to reduce the audience of 'Voice of Namibia', SWAPO's external radio station which broadcasts from independent African countries.

See also **'Voice of Namibia'**.

press (South Africa)

Commercial newspapers in South Africa are broadly divided between English-language papers and those published in Afrikaans. The former are almost all owned by two companies, the Argus Printing and Publishing Company, and South African Associated Newspapers (SAAN). While there has been intense competition between the two companies, there is a significant cross-holding between them, and the same outside group, Anglo American, also has a controlling interest in each. Partly as a result of this, the newspapers of both groups have taken largely similar political stances, usually in support of white parliamentary opposition parties.

Argus' main titles include the *Star, Argus, Daily News, Sunday Tribune, Diamond Fields Advertiser, Friend* and *Pretoria News*. In addition it publishes several titles aimed exclusively at black readers, including the *Sowetan, Post Natal, Ilanga* and *Cape Herald*. SAAN titles include the *Sunday Times, Cape Times, Eastern Province Herald, Evening Post, Business Day* and the weekly *Financial Mail*. Its best known title, the *Rand Daily Mail*, and a weekly paper, the *Sunday Express*, were discontinued amidst considerable controversy in 1985. The *Rand Daily Mail*, which had the reputation of being the most liberal of the mainstream papers, had achieved the second highest sales figures of any daily newspaper, between January and June 1984. The two papers were closed following the intensification of competition between the Argus and SAAN groups, and the launching of a new Sunday paper, the *Sunday Star*, in direct competition with the most profitable SAAN paper, the *Sunday Times*. Ex-*Rand Daily Mail* journalists subsequently set up the *Weekly Mail* as a new liberal paper. Four other English language newspapers which have remained independent are the *Natal Mercury, Natal Witness* (with the same owners), *Daily Dispatch* and *Citizen*. The last was set up by the government Information Department as an undercover project to create a pro-government English-language paper and was subsequently bought by the Afrikaans newspaper group *Perskor*.

Commercial Afrikaans-language newspapers are largely controlled by two groups, *Nasionale Pers* and *Perskor*. The former and larger controls titles which generally support the Botha government and include *Die Burger, Die Volksblad, Die Oosterlig*, and *Beeld*, and a black paper, *City Press*. *Perskor's* main titles are *Die Vaderland, Die Transvaler* and the *Citizen*, as well as *Rapport* jointly owned with *Nasionale Pers*. Some *Perskor* papers show sympathy to policies to the right of those currently pursued by the Botha government.

In addition to the commercial press, there is a growing alternative press in South Africa which includes the underground publications of the liberation movement (such titles as *Sechaba, African Communist* and *Workers Unity*), those of the independent trade union movement (including the *South African Labour Bulletin*), of student organisations (including *SASPU National*, a highly regarded paper published by NUSAS), and independent titles like *Grassroots* and *Work in Progress*.

The broadcast media in South Africa are dominated by the South African Broadcasting Corporation (SABC) which controls commercial and non-commercial radio stations and television. After the National Party victory in 1948 the SABC became increasingly pro-government in its editorial position, and after the appointment as Director General in 1958 of P. J. Meyer, former Chairman of the *Broederbond*, it became slavishly so. More recently, some commercial radio stations have been established in the 'Homelands', and short-wave broadcasts from outside South Africa, including some by the ANC, can be received. During the South African invasion of Angola in 1975, when there was a near total blackout on internal news of the operation, it was through broadcasts such as those of the BBC World Service that South Africans learned of the activities of their own government.

The South African press is subject to considerable restrictions and controls which make it impossible to operate in anything approaching a free way. Apart from legislation limiting what can be said, quoted or photographed, journalists have been arrested, searched, threatened, beaten, tortured, banned, expelled, sacked and bought off by the government at different times. In addition papers have been closed down and the Newspaper Press Union forced to accept self-censorship. Most constraining of all, however, are the restrictions imposed under the state of emergency in 1986 prohibiting the publication of any independent stories on 'unrest' or political violence in South Africa without prior approval. The result has been a dramatic fall in both internal and international coverage of struggles against apartheid, and a consequent improvement of the image of the government.

See also **ANC; Anglo American; NUSAS; state of emergency**.

Press Holdings (*Malawi*)

Press Holdings is a group of companies controlled by Hastings Banda, Malawi's president. Dr Banda is an important shareholder in Press Hold-

ings which owns some eleven companies with seventeen subsidiaries. Its main investments are in the most important areas of the Malawian economy such as the tobacco estates.
See also **Banda**.

Proclamations AG8/9 (*Namibia*) See **AG8/9**.

production councils (*Mozambique*)

Organisations of workers in factories in Mozambique: see **factory production councils**.

p

Programme of Action (*South Africa*)

The Programme of Action was adopted by the ANC in 1949. It was originally the Youth League Programme and was drawn up by the Congress Youth League. Its adoption was a considerable ideological success for the Youth League and marked the start of a period of strong influence by the Youth League on the ANC itself. The programme was a militant document stating ANC principles. Congress, it argued, should reject white leadership and segregation and struggle for freedom. It recommended a massive programme of civil disobedience and non-cooperation to achieve these ends, including boycotts and strikes. African economic power should be extended and African education improved. A national fund should be set up as well as a national press. All this was to take place under the inspiring banner of 'African Nationalism'.
See also **ANC; Congress Youth League; Defiance Campaign of 1952**.

Progressive Federal Party (*South Africa*)

In 1959 almost a third of the United Party's MPs, together with other party officials and representatives, broke away over the issue of its 'colour policies'. They subsequently formed the liberal Progressive Party which campaigned on a platform of a qualified but non-racial franchise, arguing that the alternative was revolution. The formation of the party immediately preceded the peak period of confrontation between the National Party and the Congress Alliance, at the start of the 1960s.

In the first general election after its formation the Progressive Party lost 12 of its 13 parliamentary seats. It was left with only one MP, the redoubtable Helen Suzman, for the next 13 years. In 1974 it won a further six seats, and a subsequent merger in 1975 with the Reform Party (which had split from the United Party after the 1974 general election) increased its representation to 11 MPs. At the merger congress the party's name was changed to South African Progressive Reform Party, a name which lasted only until 1977, when, after a further merger with a breakaway group from the United Party it became the Progressive Federal Party of South Africa (PFP). In the general election of that year the PFP won 17 seats and became, for the first time, the official parliamentary opposition to the governing National party. In 1981 the PFP increased its total of seats to 26, winning 18.1 per cent of the vote. In the 1987 election, however, the party lost its position as official opposition to the right wing Conservative Party, winning 20 seats and 14 per cent of the vote.

In 1983 the PFP suffered a setback when it campaigned for a 'No' vote in the 'Whites'-only referendum on the new constitution. The constitution was approved by a large majority. It suffered a further setback in 1986, when its effective and charismatic leader, Frederik van Zyl Slabbert, resigned on the grounds that the party had no realistic chance of becoming the government and that parliamentary politics were no longer the primary political locus. A former leader, Colin Eglin, who had been supplanted by Slabbert in 1979, was elected to replace him.

The spectacular growth of the PFP during the 1970s after so long in the political wilderness, reflected the polarisation of white politics which took place over that decade, and which, having destroyed the United Party, is presently tearing the National Party apart. This process affected all facets of South African society, and not least that of business. During its years as a single MP party, the Progressive Party survived because of the financial support of one man, Harry Oppenheimer. Oppenheimer, at the time the Chairman of the Anglo American Corporation, the dominant multi-national in South Africa, supported the party out of the far-sighted recognition that the state would not be able to continue forever the material and political subjugation of black South Africans. He favoured the development of a less repressive state which provided for upward mobility for a black middle-class elite and which conferred political rights on its members as they achieved set material or educational goals.

By the 1980s this view had gained sufficient currency among leading capitalists in the country to ensure that the PFP achieved a far wider base of electoral and financial support, and to contribute significantly to the pressure for change within the National Party. Despite the much-vaunted 'reformism' of the National Party under P. W. Botha, however, the PFP remained the primary political voice of liberal capitalists in South Africa.

Since the early 1970s, the party has had a warm relationship with M. G. Buthelezi, leader of the KwaZulu bantustan and the Zulu political organisation, *Inkatha YeNkululeko ye Sizwe*. Although it attracted a limited following among 'Coloured' people in the Cape during the 1960s, it is to its relationship with Buthelezi that the party has consistently pointed as evidence that its policies would be acceptable to black South Africans.

See also **Anglo American Corporation; Buthelezi; constitution of 1983; *Inkatha*; National Party; Oppenheimer; Reform Party; 'reforms'; United Party**.

Progressive Party (*Namibia*)
A constituent party of the Namibia National Front: see **National Convention of Namibia**.

Progressive Party (*South Africa*) See **Progressive Federal Party**.

Progressive Reform Party (*South Africa*) See **Progressive Federal Party**.

Progs (*South Africa*)
'Progs' is a term used for the old Progressive Party and is now sometimes

applied to the Progressive Federal Party. It is also used in the form 'Old Progs' to mean those members of the original Progressive Party still active in the Progressive Federal Party.
See also **Progressive Federal Party**.

Proposal for a Settlement of the Namibian Situation (*Namibia*)
The basis for a settlement in Namibia put forward by the Contact Group: see **Contact Group**.

p

protected villages (*Zimbabwe*)
Protected villages, or PVs, were fenced, heavily guarded areas where rural Zimbabweans were resettled to prevent them from supporting the guerrillas during the country's liberation struggle. After people had been evacuated to the PVs their property was often confiscated and their homes burned. The programme of moving people into these areas was extended as the war continued. From 1974 to 1976 some 200,000 people were forced into new government-controlled compounds, regarded by the nationalists as concentration camps. Most were in Mashonaland and had the effect of increasing support for Robert Mugabe's ZANU party which drew most of its support from Shona-speakers.

The scheme had been used in the counter-insurgency campaign in Malaya in the 1950s and its success influenced some young white officers in the rebel Rhodesian forces who had served there in the British army. The technique had also been tried elsewhere, in countries such as Vietnam, Angola and Mozambique.

PRP (South Africa)
South African Progressive Reform Party: see **Progressive Federal Party**.

PUDEMO (*Swaziland*) See **People's United Democratic Movement**.

Putterill, Sam (*Zimbabwe*)
Major-General Sam Putterill was the Rhodesian armed forces commander from 1965 to 1968. This was the period of the first guerrilla incursions. The commander was considered to hold extreme left-wing views by the rebel minority leader, Ian Smith. This was largely because of his argument that Africans, under the 1969 constitution, would have no hope of effecting any changes in the system through legitimate political action and would therefore inevitably turn to military action. The constitution, he felt, would help the guerrillas to recruit new members. Sam Putterill gave his political support to the opposition Centre Party.
See also **Centre Party; constitution of 1969**.

R

Raditladi, Leetile Disang (*Botswana*)
Founder and President of the Bechuanaland Protectorate Federal Party: see **Bechuanaland Protectorate Federal Party**.

Ramaphosa, Matamela Cyril (*South Africa*)
Cyril Ramaphosa has been the general-secretary of the NUM, South Africa's largest single trade union since its formation in 1982. He joined CUSA, the Council of Unions of South Africa, in 1981 and became an adviser in its legal department. He had previously studied law, but joined CUSA before being admitted as an attorney. He subsequently led the NUM into COSATU, the country's main trade union federation, when it was formed in 1985, and is now a leading figure in this influential new grouping. He has been detained by the South African police for his activities.
See also **COSATU; NUM**.

Rand Monetary Union (*Lesotho, South Africa, Swaziland*)
The Rand Monetary Union (in the Rand Monetary Area – RMA) was introduced in December 1974. It initially consisted of Botswana, Lesotho, Swaziland and South Africa (as well as South African-occupied Namibia). Botswana withdrew from the arrangement leaving the others as members. On 1 July 1986 the Rand Monetary Union was replaced by the Tripartite Monetary Agreement under which the rand has ceased to be legal tender in Swaziland and the authorities are left free to determine the exchange rate policy. In practice the rand remains in circulation in the country and the currencies are still linked but Swaziland now has the right to float its exchange rate if the rand falls to an unacceptable level.

The monetary union used to give South Africa considerable monetary control over the countries concerned, dramatically reducing the autonomy of their central banks. It may be seen as part of South Africa's design to dominate the region economically and financially. Any international pressure on the rand resulting from South Africa's apartheid policies was felt by the members of the union.

Under the RMA the countries received interest on rand currency circulating in their jurisdictions. By 1980 Lesotho, Botswana and Swaziland had all moved towards monetary independence by introducing their own currencies. However Lesotho's *loti* and Swaziland's *lilangeni* are still on a par with the South African rand.
See also **SADCC; Southern African Customs Union**.

Reconstituted European Advisory Council (*Swaziland*)
The Reconstituted European Advisory Council was formed in 1949 by government proclamation. It replaced the European Advisory Council and represented white interests in Swaziland. It consisted of ten elected members and seven non-voting government officials and its official purpose was to advise the resident commissioner about matters concerning whites in the country. The council was conservative and tended to represent the pro-South African sympathies of many whites. It was led by Carl Todd and attacked the colonial administration on many issues, particularly on all moves to institute a system of universal suffrage in Swaziland.

The council was keen to establish a legislative council. The idea was to share power with the conservative Swazi National Council and shut out the emerging political parties. It succeeded in 1964, when the Legislative Council replaced the Reconstituted European Advisory Council. In order to campaign for the council elections the whites formed the United Swaziland Association, winning significant representation in the vote for this multiracial body.
See also **European Advisory Council; Legislative Council; Todd**.

r

Reform Party (*South Africa*)
Not to be confused with the party of the same name formed from the white United Party, the Reform Party was formed by members of the SAIC, the South African Indian Council – a government-sponsored body – in 1976. It was seriously challenged by the campaigns against the SAIC elections in 1981 and withdrew its candidates as a result. One, however, did stand and won election to the discredited council.

The Reform Party is a member of SABA, the South African Black Alliance, chaired by Chief Buthelezi. It rejected participation in the 1983 constitution and advocated a boycott of the 'Indian' election. It is led by Yellan Chinsamay.
See also **House of Delegates; Natal Indian Congress; SABA; SAIC**.

Reform Party (*South Africa*)
Not to be confused with the party of the same name formed by members of the SAIC, the Reform Party was born in 1975 out of a series of expulsions of so-called 'Young Turks' from the United Party. Following the expulsion of Transvaal United Party Leader, Harry Schwarz, the bulk of the party's public representatives in the Transvaal left with him to form their own party. Later the same year they merged with the Progressive Party to form the South African Progressive Reform Party.
See also **Progressive Federal Party; United Party**.

'reforms' (*South Africa*)
In 1979 Dr Piet Koornhof, then the National Party government's Minister for Cooperation and Development, declared 'apartheid is dead'. The 1970s marked the start of a campaign first by Prime Minister Vorster and then taken up and developed by President P. W. Botha to 'reform' the system, stimulated by the needs of South African business for a more

settled, skilled labour force. What this involved was a restructuring of apartheid, a process which is still continuing in an attempt to retain white control while eliminating the more superficial aspects of the system. Two commissions were appointed in 1977 – the Wiehahn Commission and the Riekert Commission – to consider changes. They published reports in 1979 which influenced government policy on the 'reforms'. The response to Dr Koornhof's remarks from Archbishop Tutu, then secretary general of the South African Council of Churches was, if apartheid is dead, we 'want to see the corpse first'. Eight years later (1987) black South Africans are still living under apartheid's laws.

The central changes introduced since 1979 are the following: independent (non-racial but mainly black) trade unions were allowed to register and were brought into the official negotiating system (1979); a new constitution was introduced in 1983 incorporating 'Indians', 'Coloured' and 'Whites' in a three-chamber parliament; the Immorality Act and the Mixed Marriages Act were abolished (1985); 'job reservation' laws were repealed (1980 and, in the mines, in 1986); the pass laws were replaced with a new form of identification (1986) and South African citizenship was restored to a limited number of people from the independent 'Homelands' (1986). There is an attempt underway to give urban 'Blacks' some political participation outside the 'Homelands', probably through a system of local government and a consultative forum known as the National Council. There is also talk of replacing the Group Areas Act with something less overtly racist.

See also **apartheid; Bantu Education Act; Botha, P.W.; Group Areas Act; 'Homelands': Immorality Act; pass laws; total strategy; urban 'Blacks'**.

Reformed Trade Union Congress (*Zambia*)

The Reformed Trade Union Congress was formed after a split from the Trades Union Congress in 1960. The split took place over the policies of the TUC leader, Lawrence Katilungu, who argued strongly that unions should not become politicised and was accused of not pursuing a radical enough leadership of the congress. The Reformed Trade Union Congress was soon to become the United Trades Union Congress and the Zambia Congress of Trade Unions in 1965.

See also **Katilungu; Mineworkers' Union of Zambia; Zambia Congress of Trade Unions**.

refugees (*Botswana*)

Botswana has always been a place of political refuge for people from all over the region. Most recently the refugees have come from South Africa and Zimbabwe. In February 1979 (just before Zimbabwean independence) there were an estimated 18,000 Zimbabwean refugees in the country and those from South Africa have increased as the violence there has escalated. Even after Zimbabwean independence refugees caused political tension between Botswana and Zimbabwe. ZAPU supporters (opponents of the Zimbabwean government) have continued to flee to Botswana and in 1983 their numbers were estimated at 3–4,000 at the Dukwe camp. This led to stricter controls being imposed by the Botswana

government. Numbers of Zimbabwean refugees have decreased as the ZANU government has made its peace with ZAPU, but there will be many more South Africans seeking the sanctuary of Botswana in the years to come.
See also **Masire**.

Rehoboth Baster Vereniging (*Namibia*)
One of the small constituent parties of the Democratic Turnhalle Alliance: see **DTA**.

Rehoboth Bevrydingsparty (*Namibia*)
Small, ethnically-based party in coalition with the Rehoboth Democratic Party: see ***Bevryder Demokratiese Party***.

Rehoboth Democratic Party (*Namibia*)
Small, ethnically-based party in coalition with *Rehoboth Bevrydingsparty:* See ***Bevryder Demokratiese Party***.

Rehoboth Liberation Front (*Namibia*) See **Liberation Front**.

Rehoboth Volksparty (*Namibia*)
One of the original parties in the NCN, it later dissolved itself and joined SWAPO: see **National Convention of Namibia**.

relations with South Africa (*Lesotho, South Africa*)
Lesotho's relations with South Africa deteriorated seriously towards the end of Chief Leabua Jonathan's rule. In the latter years they had degenerated to the point of sanctions being imposed by South Africa against Lesotho, and of South African military raids into the country.

Lesotho is important to South Africa because it forms a potential natural haven for the guerrillas of the ANC of South Africa. At the same time, however, it depends on South Africa for its economic survival. The early 1980s saw South Africa constantly trying to force Lesotho to sign a security agreement pledging co-operation against the ANC. Lesotho resisted these pressures, but at considerable cost, including a military coup in 1986.

In 1978 South Africa imposed stringent border controls between Lesotho and the Transkei because Lesotho refused to recognise the Transkei (one of South Africa's 'Homelands') as an independent state. The United Nations and the EEC had to help Lesotho to survive the crisis. The South Africans subsequently launched several military raids against Lesotho in an attempt to batter the country into expelling the ANC of South Africa. The worst took place in December 1982, when a South African commando raid on Lesotho's capital city, Maseru, left 42 people dead, many of them Basotho. More border controls were imposed in 1983 leading to food shortages in Lesotho. Soon some 3,000 refugees were expelled from Lesotho following a South African ultimatum either to act or to face the economic consequences. Despite this, South Africa imposed fresh border restrictions in January 1986 demanding the expulsion of ANC cadres from the country. With Lesotho in a state of seige and fresh food running out, Chief Jonathan was deposed in a military

coup. He was replaced by a conservative soldier Major General Lekhanya, who immediately deported some 100 ANC members. There was also a noticeable decline in anti-Pretoria rhetoric, and South Africa reopened the question of a security pact with the country.
See also **Jonathan; Lekhanya; migrant workers; Moshoeshoe**.

relations with South Africa (*South Africa, Swaziland*)

Swaziland signed a secret, non-aggression pact with South Africa in February 1982, which was revealed on 30 March 1984. The terms commit both states to 'combat terrorism, insurgency and subversion individually and collectively' and to 'call upon each other wherever possible for such assistance and steps as may be deemed necessary'. Each country undertook to respect the other's sovereignty and territorial integrity and to prevent any activity within its borders which might threaten the integrity of the other. Both countries were prohibited from allowing foreign powers to establish bases in their territory from which to attack the other.

Speaking at a press conference after the pact had been revealed the South African Foreign Minister, 'Pik' Botha, compared the agreement to the Nkomati Accord which South Africa had recently signed with Mozambique. He claimed that this differed from the Swaziland agreement in two respects – South Africa and Swaziland had pledged to assist each other in the 'fight against terrorism', and had agreed not to allow any foreign troops or military bases in their respective countries without notifying the other. It was felt that only after the signing of the similar Nkomati Accord could Swaziland afford to make its relationship with South Africa public. In practice the pact is part of South Africa's plan for the region of southern Africa, and involves persuading or terrorising surrounding states into refusing help or bases to ANC activists who aim to overthrow the South African state.

Since 1984 Swaziland has increased its harassment of ANC members, deporting many and imprisoning many others. Relations with South Africa have also become more and more visible; for instance an official visit to South Africa, led by the Swazi prime minister, took place in 1984. Furthermore an agreement with South Africa to establish trade missions was signed on 27 December 1984.

None of this has prevented South Africa from launching raids into Swaziland which it claims are in order to pursue ANC guerrillas. These raids are also used to gather information on exiled South Africans as well as the ANC. For instance in 1986 an organisation which runs scholarships for South African refugees was attacked – it is widely thought by South Africa. During 1986 several raids took place including one on 12 December in which six people were kidnapped by a group of ten armed South African commandos who drove into Swaziland. Two of those kidnapped were Swiss citizens who were later returned. 'Pik' Botha expressed no regret after the raid and made it quite clear that South Africa would not hesitate to repeat its cross-border raids if it thought them necessary.

Swaziland continues to be almost totally dependent on South Africa

economically, despite its membership of SADCC, a grouping of anti-apartheid southern African states. It has been described as South Africa's most faithful ally and closest collaborator in the region.
See also **ANC (South Africa); CONSAS; Nkomati Accord; SADCC**.

Relly, Gavin Walter Hamilton (*South Africa*)
Current Chairman of the Anglo American Corporation, Gavin Relly was groomed by the man he succeeded, Harry Oppenheimer. Since taking over the reins, Relly has largely pursued similar policies to those of his predecessor. Appearing, initially, to be less ready to take overtly political positions, Gavin Relly was one of the business leaders who participated in a controversial meeting with the African National Congress of South Africa in Lusaka, Zambia, in 1986. Relly is a past president of the South Africa Foundation, and serves on the Board of Governors of the Urban Foundation.
See also **Anglo American Group; Oppenheimer**.

removals (*South Africa*)
The logic of apartheid, that 'races' should live apart, dictated a policy of 'removals' of communities which did not conform to the 'grand design'. The 'Homelands' or bantustan policy resulted in forced removals of millions of Africans from 'White' South Africa into the 'Homelands', or from one 'Homeland' to another. About a million removals of farm workers alone have taken place as a result of the mechanisation of farms in recent years.

Although the vast majority of people 'removed' are Africans other groups are also affected. A study carried out by academics in Stellenbosch University in 1986 estimated that there have been some four million removals since 1950 – excluding removals under influx control (the 'pass laws').

Often those 'removed' were 'dumped' in resettlement camps in remote areas with few facilities. One estimate calculates that nearly four million people are at present living in this way. These camps were notorious for their poverty and child mortality rates, and became known as 'the dumping grounds' after a film made about the Dimbaza camp.
See also **'Homelands'; Natives Land and Trust Act of 1936**.

RENAMO (*Mozambique*)
Resistência Nacional Moçambicana, an anti-FRELIMO force backed by South Africa: see **MNR**.

Republican Front (*Zimbabwe*)
The Rhodesian Front, the party of rebel white Rhodesians led by Ian Smith, was renamed the Republican Front in June 1981. Just before the 1985 elections it was renamed the Conservative Alliance of Zimbabwe.
See also **Conservative Alliance of Zimbabwe; Independent Zimbabwe Group; Rhodesian Front; Smith**.

Republican Party (*Namibia*)
This small party of white conservatives was formed in October 1977 and

since then has been the personal political vehicle of its leader, Dirk Mudge. Shortly after the SWANP severed its ties with the National Party of South Africa, Mudge challenged extreme right winger Abraham du Plessis for the leadership of the party. He was defeated, in a vote at the party's annual congress, by 141 votes to 135, and led a walkout by 75 delegates who formed the Republican Party a month later.

Mudge took his party into the DTA and became leader of South Africa's anti-SWAPO 'internal government' based on the DTA. In the 'ethnic election' of 1980, the Republican Party was heavily defeated by the further right SWANP which won 11 of the 18 seats in the white 'legislative assembly'. The result reflected a divide within the white population with most Republican Party support coming from German and English speaking whites and SWANP support coming from Afrikaans speakers.

After the collapse of the DTA 'internal government', the Republican Party joined South Africa's new 'Transitional Government of National Unity' as part of the MPC. While depicted as a party of white *verligtes* (or progressives), the Republican Party remains deeply conservative, implacably opposed to SWAPO and hostile to UN Security Council Resolution 435 as a basis for independence in Namibia.

See also **DTA; MPC; SWANP; Transitional Government of National Unity; UN Security Council Resolution 435; *verligte*.**

República Popular de Angola *(Angola)*

People's Republic of Angola: see **Angola**.

República Popular de Moçambique *(Mozambique)*

People's Republic of Mozambique, the official title of the country since independence in 1975: see **Mozambique**.

Republic of South Africa

The Union of South Africa became a republic on 31 May 1961 under the National Party Prime Minister, Dr Verwoerd.

See also **South Africa; Union of South Africa**.

Resistência Nacional Moçambicana *(Mozambique)*

An anti-FRELIMO force backed by South Africa: see **MNR**.

Resolution 435 *(Namibia)* See **UN Security Council Resolution 435**.

retornados *(Angola, Mozambique)*

Retornados is the Portuguese term for white settlers who returned to Portugal following the independence of its African colonies. A few subsequently re-emigrated to the colonies.

See also **settlers**.

Revolta Activa *(Angola)* See **Active Revolt**.

Revolutionary Government of Angola in Exile *(Angola)* See **GRAE**.

Rhenish Missionary Society *(Namibia)*

Founder of Evangelical Lutheran Church in Namibia: see **churches**.

r

Rhodes, Cecil John (*Zambia, Zimbabwe*)

Cecil Rhodes was one of the most famous British colonial entrepreneurs of the nineteenth century. He sent a group of some 180 pioneers across the Limpopo River to expand the British empire and search for gold. He founded the colony of Rhodesia on 12 September 1890, and it was administered for the first 33 years by his private commercial company, the British South Africa Company. A royal charter entitled the company to make laws, raise taxes, keep a police force, recruit administrators and build roads and railways. Cecil Rhodes also obtained treaties and concessions from various African chiefs north of the Zambezi River. He was unsuccessful in Katanga, which fell instead to Belgium, and became the Belgian Congo, but his agreements elswhere placed what was to become Northern Rhodesia, and later Zambia, firmly under British control.

See also **British South Africa Company**.

Rhodesia (*Zimbabwe*)

Rhodesia was the name often given to the British colony of Southern Rhodesia after the collapse of the Central African Federation in 1963, while Northern Rhodesia became Zambia the following year. Rhodesia was also the first name given to the country by Cecil John Rhodes when he established a colony there in 1890. The illegal white regime of Ian Smith declared the name of the country Rhodesia in its Unilateral Declaration of Independence. However, this was never the internationally-recognised name of the country. That remained Southern Rhodesia until the emergence of independent Zimbabwe in 1980.

See also **Central African Federation; Rhodes; Southern Rhodesia; Unilateral Declaration of Independence**.

Rhodesian Front Party (*Zimbabwe*)

The Rhodesian Front (RF) was formed in March 1962. Its programme for the election of that year was crystal clear. It sought to maintain white supremacy in the colony of Southern Rhodesia (later to become Zimbabwe). It convinced white Rhodesians that they should and could retain permanent control and that compromise with black nationalists was unnecessary.

Although the RF was formed only nine months before the election, by the time it came to the vote it had convinced the white electorate. Its leaders too were little known: Winston Field, Clifford Dupont and Ian Smith. The RF gained 35 of the 50 white seats. Soon the party became obsessed with the need for independence. It saw the remaining links with Britain as a threat to white survival.

In 1963 the Central African Federation collapsed and independence for Nyasaland and Northern Rhodesia followed shortly. The new Southern Rhodesian Prime Minister, Winston Field, began to negotiate independence with Britain. British conditions for granting independence were enumerated in five principles, known as the NIBMAR principles, 'no independence before majority rule'. On 13 April 1964 the white rebel leader Ian Smith, took over as Prime Minister from Winston Field, who had failed to advance the cause of Rhodesian independence. With 'Smithy', as he became known, at the head of the government a Unilateral Declaration of Independence (UDI) became a definite possibility. The RF got a majority of 89 per cent in a referendum on the question and in the May 1965 general election it won all 50 seats. From then on white opposition parties were to play a minor role in Southern Rhodesian politics.

On November 11 that year Rhodesia proclaimed itself independent in a Unilateral Declaration of Independence. Ian Smith countered the NIBMAR principles with five of his own. They included a statement on Rhodesia's unwillingness to accept the principle of unimpeded progress towards majority rule. Britain condemned UDI and called on the Commonwealth to help suppress the white rebellion. The United Nations Security Council in turn called on states not to recognise the illegal Smith regime and to impose selective economic sanctions. Ian Smith negotiated with the British Prime Minister Harold Wilson on two occasions with no success, and in 1969 the RF introduced a new constitution reinforcing a segregated society with no possibility of majority rule. As a sop to international opinion it introduced instead the concept of parity, a system under which black citizens would achieve equal representation with whites, although only in the distant future.

The failure of negotiations with Britain and the success Southern Rhodesia had in defying sanctions produced a right-wing reaction in the RF. This led to resolutions calling for the end of multiracialism. Ian Smith prepared to move along with this hardening political mood in his party. In 1971 Lord Goodman, an emissary from the new British Conservative government, flew into Salisbury, then the capital of Southern Rhodesia, to renew discussions. The idea of majority rule was still the problem for the Rhodesian regime, but Lord Goodman's plan envisaged a qualified kind of majority rule in the distant future. Ian Smith accepted the principle and tried to sell it to the RF. The party agreed but the country's black leaders protested that they had not been consulted. The Anglo-Rhodesian agreement was seen as a victory for the RF. All that remained was to ensure that the independence settlement was based on 'general consent'. To establish this Lord Pearce arrived in Salisbury in 1972. Much to the dismay of the RF the verdict of the Pearce Commission was a resounding 'no' from the country's black population, and the deal was off.

In 1976 Ian Smith read out terms offered by Henry Kissinger, then the US Secretary of State. For the first time he accepted the principle of majority rule, albeit in the future. The initiative proved unacceptable to the nationalists and the 'frontline' states. However, the psychological impact on the RF was substantial. Only six months earlier their leader had announced that there would never be majority rule in Rhodesia, 'never in a thousand years'. Now he had conceded the point when hundreds of RF supporters were dying in the guerrilla war. Meanwhile there was a white exodus from the country. In the first six months of 1977 an average of 1,500 whites left each month.

Reacting to the Kissinger plan, a group of RF MPs demanded more hardline action from Ian Smith, including military raids into neighbouring territories. The rebel group was eventually expelled from the RF and scores of party followers, including the chairperson, Des Frost. A new party was launched in July 1977, the Rhodesian Action party. It was a rebirth of the RF of 1962. Its leader was Ian Sandeman. To counter this Ian Smith called an election and the RF won it overwhelmingly.

During the same period guerrilla attacks on the whites were growing and RF MPs began to express discontent with the government. In 1977 Ian Smith was heckled at an RF by-election meeting in Salisbury. Bishop Muzorewa's popularity seemed to be waning and Ian Smith met Joshua Nkomo and President Kaunda of Zambia to try to involve Nkomo in his government. The meeting caused a split in the RF and when a Rhodesian plane was shot down by guerrillas from Zambia this further infuriated the party and prevented any kind of agreement. The RF retaliated with a series of morale-boosting strikes against guerrilla forces in Zambia.

After the failure of the Anglo-American agreement, Ian Smith embarked on an internal settlement, the Salisbury Agreement, signed between the RF, Bishop Muzorewa, Ndabangi Sithole and Chief Chirau in 1978. This led to a transitional government based on the 1969 constitution. The whites voted for the RF's constitution because there seemed to be no alternative. In the election that followed the transitional government in April 1979 the RF proved its popularity with the whites beyond any doubt: it won all 20 white seats.

This state of affairs did not last beyond independence, however. The RF became a minority party in Zimbabwe with little influence and most whites turned to parties more in tune with the new government. In June 1981 the RF was renamed the Republican Front and just before the 1985 general elections it became the Conservative Alliance of Zimbabwe.

See also **Anglo-Rhodesian agreement; Chirau; constitution of 1969; 'front-line' states; internal settlement; Muzorewa; Nkomo; NIBMAR; Pearce Commission; Salisbury Agreement; sanctions against Rhodesia; Sithole, Ndabaningi; Smith; Southern Rhodesia; transitional government; Unilateral Declaration of Independence**.

Rhodesian Settlement Forum (*Zimbabwe*)

The Rhodesian Settlement Forum and the African Settlement Convention were black organisations formed to give support to the 1971 Anglo-Rhodesian agreement. They enabled Ian Smith, the leader of the white Rhodesian regime, to argue that there was black support for the settlement proposals. However, the test of majority opinion in the country by the Pearce Commission found it overwhelmingly against the agreement.

See also **Anglo-Rhodesian agreement; Pearce Commission; Smith**.

Rhodesia Party (*Zimbabwe*)

The Rhodesia Party was a white liberal opposition party formed in 1972 to oppose the increasing right-wing radicalism of Ian Smith's Rhodesian

Front. It argued that only a political solution could end the war, and it was in favour of talks with nationalist groups such as the African National Council. The Rhodesian Front's victory at the polls in July 1974 indicated drastically-reduced support for the Rhodesia Party and it ceased to be regarded as a threat to the government.
See also **African National Council; Rhodesian Front**.

Riekert Commission (*South Africa*)

The Riekert Commission reported in 1979. It was set up, together with the Wiehahn Commission of Inquiry, to address the question of urban 'Blacks' and migrancy relative to the labour supply. The Wiehahn Commission was concerned with the state's response to the rapid growth of black trade unionism in South Africa. The Riekert Commission concentrated on the wider questions this raised. It proposed greater control of African industrial workers by isolating them, together with the African middle class, from the black rural population and the growing numbers of unemployed. It proposed to do this by granting some urban Africans legal status as permanent residents and, at the same time, to tighten up influx controls for all other Africans. This, it was calculated, would promote the division of Africans into a privileged minority urban group and a majority group of rural people without any rights in the townships. This would allow the state to co-opt the black middle-class as a means of extending its domination of the 'rural' black majority.

The Riekert Commission proposals had a political impact on government thinking but they were not adopted in total as the Wiehahn proposals had been. The huge levels of structural unemployment in South Africa are, however, being dealt with through the strict town-country division as recommended by the commission.
See also **'Homelands'; migrant labour; 'reforms'; urban 'Blacks'; Wiehahn Commission**.

Riruaka, Kuaima (*Namibia*)

Herero chief and leader of NUDO: see **DTA** and **NUDO**.

Rising of 1915 (*Malawi*)

The 1915 rising took place in Malawi as a protest against colonial rule. It lasted from 23 January to 4 February and was significant for its political impact, demonstrating that Africans need not accept white rule without protest. Militarily, however, it was a failure and those who took part were killed, jailed or went into exile. The rising was led by one of Central Africa's most famous early nationalists, John Chilembwe.
See also **Chilembwe**.

Rivonia Trial (*South Africa*)

On 12 July 1963 the security police captured *Umkhonto we Sizwe's* (the ANC's armed wing) most important leaders at a house called Lillieleaf Farm in Rivonia near Johannesburg. This was the underground headquarters of *Umkhonto* and documents and explosives were also found and taken by the police.

The so-called 'Rivonia Trial' began in October 1963 and Nelson

Mandela was brought from Robben Island to join the other eight accused with him. They were being tried under the General Law Amendment Act and the Suppression of Communism Act for sabotage and conspiracy to overthrow the government by revolution and by assisting an armed invasion of South Africa by foreign troops. The trial went on for eleven months and on 12 June 1964 eight of the nine accused were sentenced to life imprisonment. They were: Nelson Mandela, Walter Sisulu, Govan Mbeki, Raymond Mhlaba, Ahmed Kathrada, Denis Goldberg, Elias Motsoaledi and Andrew Mlangeni. Lionel ('Rusty') Bernstein was found not guilty and acquitted.

In a statement to the court on 20 April 1964 Nelson Mandela said he had been one of the founders of *Umkhonto we Sizwe* and continued with an eloquent and passionate attack on racial oppression in South Africa. His statement received wide international publicity.

After introducing himself as a convicted prisoner he began as follows:

> In my youth in the Transkei I listened to the elders of my tribe telling stories of the old days. Amongst the tales they related to me were those of wars fought by our ancestors in defence of the fatherland . . . I hoped then that life might offer me the opportunity to serve my people and make my own humble contribution to their freedom struggle. This is what has motivated me in all that I have done in relation to the charges made against me in this case.

The speech went on to justify the ANC's strategy and attack the injustices in South African society. 'During my lifetime I have dedicated myself to this struggle of the African people', he concluded,

> I have fought against White domination, and I have fought against Black domination. I have cherished the ideal of a democratic and free society in which all persons live together in harmony and with equal opportunities. It is an ideal which I hope to live for and to achieve. But if needs be, it is an ideal for which I am prepared to die.

See also **ANC; Mandela, Nelson; Sisulu, Walter; Treason Trial;** ***Umkhonto we Sizwe***.

RMS (*Namibia*) See **Rhenish Missionary Society**.

RNM (*Mozambique*)
Resistência Nacional Moçambicana, an anti-FRELIMO force backed by South Africa: see **MNR**.

Roach, Fred (*Lesotho*)
Fred Roach was the officer commanding the Police Mobile Unit during the troubles around the time of the suspension of the constitution in 1970. The support of this English expatriate helped Chief Jonathan's government to survive by finding and suppressing any resistance.
See also **Jonathan**.

Robben Island (*Namibia, South Africa*)
Robben Island is a bleak, rocky outcrop off the southern tip of Africa

within sight of Cape Town. Surrounded by the bitterly cold water of the South Atlantic, it was originally used for defence of the harbour of Cape Town against naval attack. It has subsequently become a maximum security prison, the most notorious in South Africa.

The list of those who have been imprisoned on 'the island', as it is called, reads like a roll-call of the political leaders of the nationalist struggles in South Africa and Namibia. Nelson Mandela, Robert Sobukwe, Walter Sisulu, Govan Mbeki and Andimba Toivo ja Toivo all had to spend many of the best years of their lives there. In March 1980 there were 53 Namibian political prisoners on Robben Island, according to the South African Minister of Prisons.

Prisoners are held under extremely rigorous conditions, and are forced to carry out hard labour in the lime quarries on the island, or to gather, process and pack seaweed washed up on its beaches. Their food is largely maize meal porridge with a small amount of meat or fish. Visits are very restricted, letters limited, censored and very slow to arrive. Academic study is discouraged and disallowed beyond high-school level.

Namibian prisoners are segregated from South Africans who are in turn further segregated according to apartheid categories. The living conditions, poor diet and hard regime have contributed to the prevalence of tuberculosis, and medical care is inadequate and often inappropriate.

Despite the harsh conditions, or perhaps because of them, Robben Island has become known among opponents of apartheid as 'the school of revolution', because of the time spent by those incarcerated there in political debate and discussion. Whether for this or other reasons, South Africa appears to be in the process of closing the island prison down and top security prisoners are gradually being moved to mainland prisons.

See also **Mandela, Nelson; Mbeki, Govan; Sisulu, Walter; Sobukwe; ja Toivo**.

Roberto, Holden (*Angola*)

An irony of Holden Roberto's position as one of the foremost Angolan nationalist leaders is that he has spent almost his entire life outside the country. He left Angola with his mother at the age of two, returned for a brief period at the age of 17, and did not live in the country for an extended period, again, until 1975, when he was 52.

Roberto followed his uncle into Kongo nationalist politics and in 1960 became the President of the UPA. In 1962 he negotiated the merger with the PDA to form the FNLA of which he has been the first and only president. He was instrumental in the formation of GRAE and in winning recognition for it from the OAU, but was unable to sustain his movement in the face of greater activism and better organisation by the MPLA.

Following the military defeat of the FNLA by the MPLA in the post-independence civil war, Roberto effectively retired to Zaire where his amicable relationship with his brother-in-law, Zairean President Mobutu, allowed him to live in safety while overseeing some residual FNLA guerrilla opposition to the government of the RPA.

See also **FNLA; GRAE; MPLA; PDA; RPA; UPA**.

Rowland, 'Tiny' (*Zimbabwe*)
'Tiny' Roland is the head of the British-based Lonrho company. He is a controversial business figure with very large commercial interests all over Africa. In Zimbabwe the Lonrho company was one of the four largest business empires; it owned about a million acres of ranches and forests, as well as gold and copper mines. It is alleged that the company was involved in breaking sanctions and keeping the rebel white Rhodesian government in funds. However, 'Tiny' Rowland was critical of Ian Smith and claimed to find the company's activities within Zimbabwe at the time of the Unilateral Declaration of Independence embarrassing. His close relationships with several African leaders made him vulnerable to their criticism, and he helped all the Southern Rhodesian nationalist movements with money and favours. He had a particularly high regard for Joshua Nkomo, the ZAPU leader.
See also **Nkomo; Sanctions against Rhodesia; Unilateral Declaration of Independence; ZAPU**.

Roman Catholic Church (*Lesotho*)
The Roman Catholic Church is a powerful institution in Lesotho. It had backed Chief Jonathan and his traditionalist and anti-communist views. However, in the last years of his rule he started to invite eastern bloc countries to establish diplomatic missions in Lesotho and this alienated the church.

In October 1986 the church came into conflict with the military regime when the Secretary General of the Lesotho Christian Council, Father Michael Worsnip, was deported after making public allegations concerning anti-ANC death squads operating in Lesotho.
See also **death squads**.

RP (*Namibia*) see **Republican Party**.

RPA (*Angola*)
República Popular de Angola: see **Angola**.

RPM (*Mozambique*)
República Popular de Moçambique: See **Mozambique**.

RSA (*South Africa*) See **Republic of South Africa**.

Ruacana hydro-electric power plant (*Angola*)
Part of the massive Cunene River Hydro-Electric Scheme: see **Cunene River Hydro-Electric Scheme**.

'Rubicon' speech (*South Africa*)
Much was expected by the international community from President Botha's so-called 'Rubicon' speech on 15 August 1985 in Durban. It was believed that he would spell out far-reaching future 'reforms' to apartheid. However, in the event, the speech did the opposite and effectively said that there would be little further change. This provided a new impetus to those demanding sanctions against South Africa, particularly in the USA.
See also **'reforms'; sanctions**.

'Russians' (*South Africa*)
The 'Russians' are a group of vigilantes who have recently re-emerged in Soweto. They are part of a growing trend towards the formation of vigilante groups in the black townships. The 'Russians' are men clad in blankets in the traditional Basotho manner and wearing Basotho-style grass hats. They appeared in Soweto during the Christmas protest campaign organised by young radicals, often known as 'comrades', against the emergency regulations imposed by the government. There were several violent clashes and some five people were battered to death. There is a suspicion that the vigilantes are being used by the government to attack the township radicals, for the 'Russians' had not been seen for thirty years since they opposed a bus boycott in the township of Evaton, south of Johannesburg.
See also **'comrades'; Crossroads; 'fathers'; townships;** ***witdoeke***.

S

SABA (*South Africa*) See **South African Black Alliance**.

***Saba Saba* meeting** (*Tanzania*)
Meeting of 7 July marking the founding of TANU in 1954: see **TANU**.

SABC (*South Africa*)
South African Broadcasting Corporation, the state broadcasting monopoly (controlling radio and television) and slavishly pro-government: see **press**.

SABRA (*South Africa*)
The South African Bureau of Racial Affairs (SABRA) was set up by the *Afrikaner Broederbond* in 1947 as an alternative to the South African Institute of Race Relations, a liberal body which campaigned for co-operation among black and white people in South Africa. SABRA was an organisation of Afrikaner intellectuals and academics who sought to provide an intellectual justification for apartheid, and who campaigned actively against what they saw as 'liberalism' in the country.

Until P. W. Botha won the leadership of the National Party, SABRA provided strong ideological and political support for the party, following closely the line determined by the *Broederbond*. After the split in the National Party, however, divisions emerged within SABRA, and it adopted an increasingly oppositionist attitude to the government. Although Botha's allies were able to regain control of the *Broederbond*, they failed with SABRA, and it actively opposed the new constitutional proposals during the 1983 referendum describing them as 'farcical and impractical'.

In 1983 the final break with the National Party came when the government withdrew financial support of R70,000 from SABRA. The decision was announced by Dr Gerrit Viljoen, Minister of National Education and a former Chairman of SABRA. It could not expect continued financial support from the state, he said, when it had placed itself in opposition to the government.
See also ***Afrikaner Broederbond*; constitution of 1983; National Party; South African Institute of Race Relations; Viljoen**.

SACC (*South Africa*)
The South African Council of Churches, outspoken representative of the

majority of Christian churches in South Africa and highly critical of the apartheid system: see **churches**.

SACLA (*South Africa*) See **South African Confederation of Labour**.

SACOL (*South Africa*) See **South African Confederation of Labour**.

SACP (*South Africa*) See **South African Communist Party**.

SACTU (*South Africa*)
SACTU, the South African Congress of Trade Unions, was formed in 1955 from unions which would not go along with the proposal by the newly formed TUCSA (Trade Union Council of South Africa) to exclude black workers from direct involvement in union negotiations. TUCSA's strategy was known as 'parallelism' and it involved a formal liaison committee with African trade unions, the white parallel unions negotiating on their behalf in exchange for their co-operation. Nineteen unions voted against this arrangement and the leading four of these – the Food and Canning Workers' Union, the Textile Workers' Industrial Union and their respective African branches – together with ten others formed a committee which finally led to the formation of SACTU.

SACTU argued that trade unionism could never be simply a struggle for rights in the workplace in South Africa, but that it had to operate in the political arena as well. It soon became part of the Congress Alliance and although it was not formally banned with the rest of the movement in the 1960s its leaders were harassed (forcing many into exile) and its activities paralysed. During the 1970s SACTU began to reorganise underground in the country and it still operates in this way, its members working within existing unions or independently. SACTU's programme links workers' demands for improved conditions with the political goal of national liberation.

At its peak in the early 1960s SACTU had 46 affiliated unions and claimed a multiracial membership of 53,323 people. Now many other independent trade unions have developed in South Africa and SACTU is no longer such a dominant force. One of SACTU's most important campaigns in the earlier years was the '£1 a-day' campaign, demanding a minimum wage of £1 a day for all workers.
See also **Congress Alliance; COSATU; Food and Canning Workers' Union; TUCSA**.

SACU (*southern Africa*) See **Southern African Customs Union**.

SADCC (*southern Africa*)
The Southern African Development Coordination Conference (SADCC) was established by the 'frontline' states in Arusha, Tanzania, in 1979 to reduce economic dependence on South Africa, to create a genuine and equitable regional economic integration, and to mobilise resources to promote national, inter-state and regional policy. When Zimbabwe became independent in 1980 it formed the last piece in the SADCC jigsaw, greatly increasing the organisation's chances of success.

SADCC consists of all the countries of southern Africa except South

Africa and South African-occupied Namibia (SWAPO has made it clear that Namibia would join, on independence). These are Angola, Botswana, Lesotho, Malawi, Mozambique, Swaziland, Tanzania, Zambia and Zimbabwe. SADCC projects are concentrated in the areas of transport, communications, industry, agriculture, labour development, fisheries and wildlife and soil conservation. Each project is coordinated by one of the member states, and those already established include, for example, the rehabilitation and upgrading of the railway line between Nakala (in Mozambique) and Malawi. This project is coordinated by the Southern African Transport and Communications Commission of SADCC, based in Mozambique. SADCC's headquarters are in Botswana.

South Africa has perceived SADCC as a major economic and political threat, and much of its programme of economic and military action against neighbouring countries has been designed to undermine it. Botswana, Lesotho, Malawi, Mozambique, Zambia and Zimbabwe have all suffered disruption of the movement of goods to and from their countries. Six SADCC countries, Angola, Botswana, Lesotho, Mozambique, Zambia and Zimbabwe, have all been subjected to direct military action, and four, Angola, Lesotho, Mozambique and Zimbabwe, have been destabilised to lesser or greater extents by South African-backed insurgency groups.

See also **Constellation of Southern African States; 'frontline' states**.

SADF *(South Africa)*

The South African Defence Force (SADF) consists of the army, air force, navy and military intelligence. It includes a core of permanent personnel (the so-called Permanent Force, about ten per cent of the total) and a majority of conscripts. All 'White' males over the age of 18 are subject to conscription. The SADF also includes a number of black contingents. Despite being termed a 'defence' force, the South African military is the main vehicle for invasion and destabilisation of neighbouring countries.

The accession of P. W. Botha to the presidency saw the military in South Africa achieve unprecedented political influence. Botha had been minister of Defence (and is now Commander-in-Chief of the Armed Forces) and has maintained his close relationship with the SADF. In 1978 he even took the unprecedented step of promoting a serving officer, General Magnus Malan (then SADF head), to Minister of Defence. The concept of 'total strategy' was first mooted in the 1977 Defence White Paper, and has since become official state policy.

Over the past 25 years direct military expenditure, as a proportion of GNP, has increased seven-fold to 5 per cent. The state is thought to have almost 500,000 personnel under arms in South Africa and Namibia at any one time, with the capacity to call up significant reserve forces (although some sources suggest a figure of 400,000 including the reserve). Since its humiliation in Angola in 1975, it has modernised much of its equipment both through secret purchases in breach of the UN arms embargo, and through local manufacture. It has developed an important local arms industry through the state company, ARMSCOR.

See also **Botha, P. W.; COSAWR; End Conscription Campaign; Malan; South African invasions (Angola); total strategy**.

SAIC (*South Africa*)

The SAIC, South African Indian Council, was set up in 1968 by the South African government. Until 1974 its members were government nominees; after this date some were indirectly elected. The SAIC was due to be directly elected in 1981 but a massive anti-SAIC campaign was launched by the Natal Indian Congress. This proved very successful and a derisory poll of 10 per cent was recorded. Members of the SAIC formed the Reform Party in 1976. Since the new constitution which provides for an 'Indian' chamber of parliament the SAIC has been incorporated into the House of Delegates.

SAIC is also the acronym of the South African Indian Congress.
See also **Reform Party; South African Indian Congress**.

SAIRR (*South Africa*)

See **South African Institute of Race Relations**.

Salim, Salim Ahmed (*Tanzania*)

Salim Ahmed Salim became Prime Minister of Tanzania in 1984 after the death of Edward Sokoine in a road accident. He had been Minister of Foreign Affairs and was the leading contender for the presidency after the resignation of Julius Nyerere. His politics are to the right of those of Julius Nyerere and if elected he would have been likely to move Tanzania away from socialism towards a more capitalist system. However, he was not elected; the presidency was won instead by Ali Hassan Mwinyi. Salim is young enough to be a possible future presidential candidate.
See also **Mwinyi; Nyerere**.

Salisbury (*Zimbabwe*)

The name of the capital city of the colony of Southern Rhodesia, renamed Harare after independence in 1980: see **Harare**.

Salisbury Agreement (*Zimbabwe*)

The Salisbury Agreement was signed on 3 March 1978. It set up a transitional government run by an executive council consisting of the white Rhodesian leader, Ian Smith, Bishop Abel Muzorewa, Eddison Sithole and Chief Jeremiah Chirau. Ian Smith retained the title of prime minister, but each member presided, in turn, over the council. The agreement marked the end of exclusive white rule in Zimbabwe.

The task of the transitional government was to arrange a ceasefire, remove racial discrimination, draft a new constitution and conduct an election towards the end of the year before handing over to a black government on 31 December 1978. Ian Smith's aim was to undercut the support for the guerrilla campaign by setting up an internal settlement. This would be based on a one-person-one-vote constitutional agreement with Bishop Muzorewa, Chief Chirau and Ndabaningi Sithole. Meetings began in December, and, at first, Bishop Muzorewa refused to attend in protest against the government's military attack on Chimoio in Mozambique. Ian Smith accused him of wasting time and the protest came to an end. Seven of Ian Smith's constitutional safeguards were agreed very quickly: the new constitution was to have a declaration of rights protecting property and pension rights; an independent judiciary and public service board; a police and security force free from political interference and kept efficient; pension guarantees for civil servants; the rights to dual citizenship for whites who wanted to keep other nationalities; and a permanent blocking mechanism to prevent a black government from amending entrenched constitutional clauses.

Bishop Muzorewa objected to the request for extra white parliamentary representatives. By February a compromise was agreed: 20 out of 100 seats would be reserved for whites and elected on a separate voter's roll; six white votes as well as 72 black votes would be necessary to change any entrenched clauses. The agreement was seen by most Zimbabweans as a victory for Ian Smith and white supremacy.
See also **Chirau; internal settlement; Sithole, Ndabaningi; Smith; Muzorewa; transitional government**.

S

SAM (*Angola*)
The *Serviço de Assistência Médica do MPLA* (SAM) was the medical wing of the MPLA.
See also **MPLA**.

SAMACO (*Malawi*) See **Save Malawi Committee**.

San (*southern Africa*)
The San people were the original inhabitants of Namibia and, with the Khoi, of much of southern Africa. They lived in small kinship groups as hunter-gatherers and were highly attuned to their environments. Conflicts with more recent settlers forced them into the most inaccessible and inhospitable parts of the sub-continent, the deserts and mountains. Many San communities were able to continue their hunter-gatherer lifestyles until relatively recently, but in Namibia, especially, most have had this disrupted as a result of their induction into the South African military to act as trackers and auxilliaries in its war against SWAPO. Surviving communities live mainly in the Kalahari areas of Namibia, Botswana and southern Angola. The term Khoisan is used to refer to the Khoi and San collectively.
See also **Khoi; SWATF**.

sanctions against South Africa (*South Africa*)
Sanctions of various sorts have been introduced against South Africa's policy of apartheid over the years. They include economic, financial and trade boycotts and corporate withdrawals. They also include actions like boycotts and divestment of shareholdings aimed at transnational corporations linked with South Africa. Withdrawal or disinvestment is the term normally used when a company leaves the country and divestment is the term often used when a shareholder sells shares in a company because of its South African links.

Calls for sanctions against South Africa go back many years. In 1959 Chief Albert Lutuli, Nobel Peace Prize winner and then the President of the ANC, appealed for action: 'The economic boycott of South Africa will entail undoubted hardship for the Africans. We do not doubt that. But if it is a method which shortens the day of bloodshed, the suffering to us will be a price we are willing to pay. In any case, we suffer already, our children are undernourished, and on a small scale (so far) we die at the whim of a policeman.' Many others have echoed his views, such as Archbishop Tutu, a prominent advocate of sanctions in the 1980s.

The National Party government has maintained a tough line about sanctions. In 1986 President Botha addressed a National Party congress with the following words: 'We do not desire sanctions, but if sanctions must come in order that our freedom, our justice, must be maintained, we will survive it. We will not just survive; we will come out stronger in the end.'

In 1946 India ended trade and diplomatic relations with South Africa – the first United Nations member state to implement sanctions. The United Nations General Assembly has passed various motions in favour of sanctions, but these have often been defeated in the Security Council as a result of vetoes from Britain, France and the United States in particular. In 1962 General Assembly Resolution 1761 (XVII) adopted on 6 November called for the breaking of diplomatic relations, the termination of trade including the supply of arms and ammunition, the closure of ports to South African vessels and the boycotting of South African goods. The General Assembly also established the Special Committee Against Apartheid to keep the situation under review. In 1963 a voluntary arms embargo was instituted by the Security Council with Britain and France abstaining. On 4 November 1977 a mandatory arms embargo, including a ban on co-operation in the manufacture and development of nuclear weapons, was imposed against South Africa under Chapter VII of the United Nations Charter, by unanimous vote in the Security Council (Resolution 418). This is the only mandatory UN sanction to be passed against South Africa to date (1987). Apart from arms the other important international embargo is the oil boycott. In 1973 most oil-exporting countries agreed to boycott South Africa; Iran joined the boycott in 1979. The oil embargo, however, is often broken.

Commonwealth countries have been in the forefront of those demanding sanctions of various sorts. In the Gleneagles Declaration of 1977, Commonwealth members committed themselves to 'taking every practical step to discourage contact or competition by their nationals with sporting organisations, teams or sportsmen from South Africa'. The Commonwealth Heads of Government Accord signed in Nassau in October 1985 banned all new government loans to the South African government and its agencies; banned the direct import of Krugerrands (a ban on all South African gold coins was introduced in May 1986); ended government funding for trade missions to South Africa and stopped trade fairs to South Africa; banned the sale of computer equipment which could be used by the security forces, the military or the police; banned new contracts for the sale of nuclear technology or materials; stopped oil sales, ended the import of arms and paramilitary equipment from South

Africa; and banned all military co-operation. In August 1986, seven Commonwealth heads of government met to consider the failure of the EPG mission to South Africa. All except Britain agreed to further measures including ending air links with South Africa.

Britain and the USA have been the most reluctant to support the growing international consensus in favour of sanctions. However, at the Commonwealth mini-summit in London in August 1986 the British government agreed to put voluntary bans on investment in South Africa and the promotion of tourism, and to accept any EEC decision to ban imported South African coal, iron and steel and gold coins. The Brussels meeting of the EEC in September 1986 agreed to ban new investment in South Africa, imports of iron and steel – but not coal – and gold coins. The US government imposed a series of sanctions in October 1986, the US Congress voting to override President Reagan's veto. These include bans on: new public and private loans and investments; imports of South African agricultural products, uranium, coal, textiles, iron, steel, gold coins, arms, ammunition and military vehicles; imports from state-owned firms; exports of crude oil, petroleum products and nuclear technology; exports of computers, software and services to the military, police, or any agency involved in administering apartheid; and aircraft landing rights.

British investment in South Africa is far more significant than that of any other country. According to figures given by the British Industry Committee on South Africa to the House of Commons Foreign Affairs Committee in July 1986, Britain has investments in South Africa worth £6,000 million – 40–45 per cent of all foreign investment there. This is 7 per cent of all British overseas investment.

Thousands of companies, trade unions, universities and other institutions have taken some kind of action against South Africa. Some of the most significant of these activities are: in 1985 the major western banks temporarily refused to renew their loans to South Africa; nearly 200 US, British and other foreign companies have withdrawn from the country since 1980 including one of South Africa's strongest supporters, Barclays Bank, which sold its subsidiary in late 1986 after a seventeen-year campaign by anti-apartheid organisations. Notable American-based companies which have withdrawn are General Motors and IBM, the world's largest computer company, both in October 1986.

See also **EEC Code of Conduct; Eminent Persons Group; Sullivan Code**.

sanctions against Lesotho (*Lesotho*) See **relations with South Africa (Lesotho)**.

sanctions against Rhodesia (*Zimbabwe*)

In 1966 the United Nations imposed selective mandatory sanctions against Southern Rhodesia after the Unilateral Declaration of Independence on 11 November 1965. These were boosted to comprehensive mandatory sanctions in 1968.

Britain began to impose sanctions against Southern Rhodesia in November 1965. The then prime minister, Harold Wilson, expelled the rebel colony from the sterling area, withdrew Commonwealth preferences on Rhodesian goods, imposed exchange controls, froze Rhodesian reserves in London, prevented access to the London money market and banned the importing of tobacco and sugar. In December reserve bank assets in London were seized and the list of banned imports extended to minerals and meat. On 17 December oil imports into Rhodesia were banned and at the end of January further measures cut off all Rhodesia's imports from and exports to Britain.

Sanctions did not succeed in crushing the rebel regime. They were imposed little by little on the country and this helped Rhodesia to organise its resistance to them. The Rhodesian leader, Ian Smith, also used sanctions as an issue to unite white citizens: 'We will never surrender to force and intimidation', was his Christmas message in 1965. As well as this, South Africa and Portugal (which controlled Mozambique at the time) refused to co-operate in imposing sanctions. This proved invaluable, and gave the Smith regime much more chance of survival. And survive it did, for a longer period than most of the world thought possible.

Sanctions against Rhodesia cost the surrounding African states dearly. Zambia, for instance, is calculated to have lost some US$500 million by mid–1976 through their operation. To apply United Nations sanctions the country had to reroute its copper away from Rhodesia to the north. This was done at considerable expense and inconvenience in 1973 after the Smith regime closed the border along the Zambezi River.

See also **Smith; Unilateral Declaration of Independence**.

SAP (*South Africa*)

The South African Police (SAP) is the main organ of internal repression in South Africa. Consisting of almost equal numbers of black and white full-timers, its strength in 1984 included over 45,000 regulars and almost as many reservists. The latter perform routine police duties when the regulars are on 'special duties' such as riot control.

As well as combating normal crime, the SAP enforces the laws providing for influx control, segregated amenities, group areas and public order which are central to the system of apartheid. It is the first line in the defence of white privilege and suppression of dissent. The most important body for the latter is the Security Police, a branch of the SAP. Sometimes called the Special Branch or SB (the Security Police was originally established as a 'special branch' on the British model), the SP is notorious for the regular use of torture to extract statements from political detainees, and for the arrest and detention of children. Administering amongst the most draconian security and public order laws in the world, the SAP may arrest or detain almost anyone, detainees may be held virtually indefinitely without trial, and there is no automatic recourse to legal advice or to medical assistance.

The SAP has become internationally known for its brutal attempts to suppress uprisings and unrest in South Africa. Scenes seen on television throughout the world of the police shooting randomly into crowds with shotguns, rifles, plastic bullets and gas, the use of CS gas in huge quantities generated by mobile 'sneeze-machines', and the use of the so-called 'Hippo' armoured vehicles have fed international opposition to apartheid. The effectiveness of such publicity has been widely credited as the reason for the clampdown on press coverage of protests which preceded the declaration of the state of emergency.

Apart from the SAP, South Africa has a multitude of other smaller police forces. The Railways Police guards and controls the rail system in the country. White municipalities throughout the country are entitled to maintain their own police forces, used mainly for traffic control and enforcement of local by-laws, including the maintenance of segregation. In 1984, 32 African local authorities were authorised to set up their own local police forces to 'maintain law and order, prevent crime and see to the implementation of city councils' decisions'. Such forces are required to support the SAP in 'riotous conditions'. In addition, each of the bantustans has its own police force which performs the duties of the SAP within the bantustan.

See also **'blackjacks'; press; state of emergency**.

S

SAP (*South Africa*) See **South African Party**.

SASO (*South Africa*)

SASO, the South African Students' Organisation, was the first black consciousness movement. It was formed in 1967 by a group of black students who broke away from the multiracial National Union of South African Students (NUSAS). The incident which led to the break took place at NUSAS's annual conference at Rhodes University in Grahamstown. There the authorities insisted on segregated social facilities and NUSAS condemned this but accepted it by continuing with the conference. Steve Biko was the moving force behind the formation of SASO which helped to develop further the ideas of black consciousness.

See also **Biko; Black Consciousness Movement; Black People's Convention; NUSAS**.

SATCC (*southern Africa*)

The SATCC, or Southern African Transport and Communications Commission, is an offshoot of the SADCC and is concerned with the development of communications and transport infrastructure in the SADCC countries. Its secretariat is based in Maputo, Mozambique.

See also **SADCC**.

Save Malawi Committee (*Malawi*)

The Save Malawi Committee (SAMACO) is the most recently established dissident Malawian movement. It was founded in Lusaka in May 1983 and claims to be a group uniting the opposition movements. Opposition parties are banned inside the country so most are formed abroad. None so far (1987) have posed a real threat to President Banda's regime.

See also **Congress for the Second Republic; Malawi Freedom Movement; Socialist League of Malawi**.

Savimbi, Jonas (*Angola*)

Jonas Savimbi and his organisation, UNITA, are notable for the alliances they have formed with a variety of both Angolan and non-Angolan forces. Despite initial financial and military support from China, a subsequent agreement with the Portuguese secret police, PIDE, led to an unannounced truce between the Portuguese and UNITA during the later stages of the liberation war in Angola. Savimbi collaborated with the Portuguese armed forces against the MPLA, and, following the withdrawal of the Portuguese, established a secret alliance with South Africa and the CIA.

Savimbi was an invited guest at the inauguration of South African President P.W. Botha, was received by US President Reagan, and is the favoured standard-bearer in Angola for the US campaign against socialist governments in Africa and South and Central America. The so-called Clarke Amendment, by which the US Congress disallowed further secret US involvement in Angola after the 1975 debacle, was overthrown at the instigation of President Reagan so that he could step up military aid to Savimbi's UNITA.

Savimbi began his political career in 1961 when he joined the UPA as its general secretary. He worked closely with Holden Roberto, UPA president, to establish the FNLA and GRAE, becoming GRAE's 'Foreign Minister'. By 1964, however, he had broken away from Roberto's FNLA to form AMANGOLA and, in March 1966, UNITA.

Savimbi formalised his secret agreement with PIDE when he negotiated an early 'ceasefire' with the Portuguese in June 1974. He subsequently campaigned energetically in Angola to build a national base for what had been an essentially regional organisation, establishing an uneasy alliance with his former colleague, Holden Roberto.

The alliance was shortlived, and following the defeat of UNITA and the FNLA by the forces supporting the MPLA, Savimbi and UNITA went back to the bush of south eastern Angola. Unlike Roberto, who never re-established the FNLA as a significant force, Savimbi was able to use his relationship with South Africa to regroup and re-arm his supporters.

See also **AMANGOLA; FNLA; GRAE; MPLA; PIDE; Roberto; UNITA: UPA**.

SB (*South Africa*)
Acronym for Special Branch, an erroneous but common name for the Security Police, originally established as a special branch of the SAP on the British model: see **SAP**.

SCA (*South Africa*) See **Soweto Civic Association**.

SDP (*Swaziland*) See **Swaziland Democratic Party**.

Secretariat (*Mozambique*)
Secretariat of the Central Committee of FRELIMO: see **FRELIMO**.

Security Police (*South Africa*)
A branch of the SAP notorious for the torture and maltreatment of detainees: see **SAP**.

S

SEIFSA (*South Africa*)
The Steel and Engineering Industries Federation of South Africa (SEIFSA) is a highly conservative body representing employers in the metal and engineering industries. Faced during the Second World War by then unprecendented levels of worker militancy, employers established an organisation both to co-opt white trade unions, and to set limits (through establishing industry-wide wage levels) on wages of black workers.

SEIFSA negotiated with white unions a series of 'racially' exclusive closed shop agreements, the basis of the rigid colour bar which existed in the industry until the mid–1970s. It has subsequently abandoned this position, but continues to be one of the employers' organisations most hostile to the independent non-racial trade unions.
See also **ASSOCOM; Chamber of Mines; FCI; job reservation**.

S

Selous Scouts (*Zimbabwe*)
The Selous Scouts was an irregular force of the Rhodesian army which was used for 'dirty tricks' operations against the liberation movements. Its forces were largely black and those who were white often blacked their faces before a military operation. One of its primary aims was to discredit the guerrilla forces of ZANLA and ZIPRA. During the war the Selous Scouts became notorious for their brutality. The guerrillas claim that many missionary murders were committed by the Scouts in order to discredit the nationalist fighters in the eyes of the church and the people. The Scouts were founded by Col. Ron Reid Daly and modelled on the brutal *flechas*.
See also ***flechas***; **MNR; SANLA; ZIPRA**.

Seoposengwe Party (Namibia)
Small, ethnically-based constituent party of Democratic Turnhalle Alliance: see **DTA**.

'separate development' (*South Africa*)
'Separate development' was an alternative name for apartheid adopted by its supporters in an attempt to indicate something positive in their policy of racial separation: see **apartheid**.

Serviço de Assistência Médica do MPLA (*Angola*) See **SAM**.

Sesotho (*Lesotho*)
The indigenous language spoken by the Basotho people: see **Basotho**.

settlers (*Angola, Mozambique*)
The initial number of European settlers in Angola was very small, the total white population numbering 44,000 in 1940, and including a large proportion of administrators temporarily working in the colony. After the Second World War, however, the Portuguese government encouraged the settlement of poor rural Portuguese in its colonies, and numbers grew to 335,000 by 1974. In Mozambique, under the government settlement

programme, numbers grew from 97,000 in 1960 to over 200,000 in 1970. By the mid–1970s Portuguese settlers dominated the commercial economy and filled the majority of the skilled jobs. At independence, for instance, only 72 of the 4,500 students at Maputo University were black and there was only one black train driver out of a total of 350.

At independence 90 per cent of the white settler populations of Angola and Mozambique opted to return to Portugal, stripping the countries in the process of the majority of their human and capital assets and plunging the economies into crisis.

See also ***retornados***.

Sewpersadh, Chanderden 'George' (*South Africa*)

George Sewpersadh is the current president of the Natal Indian Congress (NIC). He joined NIC in 1956 and played a leading role in its revival after 1971. He is vice-president of the Natal region of the UDF, the country's largest umbrella anti-apartheid organisation, and an executive member of the 'Release Mandela' committee. He has been banned and detained by the South African state for his activities. He is a lawyer and has his own attorney's practice in Verulam, Natal.

See also **Natal Indian Congress; South African Indian Congress; UDF**.

Sharpeville (*South Africa*)

Sharpeville is an African township built in 1942 near the industrial centre of Vereeniging, some 80 km from Johannesburg. On 21 March 1960 in Sharpeville police opened fire on a crowd of demonstrators killing 69 people including eight women and ten children; 180 people were wounded.

The Sharpeville demonstration was part of a PAC anti-pass campaign designed to started on 21 March (the ANC national anti-pass campaign was due to start on 30 March). The PAC used the slogan 'no bail, no defence, no fine'. On the morning of the 21st the crowds were beginning to gather and their size surprised the local police officers. At 1.15 pm some 5,000 people were gathered in Sharpeville. They refused instructions to disband. Some 300 police were present and a scuffle broke out. Stones were thrown, according to police witnesses, and the police panicked and opened fire. Most of those who were killed were shot in the back. The mass funeral took place on 30 March.

See also **ANC; PAC; pass laws**.

shebeen (*southern Africa*)

'Shebeen' is a term of Irish origin for unlicensed houses selling alcoholic liquor. In South Africa, where for many years it was illegal for Africans to drink spirits, shebeens are illegal and have become gathering places for political debate as well as drinking. Throughout the region shebeens have traditionally been cultural and social centres for black people. Most are run by women, and many supply home-brewed beer and spirits as well as commercially-produced liquor.

See also **cuca shop**.

Shipanga, Andreas (*Namibia*)

Breakaway SWAPO official who returned to Namibia to found the anti-SWAPO party known as SWAPO-Democrats: see **SWAPO-D**.

Shirazi Association (Tanzania)

The Shirazi Association was a group on the island of Pemba, one of the African political associations which had sprung up during the 1930s. It joined with the African Association on Zanzibar to form the Afro-Shirazi Union in 1957 which was closely modelled on Tanganyika's main nationalist party, TANU.
See also **Afro-Shirazi Party**.

Shona (*Zimbabwe*)

Zimbabwe's indigenous population is divided into two broad linguistic groups, the Shona or Mashona and the Ndebele or Matabele. There are about four times as many Shona as Ndebele-speakers in the country. The ruling ZANU party draws its primary support from Shona-speakers.
See also **Mashonaland; Matabeleland; Ndebele**.

S

SIDC (*Swaziland*)

SIDC, or the Swaziland Industrial Development Company, was inaugurated on 3 September 1986 after a report criticising the state-owned National Industrial Development Corporation of Swaziland (NIDCS). It will take over much of NIDCS's work especially in the area of investments. SIDC is to be a private company with the government holding 37.5 per cent of shares.
See also **National Industrial Development Corporation of Swaziland**.

Simango, Uria (*Mozambique*)

Rev. Uria Simango headed UDENAMO when it merged with other parties to form FRELIMO in 1962. He was elected Vice-President of FRELIMO at its first congress, and re-elected at its second. He was an adherent of the more conservative currents in FRELIMO, and following the assassination of FRELIMO president Eduardo Mondlane in 1969, attempted to take over the leadership. He was prevented from doing so by the Central Committee which appointed a triumverate of Simango, Samora Machel and Marcelino dos Santos, instead.

With his ambitions thwarted, Simango resigned and in November 1969 published a strong denunciation of FRELIMO policies. Simango attracted the support of a high proportion of Mozambicans outside the country, many of them students, and some of them followed him when he subsequently joined COREMO. Some of these later rejoined FRELIMO when it became clear that COREMO was not a viable alternative.
See also **COREMO; Dos Santos; FRELIMO; Machel; Mondlane; UDENAMO**.

Sinoia, Battle of (*Zimbabwe*)

On 28 April 1966, ZANLA, the liberation army of ZANU, penetrated deep into white Rhodesian territory and seven of its guerrillas were killed

at the Battle of Sinoia. Now ZANU marks 28 April as *Chimurenga* Day, the official start of the war of liberation.
See also ***Chimurenga***.

Sisulu, Nontsikelelo Albertina (*South Africa*)

Albertina Sisulu is President of the Federation of South African Women (since 1983) and Transvaal President of the UDF, South Africa's largest anti-apartheid umbrella organisation (since 1983). She joined the ANC Women's League in 1948 and was closely involved with the ANC from its earliest days. Her husband, Walter Sisulu, was general secretary for a number of years. She has been banned and imprisoned many times by the South African state and has become a leading symbolic anti-apartheid figure.
See also **ANC; Federation of South African Women; Sisulu, Walter; UDF**.

Sisulu, Walter Max (*South Africa*)

Walter Sisulu is one of the ANC's most important leaders and an activist in its military wing, *Umkhonto we Sizwe*. He was arrested, together with others, by the security police in a house in Rivonia near Johannesburg on 12 July 1963. In the ensuing 'Rivonia Trial' he was sentenced to life imprisonment for plotting to overthrow the government by violence.

He had joined the ANC in 1940 and became treasurer and a founder member of the ANC Youth League. His politics were militant and at that stage Africanist and separatist. Together with Nelson Mandela and Oliver Tambo he rose rapidly into the top leadership positions within the organisation. In 1949 he became secretary-general and was one of the main leaders in the 1952 Defiance Campaign called by the Congress Alliance against the race laws. He worked closely with the South African Indian Congress and was responsible for maintaining good relations between the ANC and democrats of other groups in the Congress Alliance. He was criticised by Africanists within the ANC for his willingness to work with non-African groups. Despite his original separatist political position, he rapidly became the motivating force behind the Congress Alliance.

Sisulu was arrested several times and banned under the Suppression of Communism Act. In December 1956 he was arrested on a charge of high treason and stood trial until his acquittal in March 1961. Finally he was arrested and imprisoned on 12 June 1964 for life. Together with Nelson Mandela he was moved from Robben Island to Pollsmoor Prison in Cape Town in April 1982.
See also **ANC; Congress Alliance; Mandela, Nelson; Rivonia Trial; Tambo; Treason Trial**.

Sithole, Edson Furatidzayi Chisingaitwi (*Zimbabwe*)

Edson Sithole was a ZANU activist who was involved in the nationalist struggle all his life. He helped to found the African National Congress in 1957 and became a leader in ZANU when that party was formed. However, he spent most of the 1960s in detention under the rebel Rhodesian regime. He became the country's second African attorney by corre-

spondence during this period. In October 1975 he disappeared from Salisbury and by 1977 was widely believed to have been kidnapped by the Rhodesian regime and killed. There has been no trace of him since his disappearance.
See also **African National Congress (Zimbabwe); ZANU**.

Sithole, Ndabaningi (*Zimbabwe*)
Ndabaningi Sithole was ordained minister of the Congregationalist Church in 1958. He became the treasurer of the National Democratic Party in 1960 and the chairperson of ZAPU, one of Zimbabwe's main nationalist organisations, from 1961 to 1963. In 1963 he led a group of ZAPU members who rebelled against Joshua Nkomo's leadership and formed ZANU. Sithole was elected ZANU's first president.

Sithole was restricted and imprisoned by the rebel Rhodesian regime and spent six years in prison with hard labour for incitement to murder its leader, Ian Smith. During that time he was deposed as leader of ZANU. He was chairperson of the African National Council in Lusaka, Zambia, for a year and founded the ANC-Sithole in 1977. In 1978 he was elected an MP in the Zimbabwe-Rhodesian parliament. Ndabaningi Sithole was a well-known and respected member of the African nationalist struggle during the 1960s and 1970s, but he lost his influence after participating in the transitional government led by Abel Muzorewa. Now he is disgraced and lives in exile, an opponent of Robert Mugabe.
See also **African National Council; Muzorewa; National Democratic Party; ZAPU; ZANU; Zimbabwe-Rhodesia**.

Sixishe, Desmond (*Lesotho*)
Desmond Sixishe was minister of information and a cabinet minister in Chief Jonathan's government. He was murdered in November 1986 together with his wife, a colleague, Vincent Makhele (former foreign minister) and his wife and a friend. He was sympathetic to the anti-apartheid ANC and his death, along with other killings, led to allegations that anti-ANC death squads were operating in Lesotho following the military coup in January 1986.
See also **death squads**.

Slabbert, Frederik van Zyl (*South Africa*)
Van Zyl Slabbert was, for seven years, the leader of the official parliamentary opposition in South Africa. In 1986, in a gesture of frustration and despair, he resigned from the leadership of the official opposition Progressive Federal Party (PFP), gave up his seat in the House of Assembly, and announced that parliamentary politics had ceased to have relevance in South Africa.

Slabbert's resignation was a great loss to the party of which he had been the star. He had joined it in 1974, after accepting nomination as the Progressive Party candidate in Rondebosch (Cape Town) in the general election. To his surprise, he was elected and forced to resign his post as Professor of Sociology at the University of Witwatersrand. Slabbert chaired the party's constitutional committee which, in the late 1970s, formulated new franchise proposals which moved away from its previous advocacy of a qualified franchise and proposed, instead, a 'consociational' system with a universal franchise circumscribed by protections for minorities and decentralisation of power.

Largely on the strength of the new policy which was adopted by the party in 1978, and an effective performance in parliament, Slabbert was elected party leader in 1979 to replace Colin Eglin. In 1981, under his leadership, the PFP increased its representation in the House of Assembly from 17 seats to 26. His sudden resignation in 1986 flabbergasted the party. Faced with the realisation that P. W. Botha had reached the limits of his 'reforms' and had no intention of abandoning apartheid, and that confrontation could only increase while government in South Africa took less and less cognisance of parliament, he decided to leave the parliamentary field and devote his energy to extra-parliamentary political initiatives. In July 1987 he headed a highly controversial delegation of Afrikaner dissidents to meet the ANC in Senegal. His place as leader of the party was taken by Colin Eglin.

See also **Eglin; Progressive Federal Party; Suzman**.

Slovo, Joe (*South Africa*)

Joe Slovo is reputed to be the chief-of-staff of the ANC's guerrilla army, *Umkhonto we Sizwe*. He is on the central committees of the ANC and the South African Communist Party. He was a founder member of the Congress of Democrats (the white branch of the Congress Alliance) and one of those arrested in 1956 and tried for treason; he was released when the charges were dropped in 1961. He has been constantly harassed by the South African state and in 1963 he left the country continuing to work for the SACP and the ANC abroad. He lived in Mozambique after the revolution brought FRELIMO to power in 1975. His wife, Ruth First, was killed in a parcel bomb explosion there in 1982 and he was asked to leave in 1984 following the signing of the Nkomati Accord between South Africa and Mozambique.

See also **Congress Alliance; Congress of Democrats; *Umkhonto we Sizwe***.

Smith, Ian Douglas (*Zimbabwe*)

Ian Smith became prime minister of the British colony of Southern Rhodesia on 13 April 1964. He was a farmer and drew much of his support from the conservative white Rhodesian farming community. He helped to found the Rhodesian Front party in 1962 and became its leader when he took office as prime minister. He is most famous for his flaunting of international opinion during the white Rhodesian rebellion against the British Crown.

On 11 November 1965 he delcared unilateral independence from Britain. This rebellion lasted fourteen years. In 1970 he introduced a new constitution and made Rhodesia a republic. After many years of guerrilla war and mounting international pressure and sanctions, Ian Smith tried to reach agreement with nationalist leaders. However, he ignored those fighting the liberation war and created Zimbabwe-Rhodesia with Bishop Abel Muzorewa as the prime minister in 1979. It was not until later that year at the Lancaster House conference in Britain that Ian Smith was forced to concede defeat and, as part of Bishop Muzorewa's delegation, agree to independence, general elections and black majority rule.

After independence Smith continued to lead the Rhodesian Front, renamed the Republican Front. Ian Smith was known by his rebel white followers as 'good old Smithy'. He had a reputation as a simple, obdurate farmer who had stepped forward reluctantly to defend his country. His hero was Winston Churchill. 'If Churchill were alive today', he once said, 'I believe he would probably emigrate to Rhodesia'. He harked back to the British spirit of Dunkirk which, he claimed, the British and the west in general, had lost. He argued that the west had become weak and no longer resisted 'communism', setting himself up as a defender of western values. Perhaps his most famous statement was made in 1976: 'I don't believe in black majority rule in Rhodesia', he said 'not in a thousand years'. Just four years later Robert Mugabe was elected the first black prime minister of an independent Zimbabwe.

See also **constitution of 1969; Lancaster House; Mugabe: Rhodesian Front; sanctions against Rhodesia; Unilateral Declaration of Independence; Zimbabwe-Rhodesia**.

S

SNASP (*Mozambique*)

The *Serviço Nacional de Segurança Popular* (National Service of Popular Security) is the Mozambican internal security organisation, formed in 1975. SNASP is charged with 'detecting, neutralising and combating all forms of subversion, sabotage and acts directed against the people's power and its representatives, the national economy or the objectives of the People's Republic of Mozambique'.

See also **FPLM**.

Soames, Christopher (*Zimbabwe*)

The British government appointed Lord Soames as a temporary governor of Southern Rhodesia to oversee the ceasefire and elections from late 1979 to 1980. The task was formidable. He had little experience of Africa or guerrilla movements and came from the most elite of British military backgrounds – Eton, Sandhurst and the Coldstream Guards. He was helped in his task by the Commonwealth Monitoring Force of 850 soldiers led by Major-General John Ackland. Its role was to monitor the guerrillas in their assembly points as well as the Rhodesia Joint Operational Commands distributed around the country. They had also to provide logistical support for the assembly points and defuse tensions between the white Rhodesians and the guerrillas. Lord Soames' operation was described by Major-General Ackland as one of the most successful in British military history. The role of 200 observers from 30 countries also served to help ensure international acceptance of the election results.

See also **Lancaster House**.

Sobhuza II (*Swaziland*)
King Sobhuza II ruled Swaziland from 22 December 1921 until his death in August 1982, aged 83. The death of each Swazi king has led to a period of political instability as various factions in the ruling monarchy vie for power. This was particularly evident after the long years of King Sobhuza's rule. Four years of political instability, following his death, were brought to an end by the crowning of the new young King Mswati III on 25 April 1986.

King Sobhuza led his country to independence in 1968 as a constitutional monarch. He represented nationalist aspirations in his attempt to regain the large amounts of land lost to the Swazi people under colonial domination. The king's policy of buying back the land won him wide popularity and this campaign dominated his early years in power.

In 1973 the king declared a state of emergency and ruled the country by decree as an absolute monarch until 1978 when he drafted a new constitution, based on the traditional power structure. Party politics were banned and the state retained all political power. By the time of the king's death the authority of the Swazi monarchy had been restored to its pre-colonial status.
See also **'constitutional crisis' of 1973; Mswati**.

S

Sobukwe, Robert Mangaliso (*South Africa*)
Robert Sobukwe was elected the first national president of the Pan-Africanist Congress on its formation in April 1959. He had been a member of the African National Congress Youth League and was closely identified with the Africanist group, becoming editor of its newspaper, *The Africanist*, in 1955. He was an intellectual who held the Africanist position with some subtlety arguing that it would be possible in future for whites to be absorbed in the 'Africa for the Africans' slogan, once they came to owe their loyalty to Africa.

On 16 March 1960 Robert Sobukwe announced a campaign against the pass laws in South Africa urging defiance of these laws under the slogan: 'no bail, no defence, no fine'. The campaign ended with the Sharpeville killings of 69 people by police. Robert Sobukwe was imprisoned and banned and he died from lung cancer in 1978 while serving a banning order in Kimberley. He was replaced by the PAC's acting president Potlake Leballo.
See also **Africanism; Congress Youth League; Leballo; PAC**.

Socialist League of Malawi (*Malawi*)
The Socialist League of Malawi (LESOMA) is one of Malawi's main opposition movements. It started a military wing in March 1980 and claims to have recruited a number of Malawians for military training in Cuba. It was formed in 1975 and is alleged to be the most left-wing of the dissident groups. Opposition parties are banned in Malawi so the various dissident groups operate beyond the country's borders and Lesoma is based in Harare, Zimbabwe. None of the groups are particularly effective, although they all claim to have wide support. In March

1983 the president of the LESOMA movement, Dr Attati Mpakati, was murdered in Harare, Zimbabwe. The new leader is Grey Kamunyembeni. See also **Congress for the Second Republic; Malawi Freedom Movement; Save Malawi Committee**.

Sokoine, Edward (*Tanzania*)

Edward Sokoine, Tanzania's prime minister at the time that President Nyerere announced his intention to retire, was widely expected to become the next president. However, in April 1984 he was killed in a road accident and succeeded as prime minister by Salim Ahmed Salim.
See also **Salim**.

Sotho (*Lesotho*)

A collective term for people speaking the Sesotho language: see **Basotho**.

South Africa

Official title: Republic of South Africa
Head of state and government: State President Pieter Willem Botha
Area: 1,221,037 sq. km
Population: 32,642,730 (1985); breakdown into official 'racial'

S

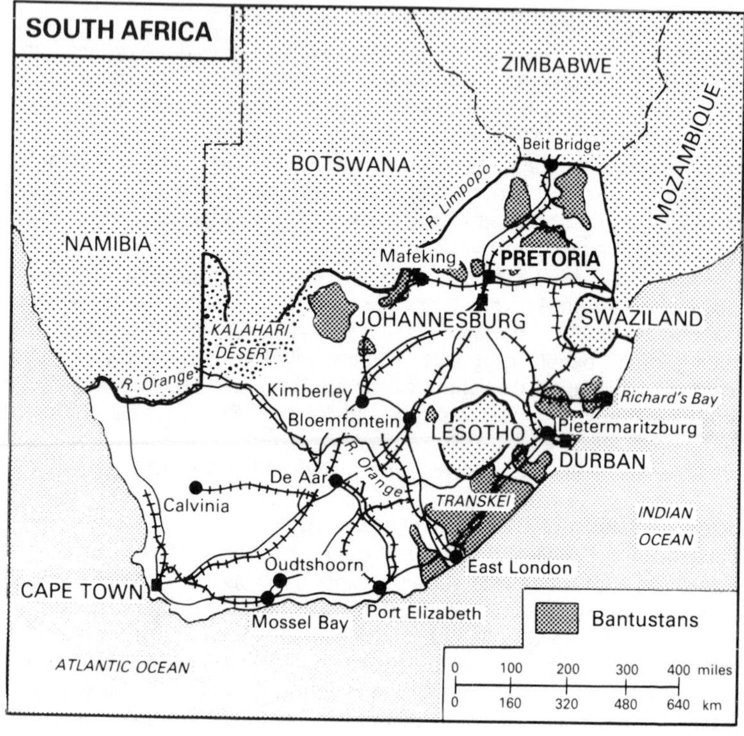

groups in mid–1984 is as follows: 'Whites': 4,818,679; 'Blacks': 24,103,458; 'Coloureds': 2,830,301: 'Indians': 890,292

Capital: Cape Town (legislative); Pretoria (administrative)

Official language: Afrikaans, English (Xhosa, Zulu, Sesotho widely spoken)

GDP per capita: US$1,666 (1985)

Major exports: gold (up to 40 per cent of total value); other minerals and raw materials eg: diamonds, uranium, silver, iron ore, copper, manganese, chrome; food; beverages; tobacco.

Currency: Rand (R)=100 cents; R2.068 per US$ (July 1987)

Main political parties: For whites these are as follows: National Party (ruling party since 1948); Progressive Federal Party; New Republic Party; Conservative Party; *Herstigte Nasionale Party;* South African Communist Party (banned 1950). For blacks: the ANC (banned 1960); the PAC (banned 1960); South African Indian Congress (banned 1960); AZAPO; Transvaal Indian Congress; Labour Party; *Inkatha*.

South Africa dominates the region of southern Africa both economically and politically. It is situated at the strategically important southern-most point of the continent with the Indian Ocean to its east and the Atlantic to its south-west. Its neighbours are Namibia to the north-west (which it occupies illegally), Botswana and Zimbabwe to the north, and Mozambique and Swaziland to the east; Lesotho is totally surrounded by the eastern part of the country. South Africa is very large with a varied climate and landscape. It contains a wealth of mineral riches and fertile land. Some of its scenery is staggeringly beautiful.

It is, however, a country torn apart by its politics. Domination by the white minority since the constitution of 1910 formed the Union of South Africa, has created an unequal society based on white privilege and exploitation of the black working population. The ruling National Party has in recent years pursued a policy designed to modify its doctrine of apartheid. These changes are in response to a new situation in which the black majority is organising and posing an ever-increasing threat to continuing white control. The 'White' state is further threatened by the gradual imposition of sanctions by the international community in an attempt to force it to come to some accommodation with the black population. Violent unrest began in the black townships in 1984 as a protest against the government's 'Bantu' education policies and has continued and escalated ever since. A state of emergency covering the whole country was reimposed on 12 June 1976 and is still in operation (1987).

President P. W. Botha introduced a new constitution on 3 September 1984, following a referendum the previous year. It provided for a three-chamber parliament to cater for 'Whites' (the House of Assembly with 178 members), 'Coloureds' (the House of Representatives with 85 members) and 'Indians' (the House of Delegates with 45 members). The President's Council is a body designed to resolve disputes among the three houses. The constitution ignored the rights of the country's majority African population, and the elections to the House of Delegates and the House of Representatives in 1984 were extensively boycotted. The government argues that Africans may vote in elections in their 'Homelands' or for municipal councils in urban areas. The large anti-apartheid umbrella organisation, the UDF, was formed in order to oppose the new constitution.

The constitution gives the state president far more power than the head of government has enjoyed hitherto. President P. W. Botha is head of state and commander-in-chief of the armed forces; he can enter into international treaties, proclaim or terminate martial law; he has enormous powers of patronage, and can, for instance, appoint members of the President's Council and intervene in its activities. He can also initiate legislation in any of the parliamentary houses.

See also **ANC; apartheid; Botha, P.W.; constitution of 1983; National Forum; National Party; 'reforms'; state of emergency; UDF; Union of South Africa.**

S

South African attacks on Botswana (*Botswana*)
These have become more frequent and serious as South Africa has sought to force Botswana to sign a 'non-aggression pact': see **Masire**.

South African Black Alliance (*South Africa*)
The South African Black Alliance or SABA was formed at the initiative of Chief Gatsha Buthelezi, the leader of the KwaZulu 'Homeland', in 1978. It is a political alliance of Chief Buthelezi's *Inkatha* movement, the *Inyandza* National Movement of the KaNgwane 'Homeland', the Indian Reform Party and the Labour Party of South Africa (which was suspended and then withdrew from SABA in 1983 because of its participation in the elections for a 'Coloured' parliament under the new constitution).

SABA is dominated by Buthelezi and is used by him when he feels he needs a broader base than the predominantly Zulu *Inkatha* movement. SABA has held protest meetings against the proposed Ingwavuma-KaNgwane transfer of land to Swaziland.

See also **Buthelezi; constitution of 1983; *Inkatha*; Labour Party; Reform Party**.

South African Broadcasting Corporation (*South Africa*) See **SABC**.

South African Bureau of Racial Affairs (*South Africa*) See **SABRA**.

South African Communist Party (*South Africa*)
The South African Communist Party (SACP) was founded in July 1921 as the Communist Party of South Africa (CPSA), and it changed its name to the South African Communist Party in 1953. Much of the history of the SACP concerned the development of its policy towards the nationalist movement. At first the party set its goal as the overthrow of capitalism,

but by 1928 most members were black and many were arguing for a more active role in the nationalist movement. The Sixth Congress of the Communist International intervened in the internal debate and the party accepted that the national question was the 'foundation of the revolution'. This meant that the party could now play an active role in the ANC. This was to lead to new problems within the ANC and provoke allegations – chiefly by the Africanists – that the movement was overly influenced by the communists. In 1950 the Suppression of Communism Act outlawed the party and in June that year it was disbanded. In 1953 it was revived and has operated underground ever since. Together with the ANC it set up *Umkhonto we Sizwe* in 1961. Joe Slovo, one of its leading members is reputed to be the present chief of staff of *Umkhonto*.

The SACP presently espouses a two stage theory of revolution, arguing that white South Africans constitute an internal colonial force oppressing the black majority. Its aim, therefore, is to 'establish an independent state of national democracy in South Africa' in the first instance (through an anti-colonial, nationalist revolution) and then to try and move this new government towards socialism (the socialist revolution). In its immediate aim of a revolutionary nationalist struggle against the 'White' state of South Africa it is allied with the ANC, and it is pledged to support the Freedom Charter. Its current chairperson is Dr Yusuf Dadoo and its general secretary Moses Mabhida.

See also **ANC; Dadoo; Fischer; Freedom Charter; Slovo**.

South African Congress of Trade Unions (*South Africa*) See **SACTU**.

South African Coloured People's Organisation (*South Africa*)
The South African Coloured People's Organisation (SACPO) was founded in September 1953 to try and gain mass support among the 'Coloureds' in South Africa, initially to oppose their removal from the common voters' roll. Later SACPO changed its name to the South African Coloured People's Congress in order to join the Congress Alliance.

See also **Coloured People's Congress; Congress Alliance**.

South African Confederation of Labour (*South Africa*)
The South African Confederation of Labour (SACLA or sometimes SACOL) was formed in September 1957. It is a group of white unions acting in the interests of the white working class. It has seen its role as protecting white workers from black workers and has traditionally supported apartheid, especially the colour bars applied to jobs. It also opposed the extension of trade union membership to black workers.

After the introduction of the Wiehahn legislation which accepted black trade unions under certain conditions, SACLA was thrown into disarray. There were disaffiliations and membership fell from 22 unions with some 179,700 members in 1980 to 14 unions with some 100,000 members in 1982. Attie Niewoudt, SACLA's former president was a member of the Wiehahn Commission and issued a minority report which reflected the

view of many SACLA members at the time. 'The racial composition of the labour force in many undertakings, industries, trades and occupations in South Africa', he wrote, 'is such that the workers who at present enjoy trade union rights would be swamped by force of numbers, should blacks be admitted to the trade union movement.' This view is held by the traditionally right wing MWU (Mine Workers' Union), led for many years by Arrie Paulus. However, SACLA's leadership is now trying to adjust to the current situation in which black trade unions play the central role in industrial relations.
See also **'job reservation'**; **Wiehahn Commission**.

South African Council of Churches (*South Africa*) See **SACC**.

South African Defence Force (*South Africa*) See **SADF**.

South African destabilisation of Lesotho (*Lesotho, South Africa*): see **relations with South Africa (Lesotho, South Africa)**.

South African destabilisation of Mozambique (*Mozambique, South Africa*)
South Africa provided moral and material support to the Portuguese troops fighting FRELIMO insurgents during the liberation war in Mozambique. Having seen the Portuguese colonies as buffer states preventing the infiltration of the ANC's *Umkhonto we Sizwe* fighters, South Africa is alleged to have been poised to intervene, in 1975, to prevent a FRELIMO takeover of the country. The alleged plan was to send troops, at the request of the Portuguese, to keep FRELIMO out of the south of the country and its capital, Maputo (then Lorenço Marques), an important port for South African goods. In the event the invitation from Portugal never came, and South Africa's troops stayed home.

In later years South Africa was less circumspect, however, and there were frequent raids on Mozambique. Lorry-loads of South African commandos attacked three houses in Matola, a suburb on the outskirts of Maputo, on 30 January 1981 killing 13 ANC and SACTU members (South African expatriates) and one Portuguese national. On 23 May 1983 South African aircraft bombed several houses and offices in Matola and another Maputo suburb, Liberdade, using fragmentation rockets designed to maximise human casualties. Six people, including three factory workers and a child playing in the street, died, and forty people were injured. Most of the damage was to a Maputo jam factory.

More important than the provocative raids by the South African military inside Mozambique, however, has been its active sponsorship and promotion of the MNR. Following the collapse of the Smith regime, South Africa took over control of the MNR, a guerrilla force set up by the Rhodesian military to attack and destabilise Mozambique and discourage FRELIMO's support for the Zimbabwean liberation movements. The MNR was subsequently built into a major guerrilla force and became the chief agency for South African-initiated destabilisation of Mozambique. South African agents were also alleged to have been the source of a parcel bomb which killed Ruth First, an academic who headed

the Centre for African Studies at Maputo's Eduardo Mondlane University.
See also **ANC; First; FRELIMO; MNR; Nkomati Accord; SACTU;** ***Umkhonto we Sizwe*****.**

South African Federated Chamber of Industries (*South Africa*): see **FCI**.

South African Foundation (*South Africa*)
A business-funded propaganda organisation which campaigns overseas to encourage investment and business confidence in South Africa, the South African Foundation was formed in 1960 shortly after the shootings at Sharpeville. Its initial efforts were focussed on rebuilding the shattered confidence of foreign investors after the post-Sharpeville collapse of share prices in South Africa. Its early propaganda and fund-raising efforts brought it into sharp conflict with the Progressive Party which viewed it as being in direct competition for business support. More recently relations have been distant but correct.

The South African Foundation maintains permanent representatives in Washington, London, Paris and Bonn and has close links with overseas-based organisations, such as the United Kingdom-South Africa Trade Association (UKSATA), which seek to promote bilateral links. Press reports associated the Foundation with the discredited South African Department of Information during the 'Information Scandal', but it denied accepting funds from the department. It campaigns most vociferously against the imposition of sanctions on South Africa, and offers free trips to the country to influential politicians prepared to accept its hospitality.
See also **Free Market Foundation; National Party; Progressive Federal Party; sanctions; Sharpeville; Urban Foundation**.

South African Indian Congress (*South Africa*)
The South African Indian Congress (SAIC) was founded in 1920 by uniting the Natal Indian Congress (founded by Mahatma Gandhi in 1894) and the Transvaal Indian Congress. In 1946 a new radical leadership was elected – Dr Yusuf Dadoo became president of the Transvaal branch and Dr G. M. Naicker became president of the Natal branch. This was followed by an agreement to co-operate with the ANC in the signing of the Dadoo-Xuma-Naicker pact in 1947. The SAIC was heavily involved in the Defiance Campaign of 1952 and many of its members were jailed. In 1955 the SAIC helped to organise the Congress of the People and became an active member of the Congress Alliance. The SAIC was never banned by the state, but its members were constantly imprisoned and harassed and it became impossible to organise effectively.

Mewa Ramgobin revived the Natal Indian Congress in 1971 and its current president is George Sewpersadh. The NIC launched a campaign against the state-imposed South African Indian Council (the Anti-SAIC campaign) and led an effective boycott of its elections in 1981, when a poll of a mere 10 per cent was recorded. The Transvaal Indian Congress

was also revived in 1983 and both it and the NIC have been prominent in the formation of the anti-apartheid organisation, the UDF.
See also **Congress Alliance; Defiance Campaign of 1952; SAIC; UDF; Xuma-Dadoo-Naicker Pact**.

South African Indian Council (*South Africa*) See **SAIC**.

South African Institute of Race Relations (*South Africa*)
The South African Institute of Race Relations (SAIRR) was founded in 1929 to promote contact and understanding between different 'races' in South Africa and to persuade government, through research and argument, to take positive steps to avoid conflict. A liberal organisation, it had some influence under the pre–1948 United Party governments, but lost this under the National Party and became part of the liberal opposition to apartheid. During the latter period it published a *Handbook on Race Relations in South Africa* in 1949 and, thereafter, annual surveys of race relations. The only such surveys available, these have become standard reference works and have become increasingly comprehensive, with time, in their coverage of events in South Africa. The latest published to date is 1985.

S

The SAIRR also publishes material on related issues on an irregular basis and administers scholarships and bursaries on behalf of other foundations and large businesses. Apart from income from membership, its publications and a number of shops selling African artifacts and craftwork, the institute's main sources of income have been liberal businesses in South Africa (to which it has also given personnel advice on occasion) and major US foundations such as the Ford Foundation. Money for its establishment in 1929, for instance, came from the Carnegie Corporation and the Phelps Stokes Fund.
See also **SABRA**.

South African invasions (*Angola, South Africa*)
The first of these occurred soon after the collapse of the Portuguese empire. In August 1975 South African forces invaded the south of the country, ostensibly to protect South African workers on the giant hydroelectric power project there. In fact, the South Africans used the opportunity to kill and expel FAPLA and MPLA forces in the area, and to support local UNITA and FNLA units which it hoped would help form a bulwark against SWAPO. On 15 October the South African Ministry of Defence admitted that it had used the invasion to launch attacks on SWAPO in southern Angola.

On 20 October South Africa launched a full-scale armoured invasion of Angola from the south in support of the FNLA/Zairean invasion in the north, and in an attempt to prevent the country becoming independence under the MPLA. By 4 November, a week before independence day, the South Africans had taken Benguela, the major port in the centre of the Angolan coastline, and seemed poised to move on Luanda. FAPLA and the recently arrived Cuban military instructors who died with the defending FAPLA soldiers, had been able to muster little more than rifles and occasional mortars to resist the South Africans. Faced by South African AFVs (Armoured Fighting Vehicles), they could hope to do little but slow up the South African advance. Following an appeal for help to Cuba, however, and the arrival of Katiusha rocket launchers, the over-extended South Africans were forced into retreat.

South African pressure on the Angolan government continued, and military incursions increased from 1981. There were major invasions by South African forces in August 1981, November 1981 and February 1982. South Africa used these operations to destroy the infrastructure in the southern-most Cunene (Kunene) province, attack SWAPO and FAPLA forces and provide military support to its surrogate UNITA forces.
See also **Cunene River Hydro-Electric Power Scheme; FAPLA; MPLA; SWAPO**.

South African Native National Congress (*South Africa*)
The South African Native National Congress – renamed the African National Congress (ANC) in 1923 – was formed on 8 January 1912 at a conference in Bloemfontein. It was the first national African political movement and the ANC is now the main South African liberation movement. The first president was John L. Dube, the leader of the Natal Native Congress, and the general secretary was Sol T. Plaatje, who had been active in the Cape-based African People's Organisation.
See also **African People's Organisation; ANC; Dube; Plaatje**.

South African Party (*South Africa*)
The original South African Party was formed in 1910, and governed the country on and off, under Louis Botha and Jan Smuts, before it merged with the National Party to form the United South African National Party (United Party). A second South African Party, small and short-lived, was formed in 1977, when six right-wing United Party MPs were expelled because of their refusal to countenance negotiations to form a united political opposition among whites to the National Party. In the general election which followed, the party, under the leadership of John Wiley, lost three of its six seats. The South African Party subsequently dissolved itself and the majority of its members joined the National Party. John Wiley became a cabinet minister in the government of P. W. Botha, but later committed suicide during the 1987 'Whites'-only election campaign.
See also **Botha, P. W.; National Party; United Party**.

South African Police (*South Africa*): see **SAP**.

South African Progressive Reform Party (*South Africa*) See **Progressive Federal Party**.

South African Students' Organisation (*South Africa*) See **SASO**.

Southern African Customs Union (*southern Africa*)
The Southern African Customs Union (SACU) is a regional arrangement which allows for unlimited free trade among its members within the area. It consists of South Africa (including South African-occupied Namibia),

Botswana, Lesotho and Swaziland. SACU began in 1910 as an agreement between South Africa and Britain (on behalf of the territories it controlled). Then it was agreed that South African currency would be legal tender, tariffs would be set by South Africa and fixed proportions of revenue would be distributed. Payments to Botswana, Lesotho and Swaziland were to be made on the basis of their share of goods imported by SACU countries, multiplied by 1.42. This extra amount is designed to compensate for a loss of freedom by these countries to conduct a completely independent economic policy. For Lesotho, revenue from SACU has been some 70 per cent of total government revenue in the 1980s; for Swaziland SACU revenue has been very variable in this period ranging from a low of 46.7 per cent in 1981–2 to a high of 64.6 per cent in 1982–3; for Botswana SACU provides some 30 per cent of total government revenue.

In recent years South Africa has tried to renegotiate the terms governing SACU. This has coincided with its attempts to force its neighbours to sign security pacts along the lines of the Nkomati Accord with Mozambique, signed in 1984. The countries of SACU are vulnerable to economic leverage from South Africa in any renegotiation of the arrangement. Swaziland and Lesotho have introduced a sales tax to try to reduce their dependence on SACU revenue. Together with Botswana they are also members of SADCC which is a wider regional organisation which seeks to create an alternative economic network not dependent on South Africa.

See also **Nkomati Accord; Rand Monetary Union; SADCC**.

Southern African Development Coordination Conference (*South Africa*) See **SADCC**.

Southern African Transport and Communications Commission (*southern Africa*)

Part of the SADCC with a secretariat based in Mozambique: see **SATCC**.

Southern Rhodesia (*Zimbabwe*)

Southern Rhodesia was the name given to the British colony founded by Cecil John Rhodes in 1893. After the collapse of the Central African Federation in 1963, it was widely known as Rhodesia, but its official name remained Southern Rhodesia until the emergence of independent Zimbabwe on 18 April 1980.

See also **Central African Federation; Rhodes; Zimbabwe**.

South West Africa (*Nambia*) See **Namibia**.

South West Africa *Herstigte Nasionale Party* (*Nambia*) See **HNP Namibia**.

South West Africa/Namibia (*Namibia*) See **Namibia**.

South West Africa National Party (*Namibia*)

The SWA National Party (SWANP) has dominated white politics in recent Namibian history. From 1950 to 1977 'White' voters in Namibia were able to elect 12 representatives to the 'White' South African Parliament, and throughout that period all 12 seats were held by the SWANP. It also held 15 of the 18 seats in the local Legislative Assembly over the same period. Its virtual monopoly of white politics was not broken until the support of many German-speaking whites was withdrawn over the period 1975–6.

The SWANP has been a staunch defender of apartheid in Namibia. It entered the MPC, where all decisions are taken by consensus, specifically to veto moves to rescind decree AG8. It had earlier opposed the DTA 'internal administration', joining with a few black politicians to form AKTUR. Although his party was participating in the so-called Transitional Government of National Unity, the leader of the SWANP, Abraham du Plessis, remained in the white Legislative Assembly arguing that it remained more important than the MPC, where the party delegation was led by his deputy, Eben van Zijl. Van Zijl was sacked in 1986, however, for supporting a move to open schools in Namibia to all children irrespective of colour. He was replaced by Jan de Wet.

The SWANP was affiliated to the National Party of South Africa until September 1977, when direct representation in the South African Parliament for white people in Namibia was ended. The SWANP split over its attitude to the so-called 'new dispensation'. In 1979 the SWANP accused P. W. Botha's National Party of surrendering the whites of the territory', and relations between the two parties have remained strained. See also **AG8; AKTUR; DTA; MPC; National Party; Republican Party; Transitional Government of National Unity**.

South West Africa National Union (*Namibia*) See **SWANU**.

South West Africa Native Labour Association (*Namibia*) See *SWANLA*.

South West Africa People's Democratic United Front (*Namibia*)
A small ethnically-based party forming part of the Democratic Turnhalle Alliance: see **DTA**.

South West Africa People's Organisation of Namibia (*Namibia*) See **SWAPO**.

South West Africa Police (*Namibia*) See **SWAP**.

South West Africa Progressive Association (*Namibia*)
An organisation of students and professionals which was dissolved on the formation of SWANU in 1959: see **SWANU**.

South West Africa Student Body (*Namibia*) See **SWASB**.

South West Africa Territory Force (*Namibia*) See **SWATF**.

Soweto (*South Africa*)
Soweto is the name of South Africa's largest African township. Its population is estimated at two million and it has been the focal point of much

political unrest in recent years. It is near Johannesburg and was named Soweto after its geographical position, South-Western Townships. After the Second World War the Africans who had been living on the outskirts of Johannesburg were sent to live in the 28 townships south-west of the city which were eventually grouped into one large township, Soweto.
See also **Soweto uprising, 1976**.

Soweto Civic Association (*South Africa*)
The Soweto Civic Association (SCA) was set up in September 1979 by Dr Nthato Motlana and the Committee of Ten. The SCA was established to broaden the base of support in the community for Soweto's self-administration. It is seen as part of the system of local democracy set up by anti-apartheid activists in the townships. The most recent attempt in this direction is the development of street committees.
See also **Committee of Ten; 'comrades'; Motlana; street committees**.

Soweto uprising, 1976 (*South Africa*)
On 16 June 1976 the black township of Soweto, just outside Johannesburg, exploded in violent upheaval. The protests were led by school students and for days pictures in the media showed children in confrontation with armed police in the streets. The bravery and political awareness of the pupils was an indication of a new kind of militancy in South Africa. The state brought police reinforcements into the township and placed army troops on standby as the unrest spread. This was the most serious and widespread uprising ever experienced in South Africa, and it started a pattern of urban protest which has not yet subsided (1987).

The immediate cause of the rebellion was the system of 'Bantu education' in South Africa. African students had been objecting to the imposition of the ruling that English and Afrikaans must be used equally as the medium of instruction from standard five onwards in secondary schools for Africans. This meant that students had to submit to being taught in two languages, one of which, Afrikaans, was widely perceived as the language of their oppressor. The first active opposition took place on 17 May after adult representations had failed to bring about any change of the hated policy. Orlando West Junior Secondary School had elected an interim pupil committee which stated they 'had tried without success to meet our circuit inspector, Mr De Beer'. Thus, they stayed away from classes and handed in a list of demands to their principal. This was just the beginning. On 13 June the South African Students' Movement met in Naledi High School and formed a Soweto Students' Representative Council (SSRC). This planned the protest march of 16 June. The march moved through Soweto and converged on Orlando West Junior Secondary School where it had all begun. The 10,000 marchers were confronted by police who fired tear gas into the crowd. In response, the children threw stones at the police. The police opened fire, first apparently with warning shots and then into the advancing children, killing at least one, Hector Peterson. Later two other adults were killed by the crowd, and all schools were officially closed.

Over the next few days the rioting spread to pupil and student groups

in several other townships copying the pattern of attacks on police patrols and symbolic buildings. Just before schools were formally re-opened on 26 June (they remained empty until the end of the year), the government dropped the ruling about being taught in Afrikaans. The pupils had won. By 1977 the official death toll from the uprising was 575 with 2,389 wounded, and many of the student leaders, including Tebello Motopayane, the first leader of the SSRC, had fled the country. The Soweto uprising spawned many community groups, like the Black Parents' Association and the Committee of Ten, as other people were politicised and drawn into the struggle.
See also **Black Consciousness Movement; Black Parents' Association; Committee of Ten; Soweto; Soweto Civic Association**.

SP (*South Africa*)
Security Police, a branch of the SAP notorious for the torture and maltreatment of detainees: see **SAP**.

Special Branch (*South Africa*) See **SB**.

SPP (*Swaziland*) See **Swaziland Progressive Party**.

SSC (*South Africa*) See **State Security Council**.

state of emergency (*Lesotho*)
Declared in 1970 to abort result of general election: see **Jonathan**.

state of emergency (*South Africa*)
The South African state president declared a state of emergency in 36 districts of the country on 21 July 1985, by the terms of the Public Safety Act of 1953. This empowers the government to issue emergency regulations which may suspend the provisions of any laws except those concerned with defence, the operation of legislatures, and industrial conciliation. By the terms of the Public Safety Act the state president may extend the areas affected at any time. The state of emergency was lifted in March 1986 and reimposed throughout the country on 12 June that year. It was renewed for up to another year on 11 June 1987. By early 1987 reported 'politically-related' deaths had fallen to one on average each day. This compares with five each day in the first half of 1986. Apart from emergencies declared in various 'Homelands', an emergency has only once before been imposed in South Africa. This was on 30 March 1960, nine days after the shooting at Sharpeville of 69 people protesting against the pass laws. It lasted until 31 August that year.

The emergency regulations confer extensive new powers on the state. Powers are given to all members of the police, army and prison service to arrest, with or without a warrant, and detain certain people for interrogation for up to fourteen days. The detention period may be renewed with the permission of the Minister of Law and Order, for an indefinite period, until the ending of the state of emergency. The Minister of Justice is empowered to make rules governing the conditions of detainees under the regulations (information about detainees under this state of emergency may only be supplied by the Commissioner of Police

– an estimated 10,000 people under eighteen years of age have been detained under the emergency so far (1987)). Police, army and prison officers have greater powers to use force – if an initial verbal warning is ignored they may apply whatever force they consider necessary. Other powers, such as curfews, control over the media (very strict in this emergency), and traffic control, may be invoked if considered necessary by the police commissioner, who must publish the details in the *Government Gazette*. Finally, the state and members of the police, army and prison service are assured in advance of indemnity against proceedings in the courts over any action they carry out under the emergency, unless it can be proved that they did not act in good faith.

The immediate background to the declaration of a state of emergency in 1985 was an increase in the level of open resistance to the government in the townships. By May the Minister of Law and Order was voicing the government's concern at the scope of the protests and their spread to small towns. When President P. W. Botha explained the re-introduction of the state of emergency on 12 June 1986, he said that one of its aims was to stop the development of what he termed 'alternative structures' within the townships. These 'alternative structures' have been a feature of black activity since 1983. Initially set up as a way of organising under conditions of repression, they have come to be seen as alternative ways of administering people's lives in a democratic way.

See also **'reforms'; Soweto uprising, 1976; street committees**.

S

State Council (*Namibia*)

The creation of the State Council was announced by South Africa in July 1983 as a replacement for the DTA-dominated Council of Ministers. Its brief was to produce 'comprehensive proposals' for an 'interim government' which were to be tested in a referendum as an alternative to UN Security Council Resolution 435. When even the DTA refused to become involved in the State Council, however, the plans were scrapped and South Africa turned instead to the MPC.

See also **DTA; interim government; MPC; Transitional Government of National Unity; UN Security Council Resolution 435**.

State Security Council (*South Africa*)

The State Security Council (SSC) was established in 1972 to bring together the various security services in South Africa with representatives of other government bodies, and to co-ordinate their work. According to a parliamentary reply by P. W. Botha in 1984, the membership of the SSC consisted of 56 per cent National Intelligence Service personnel, 16 per cent South African Defence Force, 11 per cent foreign affairs, 11 per cent Security Police, 5 per cent Railway Police and 1 per cent prison staff. It was chaired by P. W. Botha and included the ministers of defence, foreign affairs, justice, law and order, finance, constitutional development and planning, and co-operation and development. It met every 14 days and made recommendations to the cabinet.

The State Security Council has come to rival the cabinet as the most powerful body in the country. It is reported to meet the day before cabinet meetings (which are held weekly), and to determine their agendas. Through the representation of the most important ministers on the SSC, and through inter-departmental committees, it is able to liaise directly with all government departments. President Botha is reputed to have changed the role of the SSC, after becoming prime minister in 1978, from a consultative and co-ordinating one to give it responsibility for conducting the 'national strategic planning process'.

Despite denials by P. W. Botha and the secretary of the SSC, General A. P. van Deventer, most independent commentators and opposition policians accord the SSC a central role in decision making in South Africa. While the cabinet continues to take the formal decisions, they argue, these in many cases simply rubber-stamp decisions already reached by the SSC.

See also **Botha, P. W.; NIS; SADF; Security Police**.

Steel and Engineering Industries Federation of South Africa (South Africa): see **SEIFSA**.

street committees (*South Africa*)

Street committees are groups of five or six people, theoretically elected by people in each street to organise the affairs of the townships during times of unrest. They are dominated by UDF supporters and, in the reported words of its Transvaal president, Albertina Sisulu, are 'the broadest scale grassroots organisation ever'. There are reported to be street committees in Soweto, Alexandra and Sharpeville, Guguletu and Langa (in the Cape Town area) and in the townships of the eastern Cape.

In May 1986 the UDF and the Soweto Civic Association decided that the people of the townships needed protection. According to reports students and 'comrades' were sent out to go 'from house to house, from street to street' calling 'hit-and-run' meetings (necessary because meetings are illlegal under the state of emergency) and explaining the purpose of the street committees.

Dr Motlana has denied that street committees are derived from the ANC's decision in the 1950s to set up cells in every street. The street committees, he argued, are for the local community and set up by them. Eventually all committees would be represented in block committees which would send members to branch committees. Above that was the executive committee of the Soweto Civic Association which was already in place.

See also **'comrades'; Motlana; Sisulu, Albertina; Soweto Civic Association; UDF**.

Sullivan Code (*South Africa*)

In March 1977 Rev. Leon Sullivan of the Zionist Baptist Church in the USA and also a board member of IBM, co-ordinated twelve of the biggest American firms operating in South Africa to endorse a set of principles designed to end segregation and job discrimination in their plants. The six principles became known as the Sullivan Code. Ten years later, Sullivan announced that the code had not worked and called for economic sanctions and disinvestment.

The principles were: non-segregation of the races in all eating, comfort and work facilities; equal pay and fair employment practices for all employees; equal pay for all employees doing equal and comparable work for the same period; initiation of a development training programme that would prepare Africans in substantial numbers for supervisory, administrative, clerical and technical jobs; increasing the number of Africans in management and supervisory positions; improving the quality of employees' lives outside the work environment.
See also **EEC Code of Conduct; sanctions against South Africa**.

Sultan of Zanzibar (*Tanzania*)
Sayyid Jamshid ibn Abdullah, Sultan of Zanzibar from 1890 to 1964: see **Zanzibar**.

Supreme Ruling Council (*Swaziland*): see **Liqoqo**.

S

Suzman, Helen (*South Africa*)
Helen Suzman is second only to President P. W. Botha in the length of time she has served as an MP in South Africa. Born in 1917, she was elected to represent the United Party in Houghton (Johannesburg) in 1953. She joined the Progressive Party when it was formed in 1959, was the only Progressive to retain a seat in the 1961 general election, and remained the party's sole MP until 13 years later, when it won six seats in the 1974 general election.

Suzman has earned an international reputation as a redoubtable opponent of apartheid and was in many instances the only MP with the courage and insight to oppose yet further inroads into civil rights and the rule of law in South Africa. She has close links, personal and political, with the business community in Johannesburg and the USA, and has been honoured in many countries of the world for her political courage. She has been awarded seven honourary doctorates, and has twice been nominated for the Nobel Peace Prize.
See also **Anglo American Corporation; Botha, P. W.; Progressive Federal Party; United Party**.

SWA (*Namibia*)
South West Africa: see **Namibia**.

SWABC (*Namibia*)
South African controlled South West Africa Broadcasting Corporation, established in 1979: see **press**.

SWANLA (*Nambia*)
Until 1972 contract workers in Namibia were recruited, allocated employers and transported to and from their places of employment by SWANLA, the South West Africa Native Labour Association. Central to the contract labour system, and owned and run by the major employers in Namibia, SWANLA was hated by the contract workers it controlled.

It was widely felt among workers that SWANLA's monopoly control of recruitment was a major factor in holding down wages and preventing agitation against poor working and living conditions. It was disbanded after the successful challenge to it in the 1971 general strike.
See also **contract labour; general strike of 1971**.

SWANP (*Namibia*): see **South West Africa National Party**.

SWANU (*Namibia*)
SWANU, the South West Africa National Union, was formed in May 1959 by a group of students and intellectuals, most of them of Herero origin. Many of its founders had been involved in earlier organisations, among them the South West Africa Student Body (SWASB) and a subsequent organisation of students and professionals, the South West Africa Progressive Association (SWAPA). A few months after the formation of SWANU its leadership was broadened with the election of Sam Nujoma and other Ovamboland People's Organisation (OPO) members to its executive. The unity was short-lived, however, and with the formation of SWAPO, SWANU became once again a narrowly-based organisation.

Despite its failure to establish itself as a nationally-based party, SWANU's political programme remained nationalistic and radical, and this, together with the efforts of its External Council, earned it the recognition of the OAU from 1963 to 1965. Its failure to retain its status in the OAU was largely the result of its refusal to support armed struggle against the South African occupation of Namibia, and its ineffectiveness as an opposition party within Namibia. In 1966 divisions within SWANU led to the resignation of Jariretundu Kozonguizi, president of the party since its formation. Having joined SWAPO in the National Convention of Namibia in 1971, and its successor organisation, the Namibia National Convention, SWANU refused to follow other parties into SWAPO in 1976. Instead it launched the Namibia National Front.

In May 1984, at the Lusaka talks, the SWANU delegation crossed the floor and allied itself with SWAPO in the negotiations. This exacerbated major divisions within SWANU arising from its membership of the MPC. By August 1984, 14 of its 18 branches had adopted resolutions rejecting the MPC and expressing a lack of confidence in its leadership. At a congress called to resolve the issue, the President of SWANU, Moses Katjiuongua, was ousted and a new leadership elected which immediately withdrew from the MPC. Katjiuongua refused to accept the decision and remained in the MPC with a rump of supporters who continued to call themselves 'SWANU'.

The rest of the party, under its new president, Kuzeeko Kangueehi, won High Court recognition as the legal SWANU and entered into an informal alliance with SWAPO. To distinguish it from the rump led by Katjiuongua, SWANU is now sometimes referred to as SWANU (Progressives) or (Left).
See also **Lusaka talks on Namibia, 1984; MPC; National Convention of Namibia; NUDO; SWAPO**.

SWAP (*Namibia*)
The South West Africa Police (SWAP) is a white-led police force which during the 1980s developed specialised paramilitary counter-insurgency units to bear the brunt of the internal war against SWAPO and to reduce the pressure on the military. The most notorious of these units is *Koevoet*. The UN Peace Plan as contained in UN Security Council Resolution 435 envisages the continued operation of the SWAP in the interim period during which elections for an independence government will be held. The formation of *Koevoet* and other paramilitary police units has been alleged to be part of a South African plan to ensure its military dominance in Namibia after the confining to their bases of PLAN and SWATF soldiers under Resolution 435. Under the UN plan, however, the UN Special Representatives would have to verify the suitability of the police to fulfil their role in the transition, and it is likely that *Koevoet* and similar organisations would fail this test.
See also ***Koevoet*; PLAN; SWATF; UN Security Council Resolution 435**.

SWAPA (*Namibia*): see **South West Africa Progressive Association**.

S

SWAPO of Namibia (*Namibia*)
The South West Africa People's Organisation of Namibia, SWAPO, is the primary political organisation in Nambia and leads the struggle against the illegal South African occupation of that country. Recognised in 1965 by the Organisation of African Unity (OAU) as the liberation movement of the Namibian people, it was accepted as the 'authentic representative of the Namibian people' by the General Assembly of the United Nations in December 1973. It is the sole Namibian political party to have been accorded this status, and commands the support of a large majority of Namibians.

The historical roots of SWAPO lie in an organisation of Namibian contract workers, the Ovamboland People's Congress (OPC) founded by Andimba Toivo ja Toivo in Cape Town, South Africa, in 1957. The following year the OPC changed its name to the Ovamboland People's Organisation (OPO). Later that year ja Toivo was deported from South Africa as punishment for sending an appeal to the United Nations. He returned to Namibia and began organising support for the OPO in the north of the country, while Sam Nujoma and others set up branches in Windhoek and Walvis Bay. When the South West Africa National Union (SWANU) was established in 1959, OPO leaders were among those elected to its executive.

In 1959 the OPO and SWANU together organised opposition to the forced relocation of black residents of Windhoek from the Old Location to the new segregated suburbs of Khomasdal and Katutura north of the city. The protests sparked a police massacre of demonstrators in December 1959, following which OPO leaders were banished from Windhoek. The working alliance between the OPO and SWANU subsequently collapsed and a decision was made to consolidate the national support which existed for the OPO by formally setting up a national organisation. On 19 April 1960 the South West Africa People's Organisation was formed with the former president of the OPO, Sam Nujoma, as its first president. Its central objective, according to the official SWAPO history, was 'the liberation of the Namibian people from colonial oppression and exploitation'.

At its 1961 national congress in Windhoek, SWAPO accepted the principle of a joint political and military campaign for independence and began preparations for armed struggle. SWAPO members were sent abroad for training and by 1964 the first guerrillas had returned to set up training camps within Namibia. In October of the same year the Caprivi African National Union (CANU) merged with SWAPO, establishing a firm base of support in the Caprivi region. The armed struggle in Namibia was launched by SWAPO on 26 August 1966 following the failure of the International Court of Justice (ICJ) to rule on the legality of South Africa's presence in Namibia. SWAPO guerrillas attacked South African troops near Omgulumbashe in the north of the country and, a month later, destroyed a government administrative base at Oshikango.

A major re-organisation of SWAPO followed a Consultative Congress at Tanga in Tanzania held from 26 December 1969 to 2 January 1970. The SWAPO Youth League, Elders' Council and Women's Council were set up to 'give expression to the struggle against particular forms of exploitation alongside the general movement for national liberation'. SWAPO's military wing was re-organised and in 1973 was renamed the People's Liberation Army of Namibia (PLAN). Subsequent to the Tanga conference, the armed struggle was escalated, and in June 1971 SWAPO's campaign received a boost when the ICJ gave an opinion that the South African occupation was illegal. The largest churches in Namibia welcomed the decision, and in December 1971, workers in Windhoek and elsewhere launched a general strike against the contract labour system. Under pressure on both the international and domestic fronts, in 1972 South Africa was forced to send more troops into Ovamboland to suppress the rural opposition. A state of emergency was declared on 4 February 1972 under Proclamation R17 (which remains in force), and SWAPO organisers were rounded up or forced underground. In the South African organised elections to the bantustan 'legislature' the following year, a SWAPO boycott call led to a mere 2.5 per cent poll among a registered electorate which was itself only a third of those eligible.

On 1 September 1975 South Africa made a new attempt at a settlement excluding SWAPO, with the establishment of the Turnhalle Constitutional Conference. SWAPO reacted by intensifying its armed struggle. In January 1976 the UN Security Council passed a hard-hitting resolution (385) calling for South African withdrawal. The five western members of the UN Security Council came together in a so-called 'Contact Group' to defuse mounting pressure for sanctions and to facilitate negotiations for a settlement. In April 1977 they began negotiations with SWAPO and South Africa, and in February 1978 the group produced a set of proposals which was accepted in principle by both parties. Later the following year, on 29 September, the UN Security Council endorsed a report by secretary general Kurt Waldheim laying down the basis for transition to independence (UNSC Resolution 435). SWAPO accepted the programme but South Africa responded by organising elections (boycotted by SWAPO) for a 'National Assembly', later to become the basis of the DTA 'transitional government'. The elections were condemned by SWAPO and the Namibian churches as fraudulent. UN-organised negotiations continued without success, with South Africa raising new objections as each earlier sticking-point was resolved. A conference attended by SWAPO, the DTA and South Africa in Geneva failed. Momentum for a settlement then decreased with the election of conservative governments in the USA and Britain, and a South African calculation that it could rely on western support to thwart pressure for sanctions. The final sticking-point for the negotiations was a South African and US insistence on linking a settlement in Namibia to the withdrawal of Cuban troops and other personnel from Angola. A further attempt to reach agreement at talks in Lusaka in 1984 also came to grief over the question of 'linkage'. SWAPO, despite the failure to reach a settlement, continued its twin strategy of political and military confrontation and in 1984 organised major rallies in Nambia and conferences and other events abroad, to mark 100 years of resistance to foreign occupation and colonialism in Namibia.

SWAPO's 1976 constitution and political programme lays down its aims and objectives and outlines the policies a future SWAPO government would implement in Namibia. It stresses SWAPO's commitment to national unity and opposition to racism, tribalism and sexism. The constitution describes SWAPO as the 'organised political vanguard of the oppressed and exploited people of Namibia', and as 'the expression and embodiment of national unity'. SWAPO has gone a long way towards achieving these aspirations, and now enjoys support across social and colour barriers. It has expanded beyond its original class base and now includes among its members and supporters church leaders, professional people, traditional leaders and white Namibians.

Apart from Sam Nujoma, President of SWAPO, and Andimba Toivo ja Toivo, Secretary General of SWAPO, leaders include vice-presidents Hendrik Witbooi and Nathaniel Maxuilili, secretary for administration Moses Garoeb, secretary for defence Peter Mushehenge, secretary for international relations Theo-Ben Gurirab, secretary for information Hidipo Hamutenya, secretary for labour Jason Angula, and chair David Merero. Other prominent internal leaders are Nico Bessinger, Daniel Tjongarero, and SWAPO's first publicly-declared white member, Anton Lubowski.

See also **Contact Group; DTA; general strike of 1971; International Court of Justice; linkage; Lusaka talks on Namibia, 1984; Nujoma; PLAN; SWANU; SWAPO Elders' Council; SWAPO Women's Council; SWAPO Youth League; Toivo ja Toivo; UN Security Council Resolution 435**.

SWAPO-D (*Namibia*)

SWAPO-D, or SWAPO-Democrats as it is more formally known, is the small personal political vehicle of Andreas Shipanga, former SWAPO secretary for information and publicity and now 'Minister for Mines' in the MPC 'transitional government'. Shipanga was arrested in Zambia in 1976, along with some of his followers, after a dispute with the leadership of SWAPO. He had been building a personal organisation within the movement, and refused to acknowledge the authority of its president or central committee. In May 1978 Shipanga and supporters were released from detention by the Zambian authorities and after a brief stay in Scandinavia he returned to Namibia with the blessing of the South African government. Shipanga announced the formation of SWAPO-D in June 1978 while still in Sweden.

The acceptability of SWAPO-D and Shipanga to South Africa derived from his outspoken opposition to SWAPO and its president, Sam Nujoma, in particular. In addition, SWAPO-D is opposed to the armed struggle in Namibia and seeks independence on the basis of a 'reconciliation of tribal and ethnic groups'. A widely-publicised merger with the anti-apartheid NNF in 1980 failed, but SWAPO-D also resisted South African pressure to join the DTA. Internal strife during the same period led to the expulsion of two leading members, the publicity and information secretary, Kenneth Abrahams, and the secretary general, Ottile Abrahams. Following their expulsion they founded a political and economic journal, the *Namibian Review*. Despite its earlier refusal to join the DTA, SWAPO-D became part of the MPC in 1985 and participated in the Transitional Government of National Unity.

See also **DTA; MPC; NNF; Nujoma; SWAPO; Transitional Government of National Unity**.

SWAPO-Democrats (*Namibia*): see **SWAPO-D**.

SWAPO Elders' Council (*Namibia*)

The SWAPO Elders' Council was set up after the 1970 SWAPO congress at Tanga to organise and represent older people within SWAPO. It has its own constitution which, like those of the other semi-autonomous wings of the organisation, the Youth League, Women's Council, and PLAN, is subject to Central Committee ratification. The establishment of a separate Elders' wing reflects SWAPO's view that older people have an important role to play in the struggle, and its desire to be able to call on their experience and knowledge. The SWAPO Elders' Council is headed by Simon Kaukungua.

See also **PLAN; SWAPO; SWAPO Women's Council; SWAPO Youth League**.

SWAPO Pioneer Movement (*Namibia*)

SWAPO's organisation for those under 15 years old: see **SWAPO Youth League**.

SWAPO Women's Council (*Namibia*)

Set up by the 1970 SWAPO congress at Tanga, the SWAPO Women's Council (or SWC) is one of four semi-autonomous wings of SWAPO. It operates with a separate constitution (subject to approval by the SWAPO

Central Committee) and both organises women in support of SWAPO and campaigns for women's rights. Members of SWC also participate in the other wings and structures of SWAPO, including PLAN. The leaders of the SWC include Gertrude Kandanga, Pendukeni Kaulinga and Ellen Musialela.
See also **PLAN; SWAPO; SWAPO Elders' Council; SWAPO Youth League**.

SWAPO Youth League (*Namibia*)
Supporters of SWAPO between the ages of 6 and 35 are organised under the SWAPO Youth League, one of SWAPO's four specialised wings. The Youth League has its own constitution, subject to the approval of the SWAPO Central Committee, and members between the ages of 6 and 15 are organised into the SWAPO Pioneer Movement. Provision for the establishment of a separate youth wing was made at the Tanga congress of SWAPO in 1970. The National Secretary of the Youth League is Jerry Ekandjo.
See also **Namibian National Students' Organisation; PLAN; SWAPO; SWAPO Elders' Council; SWAPO Women's Council**.

S

SWASB (*Namibia*)
The South West Africa Student Body (SWASB) was a student organisation of the late 1950s, most of the members of which joined SWANU on its formation in 1959: see **SWANU**.

SWATF (*Namibia*)
The South West Africa Territory Force (SWATF) was established by South Africa in 1980 to shift the burden of the war in Namibia off the South African Defence Force (SADF) and onto Namibians. It was in part a strategy to 'Namibianise' the institutions of the country in line with plans for an 'internal settlement', and in part an attempt to reduce the political costs of deaths of white South African conscripts in the war against SWAPO. Estimates of the number of personnel under arms and fighting for South Africa vary from 50,000 to 100,000. The concentration of the bulk of these forces in the north of Namibia make this one of the most militarised areas, relative to population density, in the world.

In 1986 and 1987 three 'cultural' organisations were established, apparently by SWATF, in areas inhabited by Ovambo and Kavango speakers, and in the Caprivi Strip. Called Etango, Ezuvo and the Namwi Foundation respectively, they purported to promote the culture and 'peaceful development' of the people in each area. In fact they appear to be crude propaganda organisations used to promote 'anti-communist' and anti-SWAPO ideology and to argue against UN Security Council Resolution 435. There have also been allegations that members of Etango and Ezuvo have been involved in violent attacks on people attending SWAPO rallies.
See also ***Koevoet*; PLAN; SADF; SWAPO**.

Swazi National Council (*Swaziland*): see ***Libandla***.

Swazi Nation land (*Swaziland*)
Swazi Nation land is the land which was set aside under the 1907 Land Proclamation Act for occupation by the Swazis. It amounted to a third of the country, and was divided into 32 'native areas' by the British. The partition had a disastrous affect on the country, transforming the economy so that traditional, self-reliant Swaziland became incapable of growing enough food to sustain its population. King Sobhuza spent the first years of independence trying to reclaim the land for Swazi citizens.
See also **Land Proclamation Act of 1907**.

Swaziland

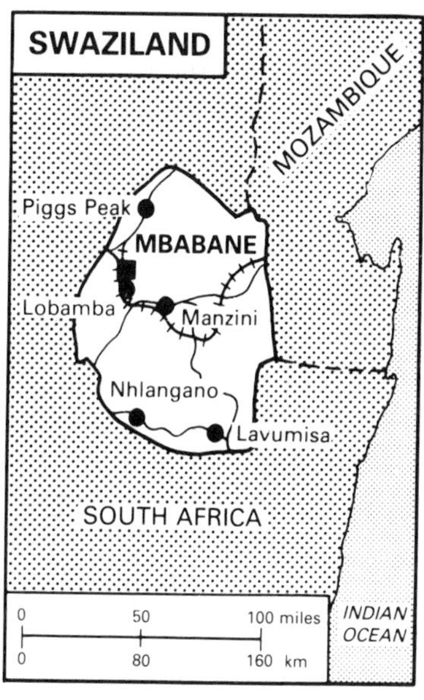

Official title: Kingdom of Swaziland
Head of State: HM King Mswati III
Area: 17,363 sq. km
Population: 706,137 (July 1986)
Capital: Mbabane
Official languages: English, SiSwati
GDP per Capita: US$790 (1984)
Major exports: sugar (40 per cent); wood pulp (20 per cent)
Currency: the *lilangeni* (plural: *emalangeni*); US$1=2.068 *emalangeni* (July 1987)
Political parties: none are permitted but activity by the Imbokodvo National Movement is allowed.

The Kingdom of Swaziland, after the Gambia, is the smallest country in Africa. It is completely landlocked, surrounded by South Africa to the north, west and south and by Mozambique to the east. It is dominated by South Africa and cannot afford to alienate its giant neighbour. It signed a non-aggression treaty with South Africa in 1982 which has drawn it more formally into that country's orbit. Economically too Swaziland is very dependent on developments in the South African economy. It is a member of the Tripartite Monetary Agreement and the Southern African Customs Union. It is a mountainous and well-watered country with four large rivers. Most of its people are engaged in agriculture and sugar is by far the largest industry and the principal export.

The country is a monarchy with a new, young king, crowned in 1986. It gained its independence from Britain in 1968 under a constitutional monarchy but since the 'constitutional crisis' of 1973 it has been ruled by

an alliance organised through the monarchy. This operates through the 1978 constitution which established a bicameral parliament comprising the Senate and the House of Assembly. The latter advises the king and its members are chosen from traditional tribal areas called *tinkundla;* the king nominates some of the members in each house. The monarch has a cabinet and a prime minister through which he runs the country. Opposition to the power of the monarchy has been suppressed over the years and there was a period of power struggles after the death of King Sobhuza II in 1982. Then the *Liqoqo*, or Supreme Ruling Council, tried to dominate the political scene. The new king has dissolved the council and is at present (1987) seeking to consolidate his power.
See also **mbokodvo National Movement; *Liqoqo*; monarchy; Mswati; *nkundla*; relations with South Africa; Sobhuza**.

Swaziland Democratic Party (*Swaziland*)
The Swaziland Democratic Party (SDP) was formed on 2 March. Its first president was Simon Sishayi Nxumalo. It was in favour of capitalism and against pan-Africanism. After initially supporting a qualified franchise it came to support universal suffrage. The party opposed the strikes of 1962–3 and spoke out against the Ngwane National Liberatory Congress for its support of the labour unrest. The party was very unsuccessful in the 1964 elections and in 1965 the SDP joined the traditionalist party of the monarchy, the Imbokodvo National Movement.
See also **Nxumalo**.

Swaziland Liberation Movement (*Swaziland*)
The Swaziland Liberation Movement was founded in 1984 by Prince Dumisa Dlamini. It was held responsible for the distribution of anti-government leaflets in the country soon afterwards. However, another opposition group, the People's United Democratic Movement, claimed responsibility for the campaign of painted slogans and pamphlets. Neither group has achieved any prominence since 1985.
See also **Dlamini, Dumisa; People's United Democratic Movement**.

Swaziland Progressive Party (*Swaziland*)
The Swaziland Progressive Party was the earliest of the country's political parties. It was founded on 30 July 1960 when the Swaziland Progressive Association transformed itself into a party. It was led by its president, J. J. Nquku and his secretary general, Dr Ambrose Zwane. It advocated a non-racial, socialist society which supported pan-Africanism. A series of splits developed in the party in 1961 which continued until 1963 and resulted in the formation of a new party, the Ngwane National Liberatory Congress. These were largely personality disputes but they succeeded in destroying the party.
See also **Ngwane National Liberatory Congress; Zwane**.

T

TAA (*Tanzania*): see **Tanganyika African Association**.

Takawira, Leopold (*Zimbabwe*)
Leopold Takawira co-founded ZANU, one of Zimbabwe's main nationalist organisations, in 1963, and was its first vice-president. He had earlier helped to found the National Democratic Party. Later he was Secretary for External Affairs of ZAPU, before the split which led to ZANU's creation. He was detained and imprisoned under the white minority regime for most of the 1960s and died in 1970 while in prison.
See also **National Democratic Party; ZANU; ZAPU**.

Tambo, Oliver R. (*South Africa*)
Oliver Tambo is the President-General of the ANC of South Africa. He has led the organisation from exile since it was banned in South Africa in April 1960. Like his friend and legal partner, Nelson Mandela, Oliver Tambo was one of the founders of the ANC Youth League in 1944 and became its national secretary and then its national vice-president. In 1949 he became a member of the national executive committee of the ANC and secretary-general in 1955. Three years later he was elected deputy president-general. During this time he was banned under the Suppression of Communism Act and arrested in 1956 on a charge of high treason; he was later released when the charges were dropped. Just before the 1960 state of emergency the ANC executive asked him to go into exile to lead the now banned organisation.

Oliver Tambo is a quiet, forceful personality and a skilled negotiator. Together with Nelson Mandela he represents the centrist political ground in the ANC between the revolutionary left and the remaining Africanists. In recent years he has devoted considerable time to diplomacy, travelling widely and meeting with senior government leaders such as the US secretary of state, George Schultz, and the British foreign secretary, Sir Geoffrey Howe, in order to discuss South Africa's political situation and to urge the imposition of economic sanctions against the country.
See also **ANC; Congress Youth League; Mandela, Nelson**.

Tanga congress (*Namibia*)
The common name for the 1970 SWAPO congress (held at Tanga) which restructured the organisation: see **SWAPO**.

Tanganyika (*Tanzania*)
Tanganyika is the name given to mainland Tanzania all through the colonial period. The country was colonised by Germany in 1885 and then passed to the British in 1919 after the First World War. The British governed Tanganyika by a system which became known throughout British colonies as 'indirect rule'. This aimed to give the traditional rulers, or others resembling these, the power to rule as agents of the colonial power. After the Second World War African nationalism began to grow and in 1954 Julius Nyerere formed TANU (the Tanganyika African National Union) to campaign for independence. In 1958 and 1959 Tanganyika's first general elections were held and on 9 December 1961 Tanganyika became independent with Julius Nyerere as prime minister.

At independence TANU seemed to have lost much of its momentum. The issue which gave it most of its problems was the Africanisation policy which sought to promote Tanganyikans into positions previously held by the colonial rulers. Here TANU was the target of resentment from the Tanganyika Federation of Labour (TFL) which wanted higher wages and more rapid change. Soon after independence Julius Nyerere resigned as prime minister in order to revitalise TANU, and nominated Rashidi Kawawa in his stead. In 1962 Julius Nyerere was elected president under a new constitution introduced by Kawawa, and in December that year Tanganyika became a republic and joined the Commonwealth. Two years later, in 1964, there was a revolution in Zanzibar. This resulted in a government proclaiming socialism led by Abeid Karume. Ties with the Tanganyikan mainland were strengthened and in October 1964 Tanganyika and Zanzibar became the United Republic of Tanzania.
See also **Africanisation; 'indirect rule'; Karume; Kawawa; Nyerere; Tanganyika Federation of Labour; TANU; Tanzania; Zanzibar**.

Tanganyika African Association (*Tanzania*)
The Tanganyika African Association (TAA) was formed in 1948 after the African Association broke its links with Zanzibar following a dispute. Soon it began to decline in popularity and in 1954 Julius Nyerere converted the organisation into TANU.
See also **African Association; TANU**.

Tanganyika African National Union (*Tanzania*): see **TANU**.

Tanganyika African Welfare Association (*Tanzania*)
The Tanganyika African Welfare Association was a student group led by Julius Nyerere, who was later to become Tanzania's president. In 1945 it converted itself into a branch of the African Association. This in turn was later renamed the Tanganyika African Association and in 1954 became TANU, Tanzania's ruling political party at independence.
See also **African Association; TANU**.

Tanganyika Federation of Labour (*Tanzania*)
On 7 October 1955 Tanganyika's trade union leaders established the Tanganyika Federation of Labour (TFL) to bring together their organisations. Its president was the Tanganyika Railway African Union's leader,

J. B. A. Ohanga, and its general secretary was Rashidi Kawawa, president of the Tanganyika African Government Servants Association (TAGSA) and later to be prime minister. Their aim was to create a general union with individual workers as members. However, the government would not register this. Instead the organisers had to form industrial unions on the British model and affiliate them to the Tanganyika Federation of Labour.

The Federation initiated an extraordinarily fast development of trade unionism in Tanganyika. By 1961 some 42 per cent of Tanganyika's workers were unionised compared with 12 per cent in Uganda and 8 per cent in Kenya. Soon after its formation a wave of strikes began which demonstrated effective support and solidarity. The Federation did not affiliate to TANU, the country's nationalist political party, but as TANU's power developed the union movement found it hard to resist its demands for loyalty. In many instances both organisations fought the colonial government, together presenting an overwhelming case for independence. In 1960 a split developed in the Federation, with Kasanga Tumbo (by then the militant leader of the railway workers) advocating independence from TANU and M. M. Kamaliza arguing for closer links. The latter won, but the tension remained. By 1962 many of the former union leaders had become cabinet ministers. However, the workers and the new Federation leaders threatened the government's plans for gradual independence by their wage demands, so the union movement was suppressed in 1962 and many of its leaders (including Kasanga Tumbo) imprisoned without trial. This process continued after independence and once TANU was in power, the party began to suppress the trade union movement and make it subordinate to the party's organisation. Rashidi Kawawa, despite his union background, led these efforts.

In 1964 the National Union of Tanganyika Workers Act established the National Union of Tanganyika Workers (NUTA) as the only legal trade union, and in 1969 workers' councils were formed to promote good industrial relations. Periodically workers complained to the government about their situation, but the state has consistently acted to limit and control the demands of industrial workers in Tanzania.

See also **Kawawa; TANU; Tumbo**.

TANU (*Tanzania*)

Julius Nyerere, later to become the President of Tanzania, was the moving force behind the creation of TANU. In 1954, at what became known as the *Saba Saba* meeting (meaning the meeting of the 7 July), Nyerere successfully relaunched the ailing African Association as the Tanganyika African National Union (TANU). The African Association had been founded in 1929 by a group of African civil servants and was the first organised nationalist movement in Tanganyika. TANU thus inherited its historical legacy as well as the loyalty of the many local groups already affiliated. TANU was also helped by the widespread use of the Swahili language throughout the country and by the absence of any predominant ethnic group.

TANU's aim during the 1950s had been to win the nationalist struggle for control of Tanganyika. The decisive confrontation between the British colonial regime and the nationalist movement came in the late 1950s. The issue was the British commitment to multiracialism. TANU's annual conference in 1958 decided to press for *madaraka* or responsible government by the end of 1959 and, failing this, to adopt positive action to pressurise the British into compliance. The conference was seriously divided over TANU's participation in the 1958–9 general elections under a voting formula devised by the British. In the end the party fought the elections and won a decisive victory. In October 1959 the young British Conservative politician, Iain Macleod, was put in charge of the Colonial Office. He recognised the realities of the situation and agreed to grant independence to Tanganyika in December 1961, after TANU had won all seats but one in the 1960 election.

Despite TANU's overwhelming victories in the pre-independence general elections, its central weaknesses soon became apparent. Democratic local government was very badly developed because of the autocratic colonial structures the party inherited, and it had to try to fill this vacuum. More significant was the clash with the workers' organisations in the country which soon developed. TANU had received its main support from the producers' co-operatives and the trade unions and after independence these began to seem like centres of opposition. The result was the suppression of the trade union movement over the years and the eclipse of the Tanganyika Federation of Labour. TANU ceased to exist in February 1977 when Julius Nyerere amalgamated it with the Afro-Shirazi Party of Zanzibar to form Tanzania's sole political party, the *Chama cha Mapinduzi*.

See also **African Association; *Chama cha Mapinduzi*; Nyerere; Tanganyika; Tanganyika Federation of Labour; Tanzania**.

Tanzam railway and road (*Tanzania, Zambia*)

The Tanzania-Zambia railway line and highway, also known as the Tanzam railway and road or Tazara, was built to provide an alternative sea outlet for land-locked Zambia. It also helps the problem of transportation to the rich Kilombero valley in Tanzania and the Iringa and Mbeya regions. The Tanzam railway was built and financed by the Chinese with an interest-free loan. The complete route was opened for regular service in October 1975. The closure of the Benguela railway from Zambia to Lobito in Angola in 1975 would have been disastrous for Zambia if not for this alternative. Nevertheless Zambia's transport problems continue because of the limited capacity of port facilities in Dar es Salaam and traffic congestion on other routes. Improvement of the railway and road as well as of the port of Dar es Salaam are priorities in the programme of SADCC (an organisation of southern African states, including Zambia and Tanzania) which seeks to develop new economic networks in the region.

See also **Benguela railway; SADCC**.

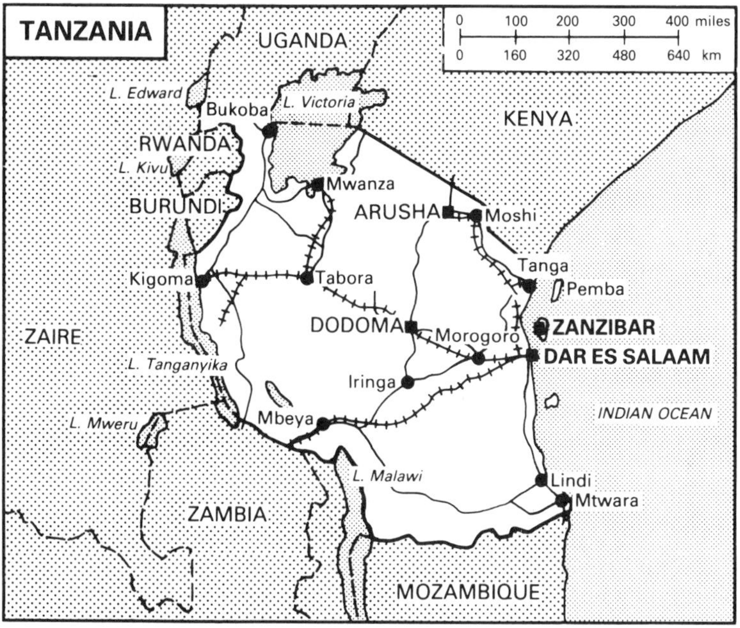

Tanzania

Official title: United Republic of Tanzania
Head of state and government: President Ali Hassan Mwinyi
Area: 945,087 sq. km
Population: 21,062,000
Capital: Dar es Salaam
Official languages: Kiswahili, English
GDP per capita: US$219 (1984)
Major exports: coffee (30–35 per cent of total value); cotton (13 per cent); cloves (10 per cent)
Currency: Shilling (Tsh)=100senti; Tsh65 per US$ (July 1987)
Political parties: *Chama cha Mapinduzi* (CCM) is the only legal political party

The United Republic of Tanzania incorporates mainland Tanganyika and the islands of Zanzibar and Pemba just off the Indian Ocean coast to the east. It includes the highest and lowest parts of Africa – Mount Kilimanjaro and the floor of Lake Tanganyika. Its neighbouring states are Kenya and Uganda to the north, Zambia to the south-west, Mozambique and Malawi to the south, and Rwanda, Burundi and Zaire to the west. The country is one of the world's least developed and poorest with few natural resources other than land. Despite its poverty Tanzania has been one of Africa's most influential independent states. President Julius Nyerere, who led the country from independence until 1985, is an intellectual, a man with controversial ideas and a vision of a particular kind of African socialism. His philosophy influenced newly developing states around him and Tanzania came to represent an example of an indigenous African route to development. Nyerere also sought for his country a part in the wider world, and Tanzania has been a leading member of the 'frontline' states in their struggle against domination by South Africa. However, the economic legacy inherited by the new president, Ali Hassan Mwinyi, when he took over in October 1985 was very bleak indeed. Exports were declining and there was a serious shortage of foreign exchange and material goods. President Mwinyi has begun to address these problems by negotiating loans from the International Monetary Fund and encouraging its aid donors with plans for an economic recovery.

The British colony of Tanganyika became independent on 9 December 1962 and united with Zanzibar to form the United Republic of Tanzania on 26 April 1964. It is a one-party state with an executive presidency. On 25 April 1977 the country's present constitution was approved by the National Assembly. It has been amended since then and, after the 1984 amendments, provides for government as follows: the National Assembly consists of 101 members directly elected from the mainland, up to 55 members directly elected from the islands and various other members indirectly elected. All candidates for election to the National Assembly must be approved by the CCM (*Chama cha Mapinduzi*), Tanzania's revolutionary party. In each constituency the official party puts forward two candidates chosen by the party membership. Members of parliament are elected for five years unless the president dissolves parliament earlier. The presidential candidate is chosen by an electoral convention of the CCM. He appoints two vice-presidents from the elected members of the National Assembly. If the president comes from the mainland the first vice-president must come from Zanzibar and vice versa. The present first vice-president is Joseph Warioba and the second vice-president and president of Zanzibar is Idris Abdul Wakil (1987). Under the recent amendments the president's powers are limited in that he is made more accountable to parliament and cannot stand for election for more than two consecutive terms.

See also **African Socialism; *Chama cha Mapinduzi*; 'frontline' states; Mwinyi; Nyerere; Tanganyika; *Ujamaa*; Zanzibar**.

TEBA (*South Africa*)

The Employment Bureau of Africa or TEBA is a centralised labour recruitment organisation established and controlled by the South African Chamber of Mines, and used to avoid wage competition for mineworkers among mining companies in South Africa.

See also **Chamber of Mines**.

Tekere, Edgar Zivania (*Zimbabwe*)

Edgar Tekere has been a prominent member of Zimbabwe's leading nationalist orgnisation, ZANU, since he helped found it in 1963. He has been its secretary-general since 1974. He was detained and imprisoned under the rebel Rhodesian regime for his commitment to the nationalist cause. After the 1980 elections he became a minister in Robert Mugabe's government. He was one of the most controversial ministers in the government and was dismissed from the cabinet by Robet Mugabe in 1980. He took a tougher stand against ZANU's rivals in ZAPU than many other members of the government and was one of the most influential leaders on the left of the party. His power-base was in the Manyika-speaking eastern region of Zimbabwe, one of the most populous of the Shona groups.

In 1980 he was found guilty of killing a white farmer but freed under a law enacted by the former Rhodesian regime which shielded ministers from criminal charges if they acted 'in good faith to suppress terrorism'. He claimed he had been warned of an assassination attempt by ZIPRA guerrillas after which he rooted them out, killing a white farmer in the process. He was removed from the position of Secretary-General of ZANU in September 1981.

See also **Mugabe; ZANU; ZAPU**.

Tembo, John (*Malawi*)

John Tembo is a member of the executive committee of Malawi's only legal party, the MCP, and he was the governor of the Reserve Bank of Malawi until April 1984. He is known to be a powerful influence on President Banda, possibly because he is the uncle of the president's Life Companion and Official Hostess, Mama Kadzamira. In 1983 there were rumours of a power struggle over the presidential succession between John Tembo and Dick Matenje, a prominent minister in the government. Later Dick Matenje and some of his colleagues died under mysterious circumstances. It was suggested by opposition politicians in exile that he had been killed because he opposed plans for John Tembo's succession. John Tembo is still regarded as a possible successor to President Banda, although he was replaced at the Reserve Bank and has not been appointed to the key vacant post of MCP secretary-general. Furthermore, one of his close allies, Mac Kamwama, lost his job as chief of police in 1987. He was reported to have resigned because of his age although he was only 51. However, in July 1987, John Tembo was appointed acting Treasurer General of the MCP.

See also **Matenje; MCP**.

Terre Blanche, Eugene Ney (*South Africa*)

Eugene Terre Blanche, a former warrant officer in the South African Police, leads the neo-Fascist *Afrikaner Weerstandbeweging* (AWB). He and six others launched the AWB in 1973 and it subsequently grew in public prominence as part of Afrikaner reaction against the policies of President P. W. Botha.

Terre Blanche believes in a non-party state and that the greatest danger to white South Africa is Islam rather than communism. His followers use violence and arm themselves for the day when the country will collapse in chaos and they will seize power and create a *volkstaat* (a people's state). In the meantime they are not above using fists and firearms against political opponents ranging from members of the ruling National Party

to activists in the anti-apartheid UDF. Terre Blanche himself has convictions for the illegal possession of arms and ammunition, and in 1982 was charged with 'terrorism' after the discovery on his brother's farm of a weapons cache including arms from the USSR, Eastern Europe, Korea and China.
See also **AWB; Botha, P. W.; National Party**.

TFL (*Tanzania*) See **Tanganyika Federation of Labour**.

TGLP (*Botswana*) See **Tribal Grazing Land Policy**.

TGNU (*Namibia*) See **Transitional Government of National Unity**.

The Employment Bureau of Africa (*South Africa*) See **TEBA**.

Tibiyo Taka Ngwane (*Swaziland*)
The *Tibiyo Taka Ngwane* fund (or 'Minerals of Ngwane', the Swazi National Development Fund) was set up in 1968 by the monarchy to give Swaziland's traditional rulers some power within the world of foreign capital in the country. *Tibiyo* was an investment undertaking based on funds derived from mineral royalties. In 1975 the system was reorganised and funds from mineral royalties were placed under a new organisation, *Tisuka Taka Ngwane*. *Tibiyo* then began to act solely as an investment corporation. Now the fund is granted substantial sums of money from the government's central budget to make its large investments. Its revenue and assets do not accrue to the finance ministry but are controlled by the traditional rulers organised through the monarchy. Its board of trustees is appointed by the monarch and *Tibiyo* is accountable neither to parliament nor to the cabinet. Its current managing director is Polycarp Dlamini, the former justice minister.

The prime minister is the *ex officio* chair of the fund which has considerable influence on the economy. In 1981 it held shares in 33 undertakings including most of the largest enterprises operating in Swaziland. Some of the main companies in which the fund holds shares are: Havelock Asbestos Mine (40 per cent), Mhlume Sugar Company (50 per cent), Royal Swaziland Sugar Corporation (32.4 per cent), Simunye Plaza (25 per cent), Swazispa Holdings (33.4 per cent), Ubombo Ranches (40 per cent). It has also acquired substantial land holdings. At the end of April 1981 its assets were valued at over 46 million South African rand.

The growing power of the fund has led to conflicts with some government officials. In 1983, for instance, *Tibiyo* requested that it be designated by the government a national development agency. This was refused. However, a month later the *Tibiyo* managing director, Sishayi Nxumalo, was appointed finance minister in a cabinet reshuffle. *Tibiyo* was certified two days later as a national development agency.

In 1979 the new prime minister, Prince Mabandla Dlamini, decided to challenge the power of the fund. He appointed a commission of inquiry into corruption and began to use this to subpoena the files of *Tibiyo*. *Tibiyo*, however, proved too powerful for him, and its managing director,

then Sishayi Nxumalo, persuaded the king to close down the commission in 1980.
See also **Dlamini, Mabandla; Nxumalo**.

***Tiger*, HMS** (*Zimbabwe*)

HMS *Tiger* was the British cruiser on which the Rhodesian rebel leader, Ian Smith, and the British prime minister, Harold Wilson, met in 1966 to discuss a settlement to the Rhodesian question. The talks took place while the ship sailed in circles in the Mediterranean off Gibraltar. The talks too seemed to go round in circles: a sixth proposal was added to the five British conditions for independence, known as the NIBMAR principles ('no independence before majority rule'). This was a concession to Ian Smith and stated that there should be no oppression of the minority by the majority or vice versa. The sticking-point was the issue of how Rhodesia should return to legality. Harold Wilson demanded that the governor of Rhodesia, Sir Humphrey Gibbs, should be given certain powers. Ian Smith returned to Salisbury, then the capital of Rhodesia, to consult his cabinet. The cabinet rejected the *Tiger* proposals, and the talks collapsed.
See also ***Fearless*, HMS; NIBMAR; Smith; Unilateral Declaration of Independence**.

t

tinkundla (*Swaziland*)

The plural of *nkundla* or *inkundla:* see ***nkundla***.

tinvuna (*Swaziland*)

The plural of *ndvuna*, the name given to a commoner chosen by the monarch to serve as an intermediary between the people and the king: see ***ndvuna***.

Tisuka Take Ngwane (*Swaziland*)

Tisuka Take Ngwane was formed in 1975 to control funds derived from mineral royalties. Previously these were invested by *Tibiyo Taka Ngwane*.
See also ***Tibiyo Taka Ngwane***.

Todd, Carl Frederick (*Swaziland*)

Carl Todd was an influential member of the white community in Swaziland for many years. He represented conservative, pro-South African sympathies in the country and was Chairman of the Reconstituted European Advisory Council for five years. He favoured a constitution which shared power between conservative whites and Swazi traditionalists and sought to resist the development of the nationalist movement in Swaziland. He helped to organise the United Swaziland Association, the white political group formed to campaign for the Legislative Council elections. However, he did not run on the Association's ticket, but instead was chosen as the first white member of the king's party, *Imbokodvo*. Carl Todd was elected on this basis to the Legislative Council and after the 1967 parliamentary elections was appointed to the Senate by King Sobhuza II whom he advised for many years.
See also **Reconstituted European Advisory Council**.

Todd, Reginald Stephen Garfield (*Zimbabwe*)

Dr Garfield Todd was a well-known critic of the rebel Rhodesian regime and its leader, Ian Smith. He had been a prominent politician in Southern Rhodesia for many years. From 1953 to 1958 he was prime minister and leader of the United Rhodesia Party. He was restricted and imprisoned by the Smith regime and acted as one of Joshua Nkomo's consultants during the Geneva Conference in 1976. He became a senator in the 1980 independence elections.
See also **Nkomo; Smith; Southern Rhodesia**.

Toivo ja Toivo, Andimba (*Namibia*)

Ja Toivo was a founder member of SWAPO in 1959 and has been active in the organisation ever since. In 1968 he was convicted along with 36 other SWAPO members at a trial under the South African Terrorism Act which had been passed the previous year and made retrospective specifically to allow it to be used against ja Toivo and his colleagues. Ja Toivo was sentenced to twenty years in prison, while nineteen of his co-defendents were given life sentences. He was held in the notorious Robben Island prison until his release in 1984.

t

After his conviction by a South African court under the Terrorism Act, ja Toivo made a statement from the dock on behalf of himself and the other accused. Addressing the Bench, he said:

> 'We find ourselves here in a foreign country, convicted under laws made by people we have always considered as foreigners. We find ourselves tried by a judge who is not our countryman and who has not shared our background.'
>
> 'You, my Lord, decided that you had the right to try us, because your parliament gave you that right. That ruling has not and could not have changed our feelings. We are Namibians and not South Africans. We do not now, and will not in the future recognise your right to govern us; to make laws for us in which we have no say; to treat our country as if it were your property and as if you were our masters.'
>
> 'We do not expect that independence will end our troubles, but we do believe that our people are entitled – as are all peoples – to rule themselves. It is not really a question of whether South Africa treats us well or badly, but that South West Africa is our country and we wish to be our own masters.'
>
> 'We have found ourselves voteless in our own country and deprived of the right to meet and state our political opinions.
>
> 'Is it surprising that in such times my countrymen have taken up arms? Violence is truly fearsome but who would not defend his property and himself against a robber. And we believe that South Africa has robbed us of our country.'
>
> 'Only when we are granted our independence will the struggle stop. Only when our human dignity is restored to us, as equals of the whites, will there be peace between us.'

Following his release from Robben Island, ja Toivo was elected Secretary General of SWAPO in August 1984, and joined its Central Committee. See also **Robben Island; SWAPO**.

Tomlinson Commission (*South Africa*)
The Tomlinson Commission was appointed by the South African government in 1950 to plan in detail the new National Party government's blueprint of applied apartheid in its 'Homelands' policy. It was to report on 'a comprehensive scheme for the rehabilitation of the Native areas ('Homelands') with a view to developing within them a social structure in keeping with the culture of the Native and based on effective socio-economic planning.'

The commission published a massive 17–volume report in 1956. It included detailed recommendations for alternative employment for Africans who were 'unproductive farmers' and in its view would have to be moved off the land. Among other things it suggested that a development corporation should develop industrial production centres at suitable sites, to help entrepreneurs. These should be located in the 'border areas' just outside the 'Homelands'. Several 'border industries' were set up but did not prove a great success. In 1968 white entrepreneurs were allowed to set up businesses in the 'Homelands' under the Promotion of Economic Development of Homelands Act.
See also **'Homelands'**.

t

Tongogara, Josiah Magama (*Zimbabwe*)
General Tongogara was a celebrated soldier and hero of the Zimbabwean struggle for independence. He joined ZANU in 1963 and soon went to Tanzania and then China for military training. In 1968 he was appointed Chief of Defence Staff and Secretary for Defence in the party's central committee. Besides being a soldier Josiah Tongogara was also an experienced politician. He had a formidable reputation as an influential personality in ZANU's central committee. His influence extended beyond ZANU and he helped to forge the joint military command of the guerrillas, called ZIPA in 1975. He died on 26 December 1979 in a motor accident in Mozambique shortly before Zimbabwean independence.
See also **ZANLA; ZANU; ZIPA**.

total onslaught (*South Africa*)
The term for the purported concerted attack on South Africa by the forces of 'world communism': see **total strategy**.

total strategy (*South Africa*)
The 'reformist' strategy of P. W. Botha's regime in South Africa is part of what he has characterised as a 'total strategy' to combat a 'total onslaught' on South Africa. National Party ideologists argue that the country is the target of a total onslaught by the forces of 'world communism' which seek to win control of South Africa for strategic reasons (the control of the Cape sea route) and in order to have access to its mineral wealth.

Total strategy involves the mobilisation of all the country's resources in a united and directed campaign to resist the total onslaught. It requires sacrifices by white South Africans, and concessions to win the support and involvement of black South Africans. It necessitates a common front among the military, business and the state in the economic, political, ideological and social-psychological fields to mobilise the country and direct its resources to the common aim. Its success would enable business to reap the benefits of a stabilised capitalist system in the country, the military would resist successfully and eventually defeat the 'communist onslaught', and white South Africans would retain their material comforts while black South Africans enjoyed new-found material advantages as well as new (though limited) political rights.

The emergence of socialist regimes in neighbouring countries, and their political and military support for the ANC, has been identified by South Africa as part of the total onslaught. As a result it has been seen as justifiable to use military, economic and political pressure on neighbouring countries to force them to end support for the ANC, and to engineer closer links and dependence. The attempt to build a Constellation of Southern African States (CONSAS) was part of total strategy, and the threat which SADCC (the group of southern African states opposed to South African domination) poses to CONSAS has made it a specific target for South Africa.

Internally, the Botha government has attempted to build a black middle class, and to co-opt it as part of its total strategy. Trade union rights have been conceded to Africans, but at the same time trade unions have been brought under tighter state control. Labour mobility for some urban black workers has been increased, both horizontally and vertically, but at the same time the bantustans have been used to tighten repression on the mass of migrant and unemployed workers. The constitution has been revised to give token representation to some black people at the centre, while real power has been concentrated in the hands of the executive president and the State Security Council.

Total strategy is a response to the growing challenges to apartheid in South Africa from inside and outside the country. It is a strategy for survival of both apartheid and white domination. It is wrapped in the language of reform, reason and abandonment of apartheid. At its centre, however, lies a revamped, reinforced and refined system of exploitation little different in its impact on the majority in South Africa from the apartheid of the past thirty years.

See also **ANC; Botha, P. W.; Confederation of Southern African States; National Party; 'reforms'; SADCC**.

townships (*South Africa*)

Townships were built exclusively for urban Africans and other black groups to live in in South Africa. They are situated nearby 'White' towns or cities and disgorge the black workforce for these towns in the mornings, sucking them back in at nightfall. Soweto is the country's biggest township with a population of about 2 million. Although African 'locations' or townships had existed previously, it was not until the 1950s and 1960s that the National Party government started enforcing the policy as part of its strategy of grand apartheid.

In 1923 the Natives (Urban Areas) Act required urban local authorities to provide segregated 'locations' or townships for African occupation. The Natives (Urban Areas) Consolidation Act, Number 25 of 1945 as amended provided that whenever the Governor-General deemed it expedient he might declare that all Africans in an urban area, unless exempted, must live in a location, African village or hostel. The Bantu Laws Amendment Act Number 76 of 1963 tightened up this legislation making it apply to all urban areas, not only those specified by the State President. Any African living in a 'white' part of town would be required to move to a township nearby. If no accommodation were available the person might be sent instead to a 'Homeland'. It also became an offence under this act for an African living in servant's quarters to share these with another African without the permission of the owner of the premises. Later this was prohibited whether the owner agreed or not. Only one full-time servant per private householder was allowed to live in 'White' areas – described by opponents as the 'White by night' policy.

The townships were internally divided along 'ethnic' lines from the 1950s and separated from towns inhabited by other 'racial' groups by industrial areas or buffer strips at least 500 yards wide. Workers were transported to and from these townships daily, often rising before dawn and returning home very late from work. The Deputy Minister of Bantu Administration stated in May 1971 that it should be feasible to transport workers daily between points up to 112 km apart, or, on a weekend basis, between points up to 640 km apart.
See also **urban 'Blacks'**.

Trades Union Congress (*Zambia*)

The Trades Union Congress was formed in 1950 and led by Lawrence Katilungu, the leader of the African Mineworkers' Union. He encouraged other industries to unionise and led the TUC for some ten years without challenge. However, the congress split in 1960 over the issue of politicisation. Some members opposed Lawrence Katilungu's anti-politicisation views and sought a more radical leadership. They left and formed the Reformed Trade Union Congress which was to become the Zambia Congress of Trade Unions in 1965.
See also **Katilungu; Mineworkers' Union of Zambia; Zambia Congress of Trade Unions**.

Trade Union Council of South Africa (*South Africa*) See **TUCSA**.

trade unions (*Botswana*)

Trade unionism is weak in Botswana. It is comparatively new and has been built up largely on the initiative of the government and with the help of the Afro-American Labour Centre and the Friedrich Ebert Foundation. The movement reflects the underdeveloped state of industry in the country and the close economic connections that Botswana has with South Africa.

The Trade Union Act of 1969 is the legislative basis for trade unionism. Unions were required to register under the act and many small unions combined into larger units. The Trade Union Education Centre was set up in 1971 to promote unionisation. By 1980 there were thirteen unions registered in the country. The Botswana Federation of Trade Unions was formed in 1977, and its current chairperson is G. U. S. Matlhabaphiri.
See also **migrant workers**.

trade unions (*Mozambique*)

Marcelino dos Santos, FRELIMO Political Economy Secretary, announced in April 1983 that trade unions would be created for the workers of Mozambique. These would be based on the factory production councils. At the same time as announcing the development, dos Santos emphasised that the party would always have primacy so that there could never be any contradiction between party and unions.
See also **Dos Santos; factory production councils; FRELIMO**.

trade unions (*Namibia*)

t

Legislation covering trade union activities in Namibia is modelled on that existing in South Africa. Until July 1978, African workers were not recognised as 'employees' by the Wages and Industrial Conciliation Ordinance of 1952 and unions with African membership were, therefore, denied the right to legal registration. After the abolition of the clause in 1978, unions were officially treated in the same way whatever their membership. Existing 'White' and 'Coloured' unions, however, were free to adopt racist clauses designed to limit or exclude African membership, a strategy of which several availed themselves.

The new legislation contained severe restrictions on political activity by trade unions, a measure apparently directed at the National Union of Namibian Workers (NUNW), a SWAPO-affiliated national, non-racial union. These made it illegal for registered unions to affiliate to, or make payments to, political organisations. Concomitantly, no registered trade unions were allowed to receive financial support from political organisations.

In 1986, when the South African National Union of Mineworkers began organising in Namibia, the Wage and Industrial Conciliation Amendment Act 1986 was rushed through by the 'internal administration' to make it illegal for non-residents of Namibia to participate in trade union activities in the country. Following the ban, independent, non-racial industrial unions began to be formed inside Namibia under the umbrella of the NUNW. The widespread support for these was shown by their burgeoning memberships.
See also **NUNW; SWAPO**.

trade unions (*South Africa*)

The central development in trade unionism in South Africa in the contemporary period has been the rapid growth and organisation of the black working class. In 1972 black unions were relatively insignificant, consisting of two former SACTU unions and some TUCSA 'parallel' unions. By 1987 most of the union movement had united into a major, largely black, trade union federation, COSATU, composed of 34 unions representing 450,000 paid-up members at its formation in November 1985 and 700,000 paid-up members by early 1987.

The sudden growth of black trade unions began after the mass strikes which broke out in 1972–3 and by the end of 1975 there were 21 unregistered unions with a membership of some 40,000. The state response was to recognise unions for the first time under an act following the Wiehahn Commission of Inquiry report in 1979. This allowed unions with black members to register officially with a registrar if they fulfilled certain conditions. The strength of black unions had forced the state to accord them a place in the country's formal industrial relations process. Previously, they had been excluded, although not made illegal, by the 1953 Bantu Labour Settlement of Disputes Act. White trade unions in South Africa, since their formation, have consistently fought to preserve the privileges of the white working class by excluding black workers.
See also **COSATU; FOSATU; NACTU; SACOL; TUCSA**.

trade unions (*Swaziland*)
1963 was the year of the most widespread strikes in Swazi history. There were six major disruptions in three and a half months, involving some five thousand workers. The first strikes took place at the Usutu Pulp Company, owned jointly by Courtaulds and the Commonwealth Development Corporation. These workers had formed Swaziland's first registered trade union, the Pulp and Timber Workers Union. There were many arrests during the strikes and British troops, the Gordon Highlanders, were brought into the country in June of that year to act against the workers. The Ngwane National Liberatory Congress supported the strikes and the party's leaders, like Dumisa Dlamini, were prominent at workers' meetings throughout the period.

Trade unions do not play a large part in Swaziland's politics largely because the monarchy has developed the traditional system to bypass unions. Although Swazis have the right to unionise (in the Swaziland Trade Union and Trade Disputes Proclamation of 1942) there are only a handful of registered trade unions in existence.
See also **Dlamini, Dumisa**.

Transitional Government (*Angola*)
The Transitional Government was set up (under the terms of the Alvor Agreement) to govern Angola from the end of January 1975 until independence elections due on 11 November that year. It was supposed to draw up a provisional constitution for the country and to oversee the elections, but proved unworkable from the start. As originally conceived, it was to have consisted of a Portuguese appointed high commissioner and a rotating premier with authority over a twelve-person cabinet. The cabinet was made up of three representatives from each of the MPLA, FNLA and UNITA and the Portuguese administration. In addition a

joint army was to have been established drawing 8,000 members from the armies of each of the nationalist movements and combined with 24,000 Portuguese troops. As open fighting among the different nationalist organisations escalated in 1975, it became clear that the Transitional Government was not viable and by July 1975 it had collapsed completely. It was formally dissolved by Portugal the following month.
See also **Alvor Agreement; FNLA; MPLA; UNITA**.

transitional government (*Zimbabwe*)

The transitional government in Zimbabwe came into being on 3 March 1978 with the signing of the Salisbury Agreement. Ian Smith, the rebel leader of the Rhodesian regime, sought to bring about an internal settlement with Bishop Abel Muzorewa as the country's new prime minister. The transitional government was to pave the way for this internal settlement.

Its task was to arrange a ceasefire, remove racial discrimination, draft a new constitution and conduct an election towards the end of the year before handing over to a black government on 31 December 1978. In fact the transitional government went on until the internal settlement came about on 1 June 1979. It was run by an executive council consisting of Ian Smith as prime minister, Bishop Abel Muzorewa, Ndabaningi Sithole and Chief Jeremiah Chirau. Each member presided in turn over the council. The transitional government did not succeed in arranging a ceasefire or removing racial discrimination, but it did draft a new constitution and conduct an election. The constitution was approved by the whites in a referendum in 1979, and Bishop Abel Muzorewa assumed office as prime minister of Zimbabwe-Rhodesia on 1 June of that year.
See also **Chirau; internal settlement; Muzorewa; Salisbury Agreement; Sithole, Ndabaningi; Smith**.

Transitional Government of National Unity (*Namibia*)

In June 1985 the South African Government established the 'Transitional Government of National Unity' (TGNU) in Namibia, based on the constitutional proposals formulated by the MPC. A 'cabinet' of eight members, three from the DTA and the rest from the other five participating parties, was nominated along with 62 representatives to an unelected 'Legislative Assembly', and a 'Constitutional Council' of 16 nominated by the assembly.

Despite the transfer of some responsibilities to the TGNU, South Africa retained direct control of crucial areas such as international affairs and security. On its formation on 17 June 1985, SWAPO estimated that the TGNU had the support of about 80,000 Namibians – 5 per cent of the population. Splits developed in the TGNU almost immediately, between those who favoured repeal of AG8 and AG9, and those wanting their retention. The latter were led by the SWANP which was determined to defend white privilege. The TGNU failed to win recognition by any sovereign government other than South Africa.
See also **AG8; AG9; MPC; SWANP; SWAPO**.

Treason Trial (*South Africa*)
In December 1956 156 people who had attended the Congress of the People in 1955 were arrested for high treason. The preparatory examination of the 'Treason Trial' began on 19 December 1956 and lasted nine months. The subsequent trial lasted until 29 March 1961, when everyone was finally acquitted. The long trial had a serious effect on the ANC and Congress Alliance as it wasted the time and energies of many of its key leaders including Walter Sisulu, Nelson Mandela, Albert Lutuli, Oliver Tambo, Ruth First, Helen Joseph, Monty Naicker and Joe Slovo. The trial led to the formation of a fund internally to raise money for legal fees. This was set up by Alan Paton, Alex Hepple and Bishop Ambrose Reeves. A similar support fund started in London by Canon John Collins was the forerunner of the International Defence and Aid Fund for Southern Africa (IDAF).
See also **Congress of the People; Mandela, Nelson; Rivonia Trial**.

Treurnicht, Andries Petrus (*South Africa*)
Andries Treurnicht was elected leader of the Conservative Party on its formation in 1982. He had previously been minister of state administration under P. W. Botha, but despite his position in the cabinet had led the opposition within the National Party to the policies of the president.

Treurnicht's power-base lay in the Transvaal National Party, of which he was leader. His differences with President Botha came to a head in 1982, when he challenged him on the question of power-sharing with black South Africans. After Treurnicht and 21 other National Party MPs refused to support Botha in a confidence vote, he was suspended as Transvaal leader. A month later the Conservative Party was formed, espousing a return to old-fashioned Verwoerdian apartheid and claiming the 'true' mantle of the ideology. In 1983 Treurnicht resigned his Waterberg (Pretoria) parliamentary seat (won in 1981 for the National Party) and re-stood as a Conservative Party candidate. He won the election to become the first MP elected on his party's ticket, and the first MP elected from a party to the right of the National Party since it came to power in 1948.

Treurnicht is a former assessor of the general synod of the *Nederduitse Gereformeerde Kerk* (NGK) and a former chair of the *Afrikaner Broederbond*. He edited the NGK newspaper *Die Kerkbode* and, later, the Pretoria daily newspaper *Hoofstad*. He was elected to his Waterberg seat for the National Party in 1971, and, before serving as Minister of State Administration, was Deputy Minister for Bantu Administration and Education, and Minister of Public Works and Tourism.
See also ***Afrikaner Broederbond*; Botha, P. W.; Conservative Party; National Party; NGK**.

Tribal Grazing Land Policy (*Botswana*)
The Tribal Grazing Land Policy (TGLP) was introduced in 1975. It is a new land tenure policy which, principally, makes provision for 50–year leases and encourages commercial ranching in Botswana. Traditionally land in the country has been treated as a communal asset, but recently

t

there has been a move towards the privatisation of land for grazing, so the TGLP is intended to regulate this process.

Tribal Trust Lands *(Zimbabwe)*
The Tribal Trust Lands (TTLs) were originally the rural areas called 'native reserves', created by a land commission formed in 1894 to look into the question of relocating the Ndebele people. They were incorporated into law in 1930 when the British and colonial governments passed the Land Apportionment Act. This divided Southern Rhodesia roughly equally between the 'European' and the 'native' population – the latter making up some 95 per cent of Zimbabwe's people.

The TTLs were where the guerrillas drew most of their support during the nationalist war. In 1970 the Land Tenure Act formalised the division of land 'for all time', and the reserves then became known as the TTLs – 165 covering 16,268,000 hectares were designated. The Land Tenure Act was the basis of the ruling white Rhodesian Front's policy for many years, and suggested changes caused an uproar in the party in 1977. The Land Tenure Act was formally dismantled in January 1979 under the transitional government and the Tribal Trust Lands became known as the communal lands.
See also **Land Apportionment Act; Land Tenure Act; Rhodesian Front**.

t

Tripartite Monetary Agreement *(Lesotho, South Africa, Swaziland)*
The Tripartite Monetary Agreement (TMA) replaced the Rand Monetary Union in July 1986. It links Lesotho, Swaziland and South Africa financially although the rand is no longer legal tender in Lesotho or Swaziland. See also **Rand Monetary Union; South African Customs Union**.

Tsumkwe Group *(Namibia)*
A small, ethnically-based party in the Democratic Turnhalle Alliance: see **DTA**.

TTLs *(Zimbabwe)* See **Tribal Trust Lands**.

TUC *(Zambia)* See **Trades Union Congress**.

TUCSA *(South Africa)*
TUCSA, the Trade Union Congress of South Africa, was formed in 1954 by uniting two other trade union groupings in order to conform with new segregationist government legislation. Its membership, in 1984, was some 420,000. It was dissolved on 2 December 1986 after a number of its affiliates had left. On 1 April 1987 the eight original TUCSA affiliates formed the National Federation of Trade Unions with a claimed membership of 150,000.

TUCSA was a union grouping which was tolerated by the state, although it was not aligned with any National Party organisations. It was multiracial but its African workers were treated differently from the 'White' and 'Coloured' workers who controlled the union. When it was first formed TUCSA put forward a policy of 'parallelism' which created subsidiary branches of its unions for African workers for whom it would

negotiate in exchange for their co-operation. After the 1979 act recommending registration of the new unions, TUCSA actively sought African members and created further 'parallel' unions. These TUCSA tried to use to compete with the rapidly-growing independent black trade unions. It tried to appeal to black workers by offering them a 'non-political' (although anti-communist) union which would campaign solely on workers' issues.
See also **COSATU; trade unions**.

Tumbo, Kasanga 'Christopher' (*Tanzania*)

Kasanga Tumbo was elected the General Secretary of the Tanganyika Railway African Union in 1959. Until then the union had a moderate leadership having begun as a staff association for clerks. In February 1960 the new leader initiated a strike, lasting a record 82 days, for more pay and as a campaign against the racism among many white railway workers. The strikers feared that if co-operative arrangements among Kenya, Tanzania and Uganda continued after independence these men would stay on in powerful positions in Tanzania.

The strike failed to break the East African co-operative arrangements, with prime minister Julius Nyerere so committed to the federation that he was even prepared to delay independence for Tanganyika if some arrangement could be made for a federation with the other two countries. Instead the union movement was suppressed. In 1963 Kasanga Tumbo was arrested by the Kenyan police and handed back to Tanzania where he remained in detention without trial for four years.
See also **East African Community; Tanganyika Federation of Labour**.

Turnhalle Constitutional Conference (*Namibia*)

A meeting of anti-SWAPO parties called to produce an internal settlement in Namibia which would exclude SWAPO: see **DTA**.

Tutu, Desmond Mpilo (*South Africa*)

Archbishop Desmond Tutu is a prominent South African clergyman and anti-apartheid spokesperson. He has an international reputation and has travelled extensively, consistently arguing the case for a free and equal South Africa. He is the Archbishop of Cape Town (from September 1986) and of the Metropolitan of the Province of South Africa, making him head of the Anglican Church in the country. He won the Nobel Peace prize in 1984 for his work against apartheid. He has held many prominent roles within the church community and was, for instance, appointed the Secretary-General of the South African Council of Churches in 1978.

Bishop Tutu is a powerful advocate of sanctions against South Africa. An interview in March 1987 quoted him as follows: 'Sanctions are for many the last possible non-violent strategy in helping us to attain the goal that we all claim to want – transition from the present vicious situation to a more just set-up'.
See also **churches; SACC; UDF**.

Twining, Edward (*Tanzania*)
Sir Edward Twining became Governor of Tanganyika in June 1949. He was appointed by Arthur Creech Jones, the British colonial secretary, and his role was to prepare Tanzania for independence. He implemented an extensive development plan in the country and modernised the constitution. However, he failed to come to terms with Julius Nyerere, the leader of the nationalist movement and future leader of Tanzania, and spent some time trying to arrest him. Sir Edward Twining served as Governor until 1958.
See also **Nyerere; Tanganyika**.

UANC (*Zimbabwe*)
In 1976 Bishop Abel Muzorewa's leadership of the African National Council was challenged and he set up the UANC (the United African National Council) The UANC was well organised and Bishop Muzorewa retained a loyal following and could draw large crowds in his support. As head of the United Methodist Church he was seen as a man of peace and he represented the voice of unity to those worried about nationalist divisions.

Ian Smith, the rebel Rhodesian leader, subsequently sought an internal settlement with the UANC, Ndabaningi Sithole and Chief Chirau. The UANC held out for more favourable terms for the country's black majority. Bishop Muzorewa hoped that the new settlement would enable more liberal whites to be elected to the government. On 15 February the Salisbury Agreement was announced after the Bishop backed down with no prospect of white seats being won by any party other than the Rhodesian Front. Soon Byron Hove, a minister sharing the law-and-order portfolio, demanded more reform from the transitional government. He was sacked and immediately became a UANC hero. The UANC threatened to pull out of the government unless he was reinstated but Bishop Muzorewa persuaded the party to stay.

The UANC was the largest black party participating in the April 1979 internal election. It gained 67 per cent of the valid votes cast and 51 of the 72 black seats in parliament. Bishop Muzorewa became the prime minister of Zimbabwe-Rhodesia on 1 June 1979. The UANC participated in the Lancaster House conference at the end of the year and fought the elections for a free and independent Zimbabwe in February 1980. The party won less than 10 per cent of the vote and three seats in the new parliament. ZAPU and ZANU won 87 per cent of the vote between them; the period of UANC influence was over.
See also **Chirau; internal settlement; Lancaster House; Muzorewa; Rhodesian Front; Salisbury Agreement; Sithole, Ndabaningi; Smith; ZANU; ZAPU**.

UDA (*Lesotho*) See **United Democratic Alliance**.

UDENAMO (*Mozambique*)
The *União Democrática Nacional de Moçambique* (UDENAMO) was established in what was then Southern Rhodesia at the start of the 1960s,

and was modelled on the National Democratic Party which had been established by Zimbabwean nationalist Joshua Nkomo. UDENAMO became part of FRELIMO when that party was formed in 1962. It was headed by Rev. Uria Simango and Adelino Gwambe.
See also **FRELIMO; National Democratic Party; Nkomo**.

UDF (*South Africa*)
The anti-apartheid UDF (United Democratic Front) was formed in Johannesburg in 1983 by 150 delegates from 30 organisations, notably the Natal Indian Congress and the Transvaal Indian Congress – both recently revived. The inspiration for its formation came from Dr Allan Boesak, President of the World Alliance of Reformed Churches, who has been a prominent spokesperson for the UDF and its elected Patron ever since. The UDF was launched nationally in August 1983 and now claims over two million members and some 700 affiliated organisations. Its *raison d'être* was to organise a boycott of elections for 'Indians' and 'Coloureds' under the constitution of 1983 devised by South Africa's President P. W. Botha. In this it was successful; the polls for these elections were low – about 20 and 30 per cent respectively of registered voters. It also campaigned against the government-sponsored town councils for African townships.

However, the significance of the UDF is far wider than these specific campaigns. It is an organisation in the tradition of the Congress Alliance and is identified with the non-racialism of the Freedom Charter, although this has not been formally adopted. Furthermore, it sports the colours of the banned ANC. Its leaders have been constantly harassed by the South African state but the organisation has not been officially banned although it was declared an 'affected organisation' in 1986, which meant that it could no longer receive overseas funds. Its local organisation is based on 'street committees' in the townships. Some of South Africa's most dinstinguished public figures are amongst its leadership. Loose organisation and many different leaders are part of its technique for survival. Some of the most prominent figures associated with the UDF are its founding presidents, Archie Gumede, Albertina Sisulu and Oscar Mpetha. Others include Allan Boesak, Winnie Mandela, Popo Molefe, (general secretary) and Patrick 'Terror' Lekote (publicity secretary).
See also **Boesak; Congress Alliance; Mandela, Winnie; Sisulu, Albertina**.

UDI (*Zimbabwe*) See **Unilateral Declaration of Independence**.

UEA (*Angola*)
The *União dos Estudantes Angolanos* (UEA) or Union of Angolan Students was the MPLA organisation for students.
See also **MPLA**.

Ujamaa (*Tanzania*)
This famous Swahili word had been adopted as the name for the brand of socialism developed by the former president, Julius Nyerere, of Tanzania. It was first used in a political context by President Nyerere in a 1962 pamphlet called *Ujamaa – the Basis of African Socialism*. *Ujamaa*

is an abstract noun that comes directly from the Swahili word for the extended family. Julius Nyerere translated it as 'familyhood'. Socialism, the president argued in this pamphlet, is a matter of distribution and 'an attitude of mind' involving moral obligations. '*Ujamaa* . . . or "familyhood" describes our socialism. It is opposed to capitalism . . . and it is equally opposed to doctrinaire socialism'.

In 1967 the idea of *Ujamaa* was combined with villagisation leading to the creation of so-called *ujamaa* villages. These were to be, according to Julius Nyerere, small groups of farmers who worked together on communal farms and used their savings to buy equipment that would benefit the group. In practice, the policy was often pushed through by force, and caused hardship and bitterness. It failed to increase production and did much to discredit President Nyerere's policy of *Ujamaa*. In June 1975 the president claimed that there were 9.1 million people living in 6,940 *ujamaa* villages accounting for 65 per cent of Tanzania's population. See also **African Socialism; Nyerere; villagisation**.

Umkhonto we Sizwe (*South Africa*)

Umkhonto we Sizwe (the Spear of the Nation) was founded in 1961 as the military wing of South Africa's main nationalist movement, the ANC. Nelson Mandela, currently in prison in South Africa for life, was its first commander-in-chief. It was formed a year after the ANC was banned and after fifty years of non-violent struggle for change by the ANC. In June 1961 the executive committee considered a proposal from Nelson Mandela on the use of violent tactics. It was agreed that the ANC itself would not change its official non-violent policy, but that ANC members who wished to follow a military path could do so. *Umkhonto we Sizwe* was open to members of all races (unlike the ANC at that point) and it was set up in conjunction with the South African Communist Party.

Umkhonto had its secret headquarters at Lilliesleaf Farm in Rivonia, northern Johannesburg. Its national high command planned a campaign of sabotage bombings which were designed to 'awaken everyone to a realisation of the disastrous situation to which the nationalist policy is leading', and thus lead to changes before 'matters reached the desperate stage of civil war'. The first attacks were in December 1961 and over 200 took place in the next eighteen months. The organisation suffered a severe setback in July 1963 when police captured the leaders in a raid on the Rivonia headquarters. The Rivonia Trial followed and many of the top leaders including Nelson Mandela, Walter Sisulu and Govan Mbeki were sentenced to life imprisonment. The focus then shifted outside South Africa where a series of regional alliances took place, for instance with Zimbabwe's nationalist movement, ZAPU. The 'Wankie campaigns' of 1967–8 tried to send guerrillas through Zimbabwe into South Africa, but these were soon ended after clashes with the forces of the Smith government in what was then Rhodesia. During this time the organisation was given material help by the Soviet Union.

The ANC itself was influenced by *Umkhonto*, and in Tanzania at the Morogoro conference of 1969 it was decided to open the ANC organis-

ation to people of all colours. The 1970s saw an upsurge in mass protest inside South Africa and the rapid growth of the trade union movement. *Umkhonto*'s strategy was to continue training and preparing to intensify the armed struggle. The 1980s have seen the start of increased activity by *Umkhonto*, and this seems likely to continue. Its present chief-of-staff is reputed to be Joe Slovo.
See also **ANC; Mandela, Nelson; Mbeki; Sisulu, Walter; Slovo; South African Communist Party**.

Umma Party (*Tanzania*)
In 1963, shortly before the pre-independence elections in Zanzibar, Abdul Rahman Babu, who led the Zanzibar National Party's (ZNP) youth wing, broke away from the ZNP together with the youth and trade union wings of the party. They formed a new party with a socialist philosophy called the Umma Party. The party was only in existence until it merged with the Afro-Shirazi Party immediately after the 1964 revolution in Zanzibar.
See also **Babu; Zanzibar National Party; Afro-Shirazi Party**.

UNAMI (*Mozambique*)
The *União Africana de Moçambique Independente* (UNAMI) was one of the three parties which united to form FRELIMO in 1962. UNAMI was established in Malawi in the early 1960s, inspired by the nationalist currents in that country.
See also **FRELIMO**.

UNCN (*Namibia*) See **UN Council for Namibia**.

UN Commissioner for Namibia (*Namibia*)
The legal head of state for Namibia is the United Nations Commissioner for the country, currently Bernt Carlsson. The first commissioner, Sean MacBride was appointed in 1973 and served until 1977, when he was succeeded by Martti Ahtisaari. Brajesh Mishra followed Ahtisaari in 1982, and was replaced by Carlsson who commenced his appointment on 1 July 1987. The UN Commissioner is also the executive officer of the UN Council for Namibia.
See also **Ahtisaari; Carlsson; MacBride; Mishra; UN Council for Namibia; UNIN**.

UN Council for Namibia (*Namibia*)
The United Nations Council for Namibia was set up in 1966 as the legal administering authority of Namibia, following the UN decision to terminate South Africa's mandate over the country. The Council for Namibia is the policy-making organ for Namibia at the UN. It gives assistance to Namibian refugees, has organised a training programme for Namibians, co-ordinates international non-governmental action to promote independence for Namibia, monitors the exploitation of Namibia's natural resources and issues travel documents. On independence, it will have the task of establishing an emergency programme to render economic and technical assistance to the country. These activities are

financed by a UN Trust Fund for Namibia set up in 1970. Its membership includes over 30 states but excludes any of the major western countries. See also **Decree No. 1; UN Commissioner for Namibia; UNIN; UN Security Council Resolution 435**.

UNFP (*Zambia*) See **United National Freedom Party**.

União Africana de Moçambique Independente (*Mozambique*)
One of the three parties which together made up FRELIMO: see **UNAMI**.

União das Populações de Angola (*Angola*) See **UPA**.

União Democrática Nacional de Moçambique (*Mozambique*)
One of the constituent parties of FRELIMO: see **UDENAMO**.

União Nacional para a Independência Total de Angola (*Angola*) See **UNITA**.

União Nacional dos Trabalhadores de Angola (*Angola*) See **UNTA**.

Unilateral Declaration of Independence (*Zimbabwe*)
Ian Smith, the rebel Rhodesian leader, made his Unilateral Declaration of Independence (UDI) from Britain, the colonial power, on 11 November 1965, renaming the colony of Southern Rhodesia, Rhodesia. The idea of UDI had been discussed for some time before the Rhodesian Front, Ian Smith's party, came to power. The business community, particularly, was anxious about its effect, predicting trade and financial problems. But independence was crucial to Ian Smith's dream of maintaining white supremacy and he rapidly persuaded the white electorate of the importance of breaking links with Britain. By 1964 he had won them over and they voted ten to one in favour of independence in a referendum on the question.
See also **Rhodesia; Rhodesian Front; Smith; Southern Rhodesia**.

UNIN (*Namibia, Zambia*)
Based in Lusaka, Zambia, the United Nations Institute for Namibia was opened on 28 August 1976 to prepare the ground for the rebuilding of Namibia after independence. UNIN is an educational institute under the direction of Hage Geingob. It provides training for civil service and administrative posts, and undertakes research into the problems of reconstruction after independence. It has initiated studies on policy options in the fields of labour, education, agriculture, health, the economy and other social services amongst others. In 1980 there were 400 Namibians in training at the institute.
See also **Decree No. 1; UNCN; UN Commissioner for Namibia; UN Security Council Resolution 435**.

UN Institute for Namibia (*Namibia*)
Lusaka based educational institution: see **UNIN**.

Union of Angolan Students (*Angola*) See **UEA**.

U

Union of South Africa (*South Africa*)

The Union of South Africa was created by the South Africa Act of 1909. In 1910 the former Afrikaner republics of the Orange Free State and the Transvaal came together with the two British colonies of the Cape and Natal to form provinces within the Union of South Africa. The Union followed the South African War which had been won by the British in 1902 when the Transvaal and Orange Free State surrendered and the Treaty of Vereeniging was signed.

At the time of Union the franchise differed in the four provinces. All male citizens of the Cape who were literate and earned £50 a year or owned fixed property to the value of £75 qualified for the franchise on the common roll. In theory the same applied to Natal except that the income or property qualification was £96. However, in practice there were so many difficulties for black voters that very few were registered. In the Transvaal and Free State only 'White' male citizens voted. The National Convention which drafted the South Africa Act decided to maintain the existing position except that it withdrew the right of Africans to sit in Parliament (implicit in the Cape) and it decided not to register any more Africans or 'Asians' in Natal. 'Coloured' men retained the franchise there until the Separate Registration of Voters' Bill removed them from the common voters' roll in the 1950s.

In 1960 National Party Prime Minister Hendrik Verwoerd set out to achieve one of the most cherished desires of Afrikaner nationalists – to make South Africa a republic. The Union of South Africa became the Republic of South Africa in May 1961. As a consequence it ceased to be a member of the Commonwealth.

See also **South Africa**.

U

Union of Tanzanian Workers (*Tanzania*)

The Union of Tanzanian Workers (JUWATA) was formed in 1978 by the National Union of Tanganyika Workers (NUTA) and Zanzibar trade unionists. It is Tanzania's sole trade union organisation and had 350,000 members in 1979. Its general secretary was Joseph Rwegasira.

See also **National Union of Tanganyika Workers**.

Union of the Peoples of Angola (*Angola*) See **UPA**.

UNIP (*Zambia*)

UNIP, or the United National Independence Party, is Zambia's sole political party. It took the country into independence in 1964 under the leadership of Kenneth Kaunda who has consolidated its power ever since. It was formed on 1 August 1959 by a merger between the African National Independence Party and the United National Freedom Party. Its aim was African unity and independence for Northern Rhodesia which was to be renamed Zambia. It sought the dissolution of the Central African Federation (Northern and Southern Rhodesia and Nyasaland), since this, it argued, worked mainly to the advantage of the Southern Rhodesian whites. To achieve this the newly-formed UNIP concentrated on party organisation and carefully planned resistance, both violent and non-violent.

The party was set up in rivalry to the Northern Rhodesia African Congress (later the ANC) which, UNIP argued, had lost its momentum and was making too slow progress in the campaign for independence. Once released from prison in 1959 Kenneth Kaunda immediately took over the leadership of UNIP and after a massive campaign of civil disobedience in 1962 the British government introduced a new constitution for Northern Rhodesia. This allowed for an African majority in the legislature and soon a coalition government was formed between UNIP and the remains of the ANC. In 1964 a new election resulted in a UNIP government and independence came in October 1964. In 1971 a rival party, the UPP (United Progressive Party) was formed by Simon Kapwepwe, the former UNIP vice-president, but many of its leaders were arrested and detained without trial. In December 1972 Zambia became a one-party state with a new constitution.

See also **African National Congress (Zambia); African National Independence Party; Central African Federation; Kapwepwe; Kaunda; Northern Rhodesia; Northern Rhodesia African Congress; United National Freedom Party; United Progressive Party**.

UNITA (*Angola*)

UNITA, or *União Nacional para a Indepêndencia Total de Angola* (the National Union for the Total Independence of Angola), founded in March 1966, was the last of the main nationalist groups in Angola to be established. It arose as a breakaway from Holden Roberto's FNLA and was formed by former UPA general secretary Jonas Savimbi out of an alliance of mainly Ovimbundu defectors from GRAE, students involved in UNEA (*União Nacional dos Estudantes Angolanos*) and other Angolan refugees based in Zambia. UNITA's first military operation, against the Benguela railway, earned the ire of its host country, Zambia, which was reliant on the line for its copper exports. Savimbi was expelled and eventually returned to Angola where he led what was the weakest of the three main nationalist organisations.

Despite its ostensibly socialist politics and its claim to be a liberation movement, by 1972 UNITA was co-operating with the notorious Portuguese secret police, the PIDE, and receiving weapons, supplies and medical assistance from the Portuguese. UNITA concentrated its efforts in the centre of Angola, attacking MPLA forces which, since 1966, had been struggling to establish a liberated zone in the east of the country.

In June 1974 UNITA signed a 'ceasefire' with the Portuguese and formed a pact with the FNLA against the MPLA. It was a party to the Alvor Agreement and, following the collapse of the Transitional Government, joined the South Africans in their efforts to overthrow the RPA through the massive 1975 invasion. UNITA subsequently became the main vehicle for South African destabilisation of Angola. It has received military and other supplies, training, logistical support and safe bases in Namibia from the South Africans in support of its efforts to

topple the government of the RPA. It has also become the major recipient of renewed US military aid to opponents of the Angolan government, and is reputed to have been supplied with highly effective Stingray ground-to-air missiles. UNITA's alliance with South Africa has ensured its hostility to SWAPO of Namibia, and it has worked with the South Africans in their attempts to establish a *cordon sanitaire* along the Angolan/Namibian border.
See also **Alvor Agreement; FNLA; GRAE; MFA; MPLA; PIDE; Savimbi; Transitional Government; UPA**.

United African National Council (*Zimbabwe*) See **UANC**.

United Democratic Alliance (*Lesotho*)
The United Democratic Alliance (UDA) was founded in 1984 and led by Phoka Chaolane. South Africa supported the new party which declared its policies to be the severence of diplomatic relations with eastern bloc countries and the banning of communism in Lesotho. The United Democratic Alliance also favours closer links with South Africa.

United Democratic Front (*South Africa*) See **UDF**.

United National Freedom Party (*Zambia*)
The United National Freedom Party (UNFP) was a small political group led by the president of one of Zambia's main trade unions. It was one of the two parties which, in 1959, united to form Zambia's main political party, UNIP.
See also **UNIP**.

United National Independence Party (*Zambia*) See **UNIP**.

United National Party of South West Africa (*Namibia*)
The precursor to the Federal Party, the United National Party of South West Africa (UPSWA) was formed in 1927. Until its dissolution in 1975, it was the main white opposition to the SWANP. Its chief support came from the minority English-speaking white community in Namibia, and it never posed a serious threat to the dominance of white politics by the SWANP.
See also **FP; SWANP**.

United Party (*South Africa*)
The United South African National Party or United Party (UP) as it was popularly known, was formed in 1934 through the fusion of the ruling National Party led by Albert Hertzog and the opposition South African Party under Jan Smuts. The United Party remained in government until 1948 when it was defeated by an alliance of Afrikaner parties.

After 1948 the United Party became the official opposition in the 'Whites'-only South African parliament, its complement of seats gradually dwindling over time as the prosperity resulting from the exploitation of black South Africans attracted more and more white support to the National Party. The UP was led by Sir de Villiers Graaff from his election as leader in 1956 until its dissolution in 1977. In that year, after decades of internal strife and repeated splits, Graaff led the party through a process of ritual self-dismemberment. It split into three parts, the right-wingers or 'Old Guard' as they had become known, forming the South African Party, the centre merging with the tiny Democratic Party to form the New Republic Party, and the left forming the Committee for a United Opposition. The last subsequently merged with the Progressive Reform Party to form the Progressive Federal Party, while the first dissolved itself into the National Party.

The United Party, by any objective standards, was a very conservative party despite its reputation in some circles as the standard-bearer of liberal opposition. It had, indeed, campaigned vigorously against the policies and actions of the National Party immediately after its 1948 defeat, but when this was repeated in 1952 by a larger margin the notion gained currency among party leaders that it could only regain its previous support by appearing more conservative than the National Party, whose apartheid programme it attempted to portray as dangerously radical. Every major split in the party after 1948 involved the departure of groups on the left frustrated at the timidity and lack of vision of the party leadership. Few bemoaned the demise of the party in 1977.

See also **Democratic Party; National Party; New Republic Party; Progressive Federal Party; South African Party**.

United Progressive Party (*Zambia*)

The United Progressive Party (UPP) was formed in August 1971 by Simon Kapwepwe, Zambia's former vice-president. The event which triggered its formation was the dismissal by Kenneth Kaunda of some Bemba-speaking ministers from his cabinet. This alienated the Bemba-speaking section of the ruling party, UNIP, led by Simon Kapwepwe.

The formation of the new party provoked a crisis in UNIP. What had hitherto been simply a faction within the party had now emerged as an opposition party with a popular leader who might challenge Kenneth Kaunda. The UPP drew its support mainly from Bemba-speakers within the Copperbelt, but the government feared that it might spread throughout the rural Bemba-speakers, the largest group in Zambia (some 34 per cent of the population). The government reacted strongly to the new threat. UPP supporters were suspended, dismissed and even threatened with losing their houses, while party organisers were detained. In 1972 the UPP was banned and those leaders not already detained, including Simon Kapwepwe, were held without trial. The next month a new party, the Democratic People's Party, was formed in protest by supporters of the UPP. In 1972 Parliament passed a law making Zambia a one-party state.

See also **Bemba; Kapwepwe; Kaunda**.

United Republic of Tanzania (*Tanzania*) See **Tanzania**.

United South African National Party (*South Africa*)

Better known as the United Party, the party of government until 1948 and subsequently the official opposition in the 'White' parliament until its demise in 1977: see **United Party**.

United Swaziland Association (*Swaziland*)
The United Swaziland Association was a political association formed by whites in Swaziland in 1963 after the Reconstituted European Advisory Council ceased to exist. Its leading member was Carl Todd, who had been head of the Advisory Council, but Willie Meyer was elected the first Chairman in 1964. The association co-operated with the Swazi traditionalists (the Imbokodvo National Movement) and sought to share power in the proposed legislative council with this group. It failed in this aim, but did win the four white seats in the council. Gradually the tranditionalists distanced themselves from the United Swaziland Association and the conservative whites became isolated. The party was unable to resist the growth of the nationalist movement and it did not contest the 1967 elections.
See also **Imbokodvo National Movement; Legislative Council; Reconstituted European Advisory Council; Todd, Carl**.

United Trade Union Congress (*Zambia*)
Successor to the Reformed Trade Union Congress, which split from the Trade Union Congress in 1960, and predecessor of the Zambia Congress of Trade Unions: see **Zambia Congress of Trade Unions**.

United Workers' Union of South Africa (*South Africa*) See **UWUSA**.

UN Peace Plan (*Namibia*)
A name for the UN proposals for ending the war in Namibia and bringing about independence: see **UN Security Council Resolution 435**.

UNSCR 385 (*Namibia*) See **UN Security Council Resolution 385**.

UNSCR 435 (*Namibia*) See **UN Security Council Resolution 435**.

UN Security Council Resolution 385 (*Namibia*)
Resolution 385 of the United Nations Security Council was passed on 30 January 1976 after it became clear that South Africa had no intention of ending its illegal occupation of Namibia. Among other things, Resolution 385 demanded an end to the bantustan system in Namibia, free elections under UN supervision, and, pending the transfer of power to the UN, the release of all political prisoners, the end of racial discrimination, and the freedom for all exiles to return, unmolested, to Namibia. South African responded by calling the Turnhalle Constitutional Conference and attempting to set up an 'internal' government which excluded SWAPO.
See also **DTA; UN Security Council Resolution 435**.

UN Security Council Resolution 435 (*Namibia*)
United Nations Security Council Resolution 435 was passed after extended negotiations between the 'Contact Group' and SWAPO and

South Africa on the best way to bring Namibia to independence. In January 1976 the UN Security Council had passed Resolution 385 calling for the unconditional withdrawal of South Africa from Namibia. On 30 March 1978 the 'Contact Group' set out its milder 'Proposal for a Settlement in Namibia', outlining the agreed basis for a way forward. Five months later, on 30 August, the UN Secretary General, Kurt Waldheim, published a report on the implementation of the Proposal for a Settlement for the Security Council, and on 29 September it passed Resolution 435 endorsing the report.

The agreed basis for a settlement in Namibia as laid out in Resolution 435 included provisions for elections based on universal suffrage, supervision of the elections by the UN, the ending of discriminatory legislation, a ceasefire and the phased withdrawal of the bulk of South Africa's troops. It provided for the establishment of the UN Transition Assistance Group (UNTAG), a joint military and civilian force, to enforce the provisions.

Implementation of Resolution 435 has never been possible due to the imposition of successive new pre-conditions by South Africa. The final sticking-point was South Africa's insistence, with the support of the Reagan administration in the US, on linkage of a settlement in Namibia to the withdrawal of Cuban troops from Angola. In July 1986, in a speech to an international conference on Namibia held in Vienna, UN Secretary General Perez de Cuellar said that all outstanding issues on the implementation of Resolution 435 had been resolved by November 1985, when agreement was reached on the electoral system. Only linkage, he said, continued to prevent a settlement.

See also **Contact Group; linkage; SWAPO; UN Security Council Resolution 385; UNTAG**.

U

UNTA (*Angola*)

The *União Nacional dos Trabalhadores de Angola* (UNTA) or National Union of Angolan Workers was the trade union wing of the MPLA and is now the official representative body of Angolan workers.

See also **FDLA; MPLA**.

UNTAG (*Namibia*)

The United Nations Transition Assistance Group (UNTAG) was provided for in UN Security Council Resolution 435 of 1978 which laid out the basis for a settlement in Namibia. UNTAG's role was to be to supervise and control the transition from South African occupation to elections and the beginning of independence. The UN Secretary General's enabling report on Resolution 435 provided for an UNTAG force of 5,000–7,500 troops and 1,000–2,260 civilian administrators and police.

See also **UN Security Council Resolution 435**.

UN Trust Fund for Namibia (*Namibia*)

A fund set up in 1970 to fund UN activities and agencies dealing with Namibia: see **UN Council for Namibia**.

UP (*South Africa*) See **United Party**.

UPA (*Angola*)
A precursor to the FNLA, the *União das Populações de Angola* (Union of the Peoples of Angola) was formed in 1958 from an emigré Kongo nationalist party, the UPNA (*União das Populações do Norte de Angola*). After the 1961 peasant uprising in northern Angola, the UPA began to organise for armed struggle against the Portuguese colonial regime. In 1962, following an initiative by Holden Roberto, UPA president, it merged with the PDA to form the FNLA.
See also **FNLA; PDA; Roberto; Savimbi**.

UPP (*Zambia*) See **United Progressive Party**.

UPSWA (*Namibia*) See **United National Party of South West Africa**.

urban 'Blacks' (*South Africa*)
The question of the urban 'Blacks' has troubled successive National Party governments for years. It became increasingly apparent that the old apartheid ideas of all urban Africans being citizens of often remote 'Homelands' and being only temporarily resident in urban townships could not endure. Nevertheless the theory of the 'Homelands' is still central to government policy so moves to introduce special provisions for urban Africans have been tentative.

The rate of urbanisation of the African population has added urgency to the government's dilemma. The percentage of the African population living in urban areas in South Africa was 31.8 per cent in 1960 and 33.1 per cent in 1970. By 1980 it had increased to 38.3 per cent, according to official figures. By the year 2,000 it will be between 60 and 75 per cent. So the government will have to accept that between 12 and 20 million more urban Africans will be added to the 1980 figures of seven million. The government has ruled out any incorporation of Africans into the three-chamber parliament. The final report of the constitutional committee of the first President's Council, published in June 1984, concluded that the incorporation of Africans into this structure would lead to 'group domination and conflict'. It was inconceivable that the process of 'Homeland independence' could be checked or reversed peacefully, the report stated.

In May 1984 President P. W. Botha said, 'we realise that millions of black people are living outside those states within the borders of the Republic', and this was why a special cabinet committee had been established. The committee, chaired by the Minister of Constitutional Development and Planning, Chris Heunis, was to investigate the position of Africans, particularly those resident outside the 'Homelands'. The government's critics rejected the committee, one suggesting that participation in its discussions, would be 'a definite betrayal of established struggles by those men and women who have been banished, exiled, or even imprisoned by the South African regime'.

The details of the government's policy were finalised during September 1984 at the height of the unrest, but were not announced formally until the State President's speech at the opening of Parliament in January 1985. The main interrelated aspects of policy concerned the issues of influx control (the pass laws), citizenship, property rights and political rights. The pass laws were repealed from 1 July 1986 and replaced with a policy of 'orderly urbanisation'. The Restoration of South African Citizenship Act restored citizenship to 1.75 million people from the 'independent Homelands' – the remaining 7.5 million were made subject to the Aliens Act and its strict controls on movement and employment. The government announced it would set up a National Council with African membership and power to make recommendations on matters of legislation and policy. However, the policy faces serious problems as even the conservative African local council representatives and the National African Federation of Chambers of Commerce have refused to participate.

A crucial part of the government's strategy is the restructuring of regional and local government. New provincial executives were introduced in July 1986 with powers previously held by central government, to administer Africans outside the 'Homelands'. In conjunction with this initiative, the 'Homelands' were given increased powers to suppress opposition, allowing them to ban organisations, restrict meetings, censor publications, among other things. The difference in the status of 'independent' and 'non-independent' 'Homelands' has narrowed as a result of this policy.

See also **apartheid; 'reforms'; townships**.

Urban Foundation (*South Africa*)

The Urban Foundation was formed in 1977 with the specific aim of organising collective intervention by multinationals and other major business enterprises in South African affairs after the 1976 uprising in Soweto. Modelled on similar initiatives by companies in the USA, it was the brain-child of Irene Menell, a leading member of the Progressive Federal Party whose husband, Clive Menell, headed a major Anglovaal company.

The Urban Foundation solicits money and loans for use in core projects designed to build a black middle class, to provide opportunities for black children, and to spur government into improving services and easing controls on black people. The bulk of money raised by the foundation has been devoted to projects on education, housing and infrastructure. Among its main achievements have been initiating the electrification of Soweto, the negotiation of long-term leasehold rights for some urban Africans, and the expansion of opportunities for further education and technical training for Africans. Apart from helping to meet specific needs of its business supporters, such as the shortage of skilled labour, the foundation has vigorously promoted their views on the need for reform and relaxation of restrictions on black South Africans.

The major donors to the Urban Foundation have included both English and Afrikaans companies, 27 of the top 100 companies in the country, and all bar one of the biggest banks. Of total donations received by 1981, 34 per cent had come from the Anglo American Corporation whose

former Chairman, Harry Oppenheimer, chairs the Foundation. His 1984 Annual Review announced that the Urban Foundation had raised R53 million in outright donations and a similar amount in loans.
See also **Anglo American Corporation; Free Market Foundation; Oppenheimer; Progressive Federal Party; South African Foundation**.

UWUSA (*South Africa*)

UWUSA, the United Workers' Union of South Africa, is attached to Gatsha Buthelezi's *Inkatha* movement and is largely based in Natal amongst the Zulu-speaking population. It was set up in May 1986 and its leader, Simon Conco, claims it has 132,000 members. It is avowedly pro-capitalist and opposes the international campaign for sanctions against South Africa. This has resulted in conflict with the independent unions, especially COSATU, and there have been many violent incidents between members of the two groups.
See also **Buthelezi; COSATU; *Inkatha*; sanctions**.

V

Vakomana (*Zimbabwe*)
This is a Shona word meaning 'the boys'. It was an affectionate term used by Zimbabweans to refer to the nationalist guerrillas who fought against white domination in their country.
See also **Shona; ZANLA; ZIPRA**.

verkrampte (*South Africa*)
Verkrampte is an Afrikaans word meaning 'ultra-conservative' or 'narrow'. It was originally applied to those on the far right of the National Party but now it generally applies to people to the right of the National Party, such as members of the official opposition, the Conservative Party.
See also ***verligte***.

verligte (*South Africa*)
Verligte is an Afrikaans word meaning 'enlightened' or 'progressive' and refers to more liberally-inclined members of the National Party and Afrikaners generally.
See also ***verkrampte***.

Verwoerd, Hendrik Frensch (*South Africa*)
Hendrik Verwoerd was born in Amsterdam, the Netherlands, in 1901 and moved to South Africa as a child with his father, a missionary. He was deeply influenced by ideas dominant in German academic circles during studies at the universities of Hamburg, Leipzig and Berlin during the 1920s, ideas on which he drew heavily in later years when he became the chief architect of apartheid. Having opposed the granting of asylum in South Africa to Jewish refugees from Germany, he subsequently, as editor of *Die Transvaler* during the Second World War, pursued a militantly anti-Semitic, and pro-Nazi line. Verwoerd stood as a National Party candidate in the 1948 general election, but was defeated. He was subsequently appointed to the Senate which he led, on behalf of the government, from 1950 to 1958, also filling the role of Minister of Native Affairs. He was elected to parliament in 1958, and became prime minister in 1959 after the death of J. G. Strydom.

Both as Minister of Native Affairs, and subsequently as prime minister, Verwoerd built a theoretical basis for apartheid, attempting to establish a moral justification (equal but separate treatment and facilities) for what had previously been (and remained) a crude system of white domination.

At the same time, however, he was responsible, with his Minister of Justice, John Vorster, for ever-increasing repression, for the shootings at Sharpeville, for the banning of the ANC and PAC, for imprisonment without trial, and for removing the last rights which some black South Africans had enjoyed in the subsequently all-white South African parliament.

Ten days after the declaration of the state of emergency in 1960, Verwoerd was shot twice in the head by a white farmer. He survived to make South Africa a republic and, in 1961, to take it out of the Commonwealth. In 1966, however, a second attempt was made on his life and he was stabbed to death by a parliamentary messenger in the House of Assembly. He was succeeded as prime minister by John Vorster.
See also **ANC; National Party; PAC; Sharpeville; Vorster**.

vigilante organisations (*Namibia*)

A number of white extreme right-wing vigilante-type organisations have emerged in Namibia to oppose perceived 'reforms' by the South African administration. The largest, *Aksie Red Blank Suidwes Afrika* (ARBSWA or Action Save White South West Africa) is a branch of the South African organisation ARBSA. Its leaders claim that it is a non-violent organisation, something which is disputed by its opponents. It is committed to the retention of white rule in Namibia and to the integration of Namibia into South Africa as a fifth province. Similar groups include the *Wit Weerstandbeweging*. (WWB or White Resistance Movement), based among German speaking residents of Namibia, and *Blank SWA* (White SWA).
See also ***Aksie Red Blank Suid-Afrika***.

Viljoen, Gerrit van Niekerk (*Namibia, South Africa*)

Cabinet Minister, ex-head of the *Afrikaner Broederbond* and National Party *verligte*, Gerrit Viljoen is a close ally of P. W. Botha and one of his party's leading intellectuals. Viljoen was born in 1926 into an academic family and was educated in South Africa, the UK and the Netherlands. A founder member of the student organisation, the ASB, he became Professor of Classics at the University of South Africa and, in 1967, Rector of the Rand Afrikaans University.

In 1974 Viljoen defeated Andries Treurnicht to become head of the *Afrikaner Broederbond*, a post he held until 1980 when he joined the cabinet as Minister of National Education. He was elected to parliament in 1981 as MP for Vanderbijlpark (Johannesburg). While serving in the education ministery Gerrit Viljoen took significant steps to bring education further under central control, but also instituted, for the first time, the principle of parity in pay between men and women. From 1978 to 1980 he served as South African Administrator General in Namibia. Viljoen's cabinet responsibilities were extended, in 1984, to include the ministry of Cooperation and Development.
See also **Administrator General; *Afrikaanse Broederbond*; ASB; FAK; National Party; Treurnicht**.

villagisation (*Tanzania*)
Villagisation was an early idea of President Julius Nyerere of Tanzania. It aimed to improve the economic life of the country but it also concerned moral development and argued that a good life was possible for those who lived in villages but not for those who lived on small family farms dispersed over wide areas. The president maintained that more modern equipment and better services could be obtained by larger groups. This lent an advantage to being in villages rather than being isolated in small family units.

However, things did not work out this way in practice and the village settlement policy was abandoned in 1966 when it became evident that only a few highly-motivated and politicised villages were succeeding. The idea of villagisation was combined with the philosophy of *Ujamaa* or 'familyhood' some years later in the concept of an *ujamaa* village. This was a village in which a philosophy of care similar to that found in an extended family operated.
See also **African Socialism; Nyerere; *Ujamaa***.

Voice of Namibia (*Namibia*)
Although all internal radio broadcasts to Namibians are controlled by South Africa through the SWABC, SWAPO has been running an externally-based service since the early 1960s, initially broadcasting from Tanzania. The Voice of Namibia is now broadcast from six African countries, Ethiopia, Congo, Tanzania, Zimbabwe, Angola and Zambia, through their national broadcasting services. It provides programmes in all Namibian languages and is judged by SWAPO as an important medium of mass communication with Namibians. It enjoys much popular support within the country.
See also **press; SWABC; SWAPO**.

Voice of the People (*Namibia*)
A component party in the National Convention of Namibia: see **National Convention of Namibia**.

volk (*South Africa*)
Volk is an Afrikaans word meaning the people or the nation. It is used in South Africa, however, to mean specifically the Afrikaner people or nation, those whites of mainly Dutch or French origin who settled the Cape and subsequently established the two Afrikaner republics of the Orange Free State and the Transvaal.
The notion of the Afrikaner *volk* was used by ideologues in the first half of the twentieth century to forge a political alliance on the basis of Afrikaner nationalism. Influenced by the national socialist ideas of the 1930s, Afrikaner nationalism was, from its start, anti-monopoly (and to some extent anti-capitalist) and racist. It was both a reaction to intense discrimination against Afrikaners by the dominant English-speaking whites, and an attempt through mutual support to escape the traps of poverty and proletarianisation to which a majority of Afrikaners were then subject. Although a significant proportion of Afrikaans-speakers

were black (or 'Coloured' in official parlance), they were excluded from the definition of *volk* because they did not fit with notions of 'racial purity' which were part of the nationalist ideology.
See also **Afrikaner; 'Coloured'**.

Voluntary Police Organisation (*Mozambique*)
An adjunct to PIDE: see **OPV**.

Voortrekker (*southern Africa*)
Originally an Afrikaans word but taken into the English language, a *Voortrekker* was an individual who took part in the Great Trek of 1835–37 or subsequent treks. Translated literally it means someone who pulls or travels before. *Trek*, another English word adopted from Afrikaans, has taken on the meaning of a journey by ox-wagon involving relocation of family and possessions. Both words originate from the Great Trek, a mass migration of settlers of Dutch extraction from the then Cape Colony to the interior of southern Africa. They were motivated by a mixture of economic factors and their resentment of British rule of the Cape which had replaced Dutch rule in 1814. Groups of such settlers, termed Boers at the time but subsequently known as Afrikaners, settled in what is now the Transvaal, the Orange Free State, Natal, and, in smaller numbers, in Namibia, Botswana, Angola, Zimbabwe and Mozambique.

The ideal of the Voortrekker as a fiercely independent, self-reliant Calvinist ready to sacrifice all for self-determination, is a potent symbol of Afrikaner nationalism. It has been invoked particularly by ultra-conservative elements, in recent years, to rally the *volk* to the standard of *ware* (genuine) apartheid.
See also **apartheid; *volk***.

Vorster, Balthazar Johannes 'John' (*South Africa*)
John Vorster was Prime Minister of South Africa from 1966 to 1979 when, for a period of eight months, he served as State President. Previously the Minister of Justice from 1961 to 1966, Vorster was elected National Party leader (and hence became prime minister) after the murder of H. F. Verwoerd in 1966.

In 1969 Vorster faced down a challenge from the right of his party and survived a split which resulted in the formation of the extreme right wing *Herstigte Nasionale Party*. He called a snap election in 1977 to capitalise on the collapse of the United Party, and achieved the largest National Party majority in the House of Assembly to that date. His own political demise followed shortly, however, when the Information Scandal forced first his resignation as prime minister and then as State President. Vorster died in November 1983, an embittered and forgotten man.
See also ***Herstigte Nasionale Party*; Information Scandal; National Party; United Party; Verwoerd**.

Walvis Bay (*Namibia*)

Walvis is an Afrikaans word meaning whale. The bay of this name is Namibia's only deep-water port. In 1878 it was annexed by Britain, along with 434 square miles surrounding the bay, and six years later it was placed under the administration of the Cape Colony. Following German acquisition of most of the rest of Namibia, an Anglo-German agreement to share the country was signed in July 1890. This secured Walvis Bay as a British possession.

At the formation of the Union of South Africa in 1911, Walvis Bay was transferred to South African control. In 1922, administration of the port became the responsibility of the mandated territory of South West Africa. On 3 August 1977, during negotiations with the United Nations on Namibian independence, South Africa unilaterally transferred the administration of Walvis Bay to the Cape Province. UN Security Council Resolution 435 demanded its re-integration into Namibia and warned South Africa against using Walvis Bay to prejudice the independence of Namibia.

90 per cent of Namibia's export trade goes through Walvis Bay and South Africa's retention of the port would allow it to control the primary trade route of an independent Namibia. South Africa has established major naval, air, infantry and armour military bases in the enclave.
See also **UN Security Council Resolution 435**.

Welensky, Roy (*Malawi, Zambia, Zimbabwe*)

Sir Roy Welensky was an influential leader of the Northern Rhodesian (later Zambian) white settler community. His driving passion was the creation of a Central African Federation incorporating Northern and Southern Rhodesia (Zambia and Zimbabwe) and Nyasaland (Malawi). However, the federation was violently opposed by emerging black nationalists in all three territories and since it was imposed against their will, it did not last long. Sir Roy Welensky was Minister of Transport in the federal government before he became Prime Minister in 1955. The Federation was dissolved in 1963.
See also **Central African Federation**.

Western Five (*Namibia*)

The five western countries, the UK, USA, France, Germany and Canada, which negotiated the basis for UN Security Council Resolution 435: see **Contact Group**.

Western settlement proposals for Namibia (*Namibia*)

The basis for a settlement in Namibia negotiated by the five western governments most involved in the country: see **UN Security Council Resolution 435**.

White Resistance Movement (*Namibia*)

Extremist white group opposed to any form of settlement in Namibia: see **vigilante groups**.

White SWA (*Namibia*)

Extremist white organisation resisting any settlement in Namibia: see **vigilante groups**.

Wiehahn Commission (*South Africa*)

The Wiehahn Commission of Inquiry was appointed in 1977 in the wake of the rapid growth of African unionisation to formulate state policy towards the new unions. It came under strong pressure from liberal business interests to find a basis for accommodation with the unions rather than simply opting for confrontation.

The commission reported in 1979, arguing that the state should insist on registration of unions as one way of submitting them to greater official control while not alienating foreign investment in the country. Prohibiting trade unionism, the report argued, would only serve to drive it underground. To become registered union constitutions had to be approved by a registrar and political activity was prohibited. A probationary period of 'provisional registration' had to be served and unions were absorbed into the bureaucratic industrial councils.

The Wiehahn Commission recommendations were accepted by the government (except that registration was not made compulsory) and enacted in the 1979 Industrial Conciliation Amendment Act (later named the Labour Relations Act). In 1981 the registrar was given power to intervene in all unions – registered or unregistered. Since then the debate among black unions for and against registration has been overtaken by events.

See also **CUSA; Food and Canning Workers' Unions; FOSATU; Riekert Commission**.

Williams, Ruth (*Botswana*)

In September 1948 in the United Kingdom Ruth Williams met and married Seretse Khama, Botswana's president from independence until his death in 1980. At the time of his marriage Khama was the hereditary chief of the important Bamangwato ethnic group. The marriage provoked a political crisis in Bechuanaland (later Botswana), and Seretse Khama was unable to return to his country with his wife until 1956, and then only because he renounced the chieftainship.

See also **Khama**.

Windhoek Massacre, 1959 (*Namibia*)
Massacre of black residents of Windhoek who were resisting relocation to an exclusively black suburb: see **SWAPO**.

wire (*Namibia*)
Among contract workers in Namibia, most of whom are recruited from the Ovambo area in the north, the system of working on contract is known as 'wire'. A 1972 denunciation of the system by striking Ovambo workers referred to it in the following way: 'because of this wrong and bad system this agreement has been changed into wire instead of the contract', and continued: 'We Ovambos do not want any improvement of or new name for wire. But we want to do away with wire, and to have a contract in the true meaning of the word.'
See also **contract labour; general strike of 1971**.

Wiriamu massacre (*Mozambique*)
One of the most notorious incidents in the Mozambican liberation war, the Wiriamu massacre involved the deaths of hundreds of peasants in Tete province. They were killed by Portuguese commandos because of their support for FRELIMO, and as an example to other rural people who might have had similar political views.
See also ***flechas***; **FRELIMO**.

witdoeke (*South Africa*)
Witdoeke is an Afrikaans term meaning 'white cloths' and refers to the 'fathers' – long-term, conservative residents of Crossroads squatter settlement near Cape Town, who wear white cloths on their arms for identification.
See also **'comrades'; Crossroads; 'fathers'; 'Russians'**.

Wit Kommando (*South Africa*)
The ***Wit Kommando*** (White Commando) is an underground, violent, right-wing organisation which specialises in the bombing and shooting of opponents of apartheid or others whom it regards as undermining apartheid, including so-called enlightened or 'verligte' members of the National Party. Among its targets was the leader of the Progressive Federal Party, Colin Eglin. Some of its members have been prosecuted and imprisoned for their activities.
See also ***Aksie Red Blank Suid-Afrika***; **AWB; Conservative Party; Progressive Federal Party**.

Wit Weerstandbeweging (*Namibia*)
An extremist white group resisting all changes in the system of apartheid in Namibia: see **vigilante groups**.

workers' councils (*Tanzania*)
Set up in 1969 with the aim of improving industrial relations in the country: see **Tanganyika Federation of Labour**.

WWB (*Namibia*) See ***Wit Weerstandbeweging***.

Xuma-Dadoo-Naicker Pact (*South Africa*)
The Xuma-Dadoo-Naicker Pact was signed in 1947 by Dr A. P. Xuma, president-general of the ANC, Dr Yusuf Dadoo, president of the South African Indian Congress and Dr G. M. Naicker, president of the Natal Indian Congress. This agreement formed the basis for political co-operation between the South African Indian Congress and the ANC, leading in the 1950s to the formation of the Congress Alliance.
See also **ANC; Congress Alliance; Dadoo; Naicker; Natal Indian Congress; South African Indian Congress**.

Yapwantha, Edward (*Malawi*)
Edward Yapwantha took over the leadership of the Malawi Freedom Movement (MAFREMO) in 1983 after the imprisonment of Orton Chirwa by President Banda. He has been active in reviving the movement and has the reported support of Zimbabwe, Mozambique and Tanzania. He studied law at the University of Zambia and at Queen's and McGill universities in Canada.
See also **Chirwa; Malawi Freedom Movement**.

Young Pioneers (*Malawi*)
The Young Pioneers evolved out of the Leage of Malawi Youth and was founded in 1963. In 1965 the movement was made part of the security services by legislative edict. This means that its members may not be arrested by the police without consultation. Members are trained in leadership, agricultural development and other skills, and all their activities are dominated by loyalty to Malawi's only political party, the MCP. A select group of the Young Pioneers is a para-military force directly controlled by President Banda. It acts as the president's bodyguard.
See also **MCP**.

Z

Zambezi River (*southern Africa*)
The Zambezi River is one of Africa's greatest rivers. It rises at Kalene Hill, Zambia and flows east to the Indian Ocean passing through the frontiers or territories of Angola, Zambia, Namibia, Botswana, Zimbabwe and Mozambique, some 2,200 miles. It forms the boundary between Zimbabwe and Zambia. Along the Zambezi lies Lake Kariba, the Kariba Dam and the Victoria Falls.
See also **Kariba**.

Zambezi Project (*Botswana*)
The Zambezi Project is a scheme to build a 1,200 km aqueduct to link the Zambezi River to the Witwatersrand in South Africa. The idea was put forward by Professor Gubrer Borchert of Hamburg University and it would tap the river at Kazangula, the point at which Zimbabwe, Zambia, Namibia and Botswana meet. All the pumpworks and the aqueduct would be in Botswana which would be able to use a third of the water along the way. This scheme, together with the Highland Water Project (involving Lesotho), would be of considerable benefit to South Africa which has great need of water, particularly in the Witwatersrand industrial area.

The Zambezi Project is bound to be a matter of political controversy as it represents another economic link with South Africa – just the kind of economic pattern that the countries of SADCC (the Southern African Development Coordination Conference), including Botswana and Lesotho, are trying to break.
See also **Highland Water Project**.

Zambia

Official title: Republic of Zambia
Head of state and government: President Kenneth David Kaunda
Area: 752,614 sq. km
Population: 6,650,000 (1985)
Capital: Lusaka
Official languages: English (Bemba, Lozi, Lunda, Tomga, Nyanja also spoken)
GDP per capita: US$269 (1985)
Major exports: copper (80–90 per cent of total value)

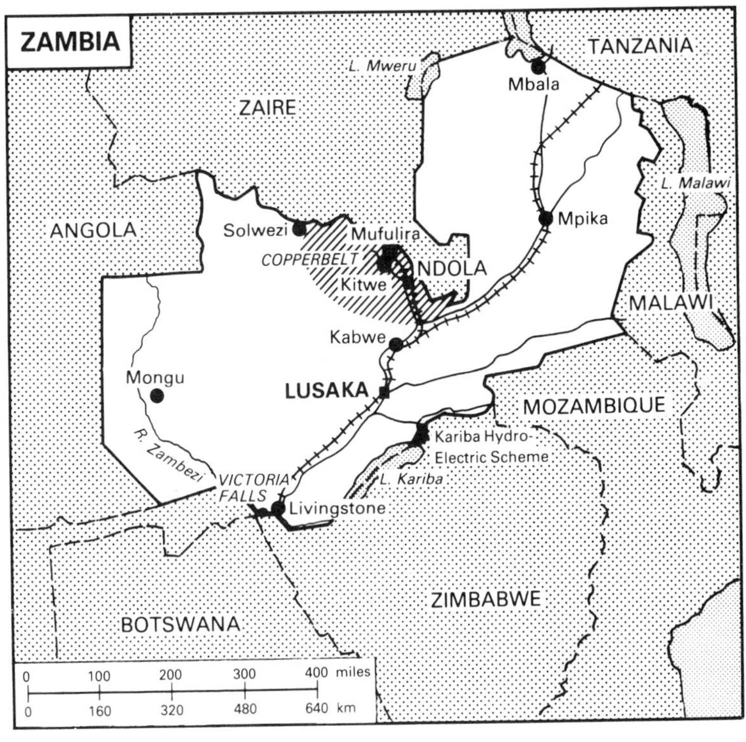

Currency: Kwacha (K)=100ngwee; K8 per US$ (July 1987)
Political parties: United National Independence Party (UNIP) – the sole legal party since 1972.

Zambia lies in the heart of south-central Africa, a land-locked country of irregular shape. The pedicle of Zaire intrudes into its centre from the north dividing the Copperbelt from the north-east of the country. Zambia's other neighbours are Tanzania to the north-east, Malawi and Mozambique to the east, Angola to the west, Namibia's Caprivi Strip, Botswana and Zimbabwe to the south. The centre of activity in the country surrounds the railway line extending south through the Copperbelt, Lusaka and to the Victoria Falls on the border with Zimbabwe. Zambia is very much a 'frontline' state in that it is dependent on South African-dominated road and rail links to the sea for its imports and exports. The 'frontline' states are trying to construct alternatives through the Southern African Development Coordination Conference (SADCC). Zambia's only major export is copper and the economy has been badly affected by falling prices in recent years.

At present (1987) Zambia is in the grip of an economic crisis owing to the dramatic fall in commodity prices in recent years. These are at an historic low and Zambia is almost totally dependent on copper and cobalt which make up some 95 per cent of its export earnings. The government has introduced an austerity programme under the auspices of the International Monetary Fund and there have been serious riots as a result of attempts to end basic food subsidies.

Zambia – formerly the British colony of Northern Rhodesia – became independent on 24 October 1964. The country has been led by President Kenneth Kaunda ever since. His ruling United National Independence Party is the sole legal political party. President Kaunda is one of Africa's most respected leaders and Zambia plays an important role in African politics. For instance, it supported the guerrillas of ZAPU, one of Zimbabwe's nationalist movements, during the long liberation war at considerable economic cost to itself as it incurred the wrath of the white-ruled south. It has suffered since for its support of SWAPO and the ANC of South Africa. In May 1986 South Africa launched an air attack on an alleged ANC guerrilla base near Lusaka resulting in two deaths, and a year later infiltrated irregular troops who killed five Zambians alleging that they were ANC guerrillas. The president has been arguing strongly for the imposition of economic sanctions against South Africa for a number of years.

Zambia's present constitution was introduced on 25 August 1973. The President is Head of State and Commander-in-Chief of the armed forces and is elected at the same time as the legislature, the National Assembly, for a five-year term. In October 1983 President Kaunda was elected unopposed, gaining 93 per cent of the votes cast compared with 81 per cent in 1978. The National Assembly consists of 135 members, 125 of whom are elected and ten of whom are chosen by the president. Some 766 candidates stood in the 1983 elections making these the most heavily contested since Zambia became a one-party state. There is a Central Committee for UNIP, the only legal party, of 25 members which has more powers than the cabinet. There is also a House of Chiefs which is an advisory body.

See also **austerity programme; Kaunda; UNIP**.

Zambia African National Congress (*Zambia*)

Kenneth Kaunda formed the Zambia African National Congress (ZANC) in 1958 after walking out of a meeting of the Northern Rhodesian African Congress (later the ANC) together with some colleagues. The split was a challenge to the leadership of Harry Nkumbula. Kenneth Kaunda, Simon Kapwepwe and Manu Kayambwa Sipalo, the ZANC's main leaders, felt that a more disciplined and highly organised nationalist movement led by people of great commitment was necessary to achieve independence.

The ZANC aimed to supplant the ANC as the main nationalist party in the country, and Kenneth Kaunda made a good impression on other African leaders in Ghana at the All-African People's Conference. Harry Nkumbula, in contrast, was often isolated at the conference and seemed to many there to be less radical than the ZANC leaders. The ZANC also boycotted an election supported by the ANC in 1959, and, as the ZANC's

campaign became effective, the party was banned, and Kenneth Kaunda and other leaders put in prison. However, the government was unable to suppress the militant support the party had achieved and the ZANC continued under another name, UNIP. UNIP was soon to become Zambia's main nationalist party and was led to independence and government by Kenneth Kaunda.
See also **Kapwepwe; Kaunda; Nkumbula; Northern Rhodesian African Congress; UNIP**.

Zambia Congress of Trade Unions (*Zambia*)
The Zambia Congress of Trade Unions (ZCTU) was formed to take over from the disintegrated United Trades Union Congress by an act of parliament in 1964, and the government hoped all trade unions would join. Most of the small unions were happy to oblige but the large mineworkers' union (now the Mineworkers' Union of Zambia) was wary of government control over its affairs, and it did not affiliate until 1966.

There was considerable potential for disagreement between governments and trade unions in the post-independence period. Governments were concerned to build and develop nations, and the trade unions were concerned above all to negotiate well on behalf of their members. This clash of interests persists in many countries, Zambia included, and in 1981 several ZCTU members were suspended from UNIP and its chair, Frederick Chiluba, was detained for some months. The government has, however, been trying to achieve a closer relationship with the trade union movement, recognising it as a potential alternative power base in the country. The ZCTU has 18 affiliated unions with some 400,000 members. Its chairperson is Frederick J. Chiluba.
See also **Chiluba; Mineworkers' Union of Zambia; Trade Union Congress; UNIP**.

Zambia Industrial and Mining Corporation (*Zambia*) See **ZIMCO**.

ZANC (*Zambia*) See **Zambia African National Congress**.

ZANLA (*Zimbabwe*)
ZANLA (the Zimbabwe African National Liberation Army) was the military wing of ZANU, one of Zimbabwe's main nationalist parties. Its first group of five guerrillas went to China for training in September 1963. Most of ZANLA's international support came from China in contrast to the Soviet Union's support for ZIPRA, the armed wing of ZAPU, Zimbabwe's other main nationalist group. A ZANLA incursion into Southern Rhodesia (later to become Zimbabwe) on 28 April 1966 marked the official start of the war and is now commemorated in Zimbabwe as *Chimurenga* or 'freedom struggle' day. At this Battle of Sinoia seven guerrillas died in a fierce encounter with white Rhodesian troops. Afterwards ZANLA's strategy was reconsidered and any conventional fighting shunned as the guerrillas sought to move among the masses, politicising them and gaining their support. In 1972 an important new stage of the war began with a ZANLA attack on the white Altena Farm in Rhodesia;

the first white farm to be attacked since 1966. But the war did not really escalate until 1976 when major guerrilla incursions from newly-independent Mozambique began. ZANLA's relationship with Mozambique's FRELIMO government was vital to its success in the Zimbabwean war.

To cement the political alliance of the Patriotic Front formed between ZANU and ZAPU in 1976, a military alliance called ZIPA (the Zimbabwe People's Army) was formed, instigated by President Julius Nyerere of Tanzania. This was led by Rex Nhongo, the most senior ZANLA fighter at liberty. However, problems developed and some ZIPA leaders started to assume a political role. They were effectively stopped by Mozambique's president, Samora Machel, who later detained some of them in Mozambique. ZIPA subsequently declined in importance.

ZANLA's success as a fighting force is widely seen as the main reason for Robert Mugabe's victory in the 1980 elections. It played a more active role than ZIPRA in the war and controlled the crucially-important Mozambique front. After independence in 1980, ZIPRA and ZANLA as well as the other armies were combined to form the Zimbabwean national army.

See also ***Chimurenga*****; Nhongo; Patriotic Front; Tongogara; ZANU; ZAPU; ZIPA; ZIPRA**.

ZANU (*Zimbabwe*)

In August 1963 Ndabaningi Sithole, Leopold Takawira and Robert Mugabe formed ZANU (the Zimbabwe African National Union). They took this decision following a split with ZAPU (the Zimbabwe African People's Union) and its leader, Joshua Nkomo. The split was based on a disagreement with Joshua Nkomo's strategy of international diplomacy and on what they saw as a willingness by ZAPU's leader to compromise too readily on his principles. They also did not agree with Joshua Nkomo's decision to move ZAPU's headquarters out of Zimbabwe.

ZANU's aims were the same as ZAPU's – majority rule. And although their support came from different areas (ZANU's from the south, east, midlands and many intellectuals) neither was ethnically based. Nevertheless months of faction fighting in Zimbabwe followed the split, and in 1964 ZANU was banned and Ndabaningi Sithole and Robert Mugabe arrested. ZANU established itself in exile in Lusaka, Zambia, and began to train its guerrillas. Its army, ZANLA, started the first serious guerrilla incursion into white Rhodesia in April 1966. But these raids were not very successful, and a specially-commissioned study showed that more training and political preparation for armed struggle were needed. ZANU focussed on this in the early 1970s.

In 1968 Ndabaningi Sithole's position as president of ZANU was challenged. He urged his supporters in a letter from prison to assassinate Ian Smith, the rebel Rhodesian leader, and two of his ministers. The letter was intercepted and Ndabaningi Sithole put on trial. He then issued a statement denouncing violence which his fellow-detainees and ZANU guerrillas saw as treachery. A struggle for leadership developed among the leaders in Que Que prison in Rhodesia and in 1970 six executive members voted for Robert Mugabe as the new secretary-general. Herbert Chitepo had been given full powers by the ZANU central committee to run the party's external operations and in 1966 he was named as the permanent head of the War Council.

ZANU was influenced by Maoist ideology and helped by Chinese supplies of arms. In 1974, under pressure from the 'frontline' states, ZANU signed a unity agreement with ZAPU, FROLIZI (another nationalist group) and the African National Council led by Bishop Muzorewa, in which all agreed to co-operate while each party retained its own identity.

In 1975 Herbert Chitepo was assassinated in Lusaka. This caused serious dissension within the nationalist movement and between ZANU and the 'frontline' states. The Zambians arrested many ZANU officials and closed the ZANU offices in Lusaka, blaming party rivalry for his death. It was decided that ZANU leaders should disperse and Josiah Tongogara and four officials went to Mozambique, Rex Nhongo and others to Tanzania. ZANU then operated from Mozambique sending its ZANLA guerrillas into action against the Rhodesian regime, and opening a new phase of the war.

In 1976 ZANU and ZAPU formed the Patriotic Front, an alliance fostered by the 'frontline' states, to represent the nationalists at the Geneva Conference in October of that year. Both ZAPU and ZANU opposed the internal settlement and Bishop Muzorewa's prime ministership in 1979. They fought the 1980 elections for a free and independent Zimbabwe separately and ZANU won a majority of the votes. Robert Mugabe became prime minister with 57 seats in Parliament. ZANU again fought and won an election in 1985. Regional rivalries in Zimbabwe have become more marked since independence and consolidated around ZANU and ZAPU, with ZANU having mainly Shona support and ZAPU support from Ndebele-speakers. This has led to violence and allegations of brutality and intimidation against the ruling ZANU. Robert Mugabe had stated his intention to move Zimbabwe towards a one-party state and in 1987, after years of difficult negotiations, the two parties united under the name ZANU-PF.

See also **African National Council; Chitepo; FROLIZI; 'frontline' states; Geneva conference; Mugabe; Muzorewa; Ndebele; Nhongo; Nkomo; Patriotic Front; Shona; Sithole, Ndabeningi; Takawira; Tongogara; ZANU-PF; ZANLA; ZAPU**.

ZANU-PF *(Zimbabwe)*

Following the decision by ZANU to fight the 1980 elections separately from ZAPU, its partner in the Patriotic Front, it was renamed ZANU-PF to distinguish it from the very small, breakaway ZANU-Sithole party: see **ZANU**.

Zanzibar *(Tanzania)*

Zanzibar is an island off the Tanganyika coast which, together with the island of Pemba, joined with Tanganyika to form the United Republic

Z

of Tanzania in 1964. Its population in the 1978 census was estimated at 110,669.

Zanzibar became a British protectorate in 1890 under the Sultan of Zanzibar, Sayyid Jamshid ibn Abdullah. The Zanzibar Sultanate included the island of Pemba and several islets. Most of the land was owned by Arabs while the African majority held plots or tended to work on Arab plantations. In 1960 a new constitution was approved for the country but conflicts developed among the main political parties, the Afro-Shirazi Party (ASP), respresenting mainly Africans, the Zanzibar and Pemba People's Party (ZPPP), representing mainly Zanzibari Arabs, and its offspring, the Zanzibar Nationalist Party (ZNP). In 1963 the ZNP-ZPPP coalition won an election and later that year Zanzibar became independent. In 1964 a revolutionary group under John Okello overthrew the Sultan's government. Later that year, after some internal upheaval, a union was signed between Zanzibar's president, Sheik Abeid Karume and Tanganyika's president, Julius Nyerere.

Despite the union and sharing the same political party, *Chama cha Mapinduzi*, the islands have separate elections and Zanzibar has its own president and revolutionary council concerned specifically with its affairs. A new constitution for Zanzibar was introduced in 1985. On 15 October 1985 Idris Abdul Wakil was elected unopposed as President of Zanzibar and thus as Second Vice-President of Tanzania, obtaining 61 per cent of the votes. A number of islanders, particularly in Pemba, were reportedly in favour of a different candidate, Seif Shariff Hamad, and this accounted for the relatively low percentage of votes cast for the new president.
See also **Afro-Shirazi Party; Karume; Nyerere; Okello; Pemba; Tanganyika; Tanzania; Zanzibar and Pemba People's Party; Zanzibar Nationalist Party**.

Zanzibar and Pemba People's Party (*Tanzania*)

The Zanzibar and Pemba People's Party or the ZPPP was formed in 1959 by a group of young Africans who left the Afro-Shirazi Party accusing it of corruption. It remained a minority party although at first its membership grew rapidly.
See also **Afro-Shirazi Party; Zanzibar; Zanzibar National Party**.

Zanzibar National Party (*Tanzania*)

The Zanzibar National Party or the ZNP was founded in 1956 with the support of a number of Arab intellectuals and some African peasants. Its leadership was mainly Arab and it sought self-government as soon as possible. Its rival, the Afro-Shirazi Party, opposed this, wanting to allow time for Zanzibar's African majority to develop further. Nevertheless, in 1963 Zanzibar became independent under a ZNP-ZPPP coalition government. One of the leading figures in the ZNP was Abdul Rahman Babu, its general secretary; Ali Muhsin was its president.
See also **Afro-Shirazi Party; Babu; Zanzibar and Pemba People's Party**.

ZAPU (*Zimbabwe*)

In December 1961 the National Democratic Party was banned and ZAPU

(Zimbabwe African People's Union) formed in its place. The party was led by Joshua Nkomo and had the same aims as its predecessor – majority rule. ZAPU was also banned, in 1962 and many of its leaders detained. Loyal supporters set up the People's Caretaker Council inside the country and this continued to be known as ZAPU.

Joshua Nkomo at this time was travelling extensively trying to persuade Britain to intervene and impose majority rule. Eventually he returned to Zimbabwe and was restricted for three months to a village south of Bulawayo. During this period he decided that a government-in-exile would be more effective so he moved the ZAPU executive to Dar es Salaam in Tanzania. Ndabaningi Sithole, who had been outside Zimbabwe at the time ZAPU was banned, had already set up emergency headquarters there. Many ZAPU members disagreed with their leader's decision to move the party outside the country and believed the struggle should be based inside Zimbabwe. Joshua Nkomo suspended his critics – Ndabaningi Sithole, Robert Mugabe and Herbert Chitepo – from the party. Sithole returned to Zimbabwe and set up ZANU as a rival organisation.

ZAPU and ZANU had very much the same aims. Both sought majority rule, external support and lobbied the British government. Both set up external bases in Africa and recruited members for guerrilla training outside the country (ZAPU's military wing was ZIPRA). At this stage both parties drew support from the same Shona-speaking areas. ZAPU's main support came from Salisbury (then the capital) and Bulawayo. However, rivalry between ZANU and ZAPU grew into violence. Both the People's Caretaker Council and ZANU were banned on 26 August 1964 and prominent nationalists were detained. Joshua Nkomo was sent in April to a camp where he remained for ten years. From this date onwards the parties continued their operations from Lusaka, Zambia. James Chikerema became acting-president of ZAPU during Joshua Nkomo's detention. The party worked together with the African National Congress of South Africa, training guerrillas.

In 1971 the African National Council (ANC) was formed by ZAPU and ZANU and Bishop Abel Muzorewa appointed leader. It was set up with a single purpose – to mobilise African opposition to the proposals of the Anglo-Rhodesian agreement, designed to settle the question of Rhodesia's rebellion against the British Crown. The ANC was successful and Zimbabweans decisively rejected the proposals.

In 1974 Ian Smith released many nationalist leaders from detention. A unity agreement was signed between them in Lusaka, under an enlarged ANC consisting of ZAPU, ZANU, FROLIZI and the UANC. However, nationalist rivalries persisted much as before. There was an attempt to form a ANC army to use as a stronger background to talks with the Smith regime which were taking place at this time. Ian Smith favoured reaching some separate agreement with Joshua Nkomo, but negotiations failed.

On 9 October 1976 Joshua Nkomo and Robert Mugabe formed the Patriotic Front (PF), an alliance between ZAPU and ZANU, encouraged and fostered by the 'frontline' states. The PF was to represent the nationalists at the Geneva Conference in October 1976. However, ZAPU and ZANU maintained their separate identities and armies within the alliance. They both opposed the internal settlement which led to the Zimbabwe-Rhodesian government of 1979, and fought the 1980 elections for a free and independent Zimbabwe as two separate parties. ZAPU lost these elections and became the main opposition party inside Zimbabwe. It won 20 seats in the new parliament.

Subsequently, hostility between the two parties grew and ZAPU drew its support increasingly from the south-west of the country. In the 1985 elections it won 15 seats. Robert Mugabe began moving the country towards a one-party state and there were protracted negotiations for a merger between the two parties. This was achieved in 1987, and in 1988 ZAPU rejoined the government.

See also **African National Council; Anglo-Rhodesian agreement; Mugabe; Muzorewa; National Democratic Party; Nkomo; Patriotic Front; People's Caretaker Council; ZANU; Zimbabwe-Rhodesia; ZIPRA**.

ZCTU (*Zambia*) See **Zambia Congress of Trade Unions**.

Zimbabwe

Official title: Republic of Zimbabwe
Head of state: President Robert Mugabe
Area: 390,759 sq. km
Population: 8,174,892 (1985)
Capital: Harare
Official languages: English (Chisona, Sindabele widely spoken)
GDP per capita: US$595 (1984)
Major exports: tobacco (19–20 per cent of total value); ferro-alloys (10–11 per cent)
Currency: Zimbabwe Dollar (Z$)=100 cents; Z$1.69 per US$ (July 1987)
Political parties: The main political parties are ZANU-PF (the ruling party); ZAPU; United African National Council; Republican Front

Zimbabwe is a landlocked country on the plateau between the Zambezi and Limpopo rivers. It is bordered by Zambia to the north-east, Mozambique to the north-west and west, Botswana to the south-west and South Africa to the south. Most of the countryside is undulating with some dramatic rock features like the Great Dyke, but the eastern highlands along the border with Mozambique are moutainous. Zimbabwe has rich mineral deposits, variable soil and poor rainfall. Most of its population is still engaged in agriculture but the country had a higher level of industrialisation than any other black African country at independence.

Zimbabwe had a turbulent transition to majority rule. The British colony of Southern Rhodesia was ruled by the rebel white prime minister, Ian Smith, for fifteen years after a unilateral declaration of independence on 11 November 1965. During this period the liberation armies fought a

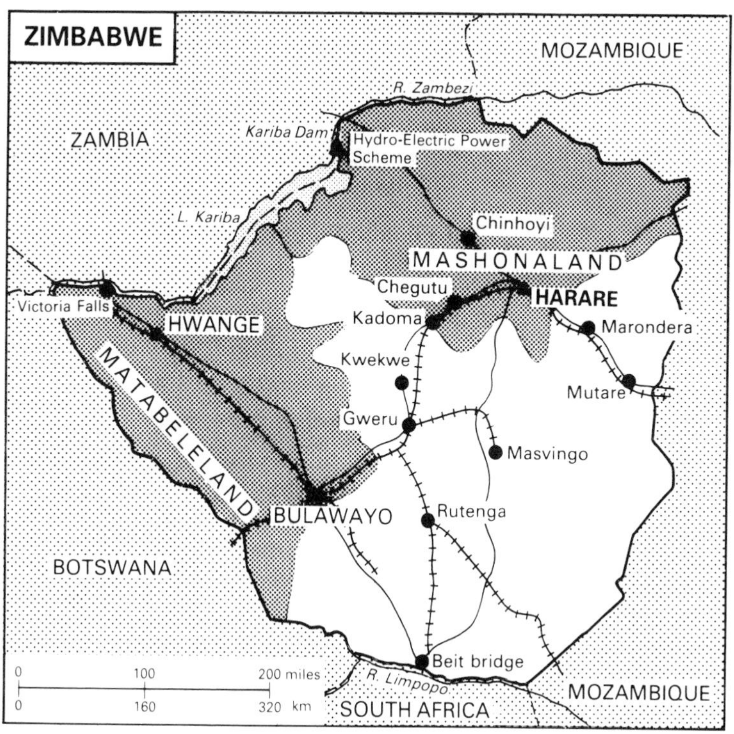

war of independence which they won on 18 April 1980 in the independence election. Robert Mugabe, the leader of ZANU-PF, became the first Prime Minister of the Republic of Zimbabwe.

The Zimbabwean constitution provides for a president who is also the head of state and the commander-in-chief of the armed forces. The government is run by a prime minister who can command the support of the majority of members of the House of Assembly. He chooses his cabinet which is then appointed by the president. Parliament consists of a 40-member Senate and a 100-member House of Assembly. The Senate is elected indirectly in various ways and the House of Assembly by univeral adult suffrage. There were special provisions for protecting white voting rights in the 1980 constitution. These provided for ten senators to be elected by 20 members of the House of Assembly who had themselves been elected on the separate white electoral roll. The government abolished the white minority seats as soon as it was able to do so legally, in 1987. A new constitution was due to come into effect in 1988, establishing an executive presidency.

Since independence Robert Mugabe's ZANU-PF party has won elections with overwhelming majorities. In 1985 it won 63 seats of the total 80 black seats in the House of Assembly. There was some conflict with Joshua Nkomo's Patriotic Front (still commonly known as ZAPU) but

this now seems to be resolved and the parties united in December 1987. The aim is to introduce a one-party state sometime in the future.
See also **Mugabe; Muzorewa; Nkomo; Patriotic Front; Republican Front; Smith; UANC; ZANU; ZAPU**.

Zimbabwe African National Liberation Army (*Zimbabwe*) See **ZANLA**.

Zimbabwe African National Union (*Zimbabwe*) See **ZANU**.

Zimbabwe African People's Union (*Zimbabwe*) See **ZAPU**.

Zimbabwe Congress of Trade Unions (*Zimbabwe*)
The inaugural congress of the ZCTU (Zimbabwe Congress of Trade Unions) took place in February 1981. It was formed as a new umbrella organisation designed by the ruling party, ZANU-PF, to bring the divided trade union movement together. The party hoped that it would gain more influence over the union movement and the unofficial workers' committees. During 1980 Zimbabwe had experienced the biggest strike wave since the early 1960s in terms of lost days of production. The strikes represented a build up of frustration by the country's workers who sought to bypass the industrial conciliation machinery and protest against reactionary, often racist unions with direct appeals to government. In the process union leaders had lost control over many of their members, so it was hoped that the ZCTU would restore some of their influence.

The new organisation was led by the late Albert Mugabe, the prime minister's brother. Unions affiliated to previous federations such as the ZTUC (Zimbabwe Trade Union Congress), the ATUC (the African Trade Union Congress), the ZFL (the Zimbabwe Federation of Labour), the NATUC (the National Trade Union Congress), the ZACU (the Zimbabwe African Congress of Unions) and the Trade Union Congress of Rhodesia came together to form the ZCTU. Since its formation almost all industrial trade unions in the country have become affiliated to the ZCTU. Its current president is Jeffrey Mutandare and its secretary-general, Anselm Chitehwe.

Before independence three types of unions existed: the white-dominated, multiracial unions, like the miners and the printers, the African industrial unions like the clothing and textile unions and the white staff associations. The ZCTU created a new kind of union movement under black control in which some of the existing unions were strengthened and some new unions emerged, like domestic workers and agricultural workers. In 1980 at the time of independence there were some 85,000 union members including staff association members and other non-industrial workers; the number of industrial workers at that time is estimated at about 30–40,000. By 1985 the membership of industrial trade unions was estimated as 150,000. If white-collar workers such as the nurses were included, the number would be nearer 200,000 members, or 18 per cent of the waged work force.
See also **ZANU**.

Zimbabwe People's Army (*Zimbabwe*) See **ZIPA**.

Zimbabwe People's Revolutionary Army (*Zimbabwe*) See **ZIPRA**.

Zimbabwe-Rhodesia (*Zimbabwe*)
Zimbabwe-Rhodesia was the name given to the colony of Southern Rhodesia by Ian Smith in 1979, following the Salisbury Agreement signed on 3 March 1978. This created a transitional government led by Ian Smith, the leader of the Rhodesian regime, Bishop Abel Muzorewa, Ndabaningi Sithole and Chief Jeremiah Chirau. On 1 June 1979 Bishop Muzorewa became the Prime Minister of Zimbabwe-Rhodesia.
See also **Chirau; internal settlement; Muzorewa; Salisbury Agreement; Sithole, Ndabaningi; Smith; transitional government**.

Zimbabwe United People's Organisation (*Zimbabwe*)
In January 1977 Ian Smith, the leader of the rebel Rhodesian regime, advised Chief Ndiweni and Chief Chirau to resign from the internal government. He wanted them to form a new political party to create support for a negotiated settlement with the Rhodesian regime. Thus the Zimbabwe United People's Organisation (ZUPO) was an attempt by Ian Smith to undercut the nationalists.

The party claimed to support majority rule and offered a programme of reform, but Chief Chirau stated that he did not believe in 'merely the counting of heads'. ZUPO did not get much support because of its suspected connections with Ian Smith. It was one of the four black parties taking part in the internal elections in 1979. By then Chief Ndiweni had broken away to form the United National Federal Party after complaining of unfairness to the Ndebele people. The United National Federal Party won nine seats and ZUPO lost badly in the election winning no seats at all.
See also **Chirau; Smith**.

ZIMCO (*Zambia*)
ZIMCO (the Zambia Industrial and Mining Corporation) was set up by the government in 1973 and took over control of the copper mines from the South African multinational Anglo American Corporation. It was designed to sell the copper produced in the Copperbelt. ZIMCO is the largest company in independent black Africa with net assets of over K2,000 million. Following reorganisation in December 1978, ZIMCO is the holding company for many Zambian companies as well.
See also **Anglo American Corporation**.

ZIPA (*Zimbabwe*)
ZIPA (the Zimbabwe People's Army) was formed in January 1976. It included cadres from ZANLA and ZIPRA, the military wings of the two main nationalist organisations, and was led by an 18–member military committee divided between the two forces. The overall commander was Rex Nhongo (the most senior ZANLA fighter at liberty) and the second

position of political commissar was filled by Alfred 'Nikita' Mangena, the ZIPRA commander.

ZIPA was formed largely under pressure from the 'frontline' states who also helped bring about the Patriotic Front, the political alliance between the two main Zimbabwean parties, ZANU and ZAPU. They were concerned to unify the guerrillas before the negotiating process on the future of the country. By the end of 1975 Ian Smith's attempts at detente with Joshua Nkomo had dramatically slowed down the war, and it was ZIPA which renewed the fighting. However, a crisis developed when those in command of ZIPA sought to assume a more political role. They were effectively stopped by President Samora Machel of Mozambique who later detained several of the leaders in Mozambique. Robert Mugabe, ZANU's leader, sent Josiah Tongogara from the Geneva Conference on the future of Southern Rhodesia to take over control of ZIPA.

See also **'frontline' states; Mugabe; Nhongo; Nkomo; Patriotic Front; Smith; Tongogara; ZANLA; ZANU; ZAPU; ZIPRA**.

ZIPRA (*Zimbabwe*)

ZIPRA was the military wing of ZAPU, one of Zimbabwe's main nationalist organisations. It received most of its weapons and training from the Soviet Union and was based in Zambia. Throughout the guerrilla war against the rebel Rhodesian regime ZIPRA operated mainly from Zambia and ZANLA from Mozambique. In May 1965 ZIPRA guerrillas launched their first raid into Rhodesia. Many of ZIPRA's fighters were killed in raids during the 1960s. However, overall, ZIPRA was less involved in the fighting than ZANLA, the military wing of ZANU.

Differences in ZIPRA emerged in the late 1960s between James Chikerema, the head of defence, leader-in-exile and vice-president, and Jason Moyo, the second-in-command in the defence department. The disagreement was mainly over the military failures of the late 1960s when many ZIPRA soldiers died in unsuccessful ventures, and over the autocratic style of James Chikerema's leadership. James Chikerema left ZAPU and formed a new party, FROLIZI, in 1972.

In 1976 major guerrilla raids from Mozambique marked a new phase in the war. At the same time the Patriotic Front, an alliance between ZAPU and ZANU, was formed. To cement this there was an attempt to create an alliance between the Zimbabwean guerrilla armies, ZANLA and ZIPRA, called ZIPA. The first ZIPA military committee was appointed on 12 November 1975; ZIPRA and ZANLA units remained separate within the new umbrella force. ZIPA declined as a significant force after divisions and the imprisonment of some of its leaders in Mozambique. After independence in 1980 the guerrilla armies of ZIPRA and ZANLA, as well as the white Rhodesian army, were amalgamated to form the new Zimbabwean national army.

See also **Chikerema; Moyo; Nhongo; ZANLA; ZAPU; ZANU; ZIPA**.

ZNP (*Tanzania*) See **Zanzibar Nationalist Party**.

Zomba (*Malawi*)
Zomba was the capital and administrative centre of the British protectorate of Nyasaland. It remained the capital of Malawi until 1975, when the seat of government was moved to Lilongwe.
See also **Lilongwe**.

ZPPP (*Tanzania*) See **Zanzibar and Pemba People's Party**.

ZUPO (*Zimbabwe*) See **Zimbabwe United People's Organisation**.

Zvobgo, Eddison Jonas Mudadirwa (*Zimbabwe*)
Dr Eddison Zvobgo was a minister in Zimbabwe's government from independence until recently. His final position was Minister of Justice, Legal and Parliamentary Affairs. He was a member of the ZANU-PF Central Committee and the Chairman of the ZANU-PF Masvingo Province for many years. In December 1986, however, he was dismissed from his party position as chair after the prime minister said that Dr Zvobgo had used 'obscene' and 'tribalistic' language to a superior in the party at a funeral earlier in the year.

He was founder member of ZANU and its deputy secretary-general until 1977. He is one of the country's most outspoken and popular politicians, and was on the left of the political spectrum within the ruling party.
See also **ZANU**.

Zwane, Ambrose Phesheya (*Swaziland*)
Dr Zwane was the head of the Ngwane National Liberatory Congress (NNLC). He has led the party from its formation after a split from the Swaziland Progressive Party in 1963. The NNLC, along with all other parties, was banned by the king in 1973.
See also **'constitutional crisis' of 1973; Ngwane National Liberatory Congress; Swaziland Progressive Party**.

List of entries by country

Angola
Active Revolt
aldeamentos
ALIAZO
Alves, Nito
Alvor Agreement
AMANGOLA
Angola
assimilado
ASSOMIZO
B52s
bairro
Benguela railway
Cabinda
Chipenda, Daniel
churches
CIR
CONCP
contratados
coup attempt, 1977
Cuban troops
Cunene River Hydro-Electric Scheme
DIAMANG
DISA
Do Nascimento, Lopo
Dos Santos, José Eduardo
Eastern Revolt
ELNA
EPLA
FAPLA
FDLA
FLEC
FNLA
forced labour
GRAE
indigena
JMPLA
Kassinga massacre
Katiusha rocket
Kifangondo marshes
Lára, Lúcio
linkage
Lusaka Agreement between South Africa and Angola, 1984
MDIA
mestiço
MFA
MIA
MINA
Mingas, Saydi
MNA
MPLA
MPLA–PT
musseques
Nakuru Agreement
Neto, Agostinho
Ngwizako
Nitistas
ODP
OMA
PDA
PIDE
Pinto de Andrade, Joaquim and Mario
PLUA
Poder Popular
retornados
Roberto, Holden
Savimbi, Jonas
settlers
South African invasions
Transitional Government
UEA

Mozambique

Namibia

SWAPO of Namibia
SWAPO-D
SWAPO Elders' Council
SWAPO Women's Council
SWAPO Youth League
SWATF
Toivo ja Toivo, Andimba
trade unions
Transitional Government of National Unity
UN Commissioner for Namibia
UN Council for Namibia
UN Security Council Resolution 385
UN Security Council Resolution 435
UNIN
United National Party of South West Africa
UNTAG
vigilante organisations
Viljoen, Gerrit van Niekerk
Voice of Namibia
Walvis Bay
wire

South Africa

Abdurahman, Abdulla
Africanisation
Africanism
African National Congress of South Africa
African People's Organisation
Afrikaanse Handelsinstituut
Afrikaner Broederbond
Afrikaner nationalism
Afrikaner Party
Afrikaner Volkswag
Aggett, Neil
Aksie Eie Toekoms
Aksie Red Blank Suid–Afrika
Anglo
Anglo American Corporation
apartheid
ASB
ASSOCOM
AWB
AZACTU
Azania
AZAPO
AZASM
AZASO
baasskap
banning order
Bantu Education Act
bantustans
Biko, Stephen
'Black'
Black Consciousness Movement
'blackjacks'
Black Parents' Association
Black People's Convention
Black Sash
'black spots'
Bloedsap
Boesak, Allan
BOSS
Botha, Pieter Willem 'P.W.'
Botha, Roelof Frederik 'Pik'
Botha, Thozamile
Buthelezi, Mangosuthu Gatsha
Camay, Phirowshaw
Chamber of Mines
churches
'Coloured'
Coloured People's Congress
Committee of Ten
Committee on South African War Resistance
'comrades'
Congress Alliance
Congress of Democrats
Congress of the People
Congress Youth League
conscription
Conservative Party
Constellation of Southern African States
constitution of 1983
COSAS
COSATU
Crossroads
Cunene River Hydro-Electric Scheme
CUSA
CUSA – AZACTU

southern Africa

Swaziland

Tanzania

Zambia

Zimbabwe

ZANU–PF
ZAPU
Zimbabwe
Zimbabwe Congress of Trade Unions
Zimbabwe–Rhodesia
Zimbabwe United People's Organisation
ZIPA
ZIPRA
Zvobgo, Eddison Jonas Mudadirwa

Further Reading

Angola

Martin, P. M. *Historical Dictionary of Angola* (London and Metuchen, The Scarecrow Press Inc., 1980)
Wolfers, M. and J. Bergerol *Angola in the Front Line* (London, Zed Press, 1983)

Botswana

Munger, E. S. *Bechuanaland* (London, Oxford University Press, 1965)
South African Labour Bulletin, 'Focus on Botswana', Jan. 1980, Vol. 5, No. 5.

Lesotho

South African Labour Bulletin, 'Focus on Lesotho', Nov. 1980, Vol. 6, No. 4.

Malawi

Crosby, C. A. *Historical Dictionary of Malawi* (London and Metuchen, The Scarecrow Press Inc., 1980)
McMaster, C. *Malawi: Foreign Policy and Development* (London, Julian Friedmann Publishers Ltd, 1974)

Mozambique

Hanlon, J. *Mozambique: The Revolution Under Fire* (London, Zed Press, 1984)

Namibia

Catholic Institute for International Relations *Namibia in the 1980s* (London, CIIR, 1980), and subsequent books in the series.
International Defence and Aid Fund *Namibia: The Facts* (London, IDAF, 1980), and other books by IDAF.
Moleah, A. T. *Namibia: The Struggle for Liberation* (Wilmington, Disa Press, 1983)
SWAPO *To Be Born a Nation: The Liberation Struggle for Namibia* (London, Zed Press, 1981)

South Africa

Davies, R., D. O'Meara and S. Dlamini *The Struggle for South Africa* (2 vols) (London, Zed Books, 1984)
Hanlon, J. and R. Omond *The Sanctions Handbook* (Harmondsworth, Penguin, 1987)
Lodge, T. *Black Politics in South Africa Since 1945* (London and New York, Longman, 1983)
Marks, S. and S. Trapido (Eds) *The Politics of Race, Class and Nationalism in Twentieth Century South Africa* (London, Longman, 1987)
Murray, M. *South Africa: Time of Agony, Time of Destiny* (London, Verso, 1987)
Omond, R. *The Apartheid Handbook* (Harmondsworth, Penguin, 1985)

Southern Africa

Africa Review (Saffron Walden, World of Information, published annually)
Africa South of the Sahara (London, Europa Publications, published annually)
Africa Who's Who (London, Africa Journal Ltd, 1981)
Cawthra, G. *Brutal Force: The Apartheid War Machine* (London, IDAF, 1986)
de Bragança, A. and I. Wallerstein (Eds) *The African Liberation Reader* (London, Zed Press, 1982)
Hanlon, J. *Apartheid's Second Front* (Harmondsworth, Penguin, 1986)

Swaziland

Davies, R., D. O'Meara and S. Dlamini *The Kingdom of Swaziland: A Profile* (London, Zed Books, 1985)
Grotpeter, J. J. *Swaziland* (London and Metuchen, The Scarecrow Press Inc., 1975)

South African Labour Bulletin 'Focus on Swaziland' (Apr. 1982, Vol. 7, No. 6)

Tanzania

Coulson, A. *Tanzania, A Political Economy* (London, Oxford University Press, 1979)
Iliffe, J. *A Modern History of Tanganyika* (London, Cambridge University Press, 1979)

Zambia

Hall, R. *The High Price of Principles* (Harmondsworth, Penguin, 1973)
Mulford, D. C. *Zambia: The Politics of Independence, 1957–64* (London, Oxford University Press, 1967)
Pettman, J. *Zambia, Security and Conflict* (Lewes, Julian Friedmann, 1974)
Tordoff, W. *Politics in Zambia* (Manchester, Manchester University Press, 1974)

Zimbabwe

Astrow, A. *Zimbabwe: A Revolution that Lost its Way* (London, Zed Press, 1983)
Callinicos, A. and J. Rogers *Southern Africa after Soweto* (London, Pluto Press, 1977)
Frederikse, J. *None But Ourselves: Masses vs Media in the Making of Zimbabwe* (Johannesburg, Raven Press, 1982)
Martin, D. and P. Johnson *The Struggle for Zimbabwe: The Chimurenga War* (London, Faber and Faber, 1981)
Meredith, M. *The Past is Another Country, Rhodesia 1890–1979* (London, Andre Deutsch, 1979)
Report of the Constitutional Conference, Lancaster House, London September–December 1979 (London, HMSO, Jan. 1980)